D1753687

Mario Birkholz
**Thin Film Analysis
by X-Ray Scattering**

Related Titles

Messerschmidt, A.

A Practical Guide to X-Ray Crystallography of Biomacromolecules

Principles and Applications

approx. 350 pages with 8 figures
Hardcover
ISBN 3-527-31396-6

Bubert, H., Jenett, H. (eds.)

Surface and Thin Film Analysis

A Compendium of Principles, Instrumentation, and Applications

353 pages with 200 figures and 8 tables
2002
Hardcover
ISBN 3-527-30458-4

Talman, R.

Electron Accelerators as X-Ray Sources

Especially the Energy Recovery Linac

approx. 470 pages with approx. 150 figures and approx. 25 tables
Hardcover
ISBN 3-527-40590-9

Janssens, K. H. A., Adams, F. C. V., Rindby, A. (eds.)

Microscopic X-Ray Fluorescence Analysis

433 pages
2000
Hardcover
ISBN 0-471-97426-9

Tsuji, K., Injuk, J., Van Grieken, R. (eds.)

X-Ray Spectrometry

Recent Technological Advances

616 pages
2003
Hardcover
ISBN 0-471-48640-X

Lifshin, E. (ed.)

X-ray Characterization of Materials

277 pages with 142 figures and 19 tables
1999
Hardcover
ISBN 3-527-29657-3

Thin Film Analysis by X-Ray Scattering

Mario Birkholz

With Contributions by

Paul F. Fewster
Christoph Genzel

WILEY-VCH

WILEY-VCH Verlag GmbH & Co. KGaA

The Author(s) of this Book

Dr. Mario Birkholz
IHP Microelectronics GmbH
Im Technologiepark 25
15236 Frankfurt (Oder)
Germany
E-mail: birkholz@ihp-microelectronics.com

Thin Film Analysis and Technology
www.thinfilm-at.com

With Contributions by

Prof. Paul F. Fewster
PANalytical Research Centre
Sussex Innovation Centre
Science Park Square
Falmer, Brighton
BN1 9SB
UK
E-mail: Paul.Fewster@PANalytical.com

PD Dr. rer. nat. Christoph Genzel
Hahn-Meitner-Institut Berlin
Bereich Strukturforschung
c/o BESSY
Albert-Einstein-Straße 15
12489 Berlin
Germany
E-mail: genzel@hmi.de

Cover
G. Schulz, Fußgönheim

■ All books published by Wiley-VCH are carefully produced. Nevertheless, authors, editors, and publisher do not warrant the information contained in these books, including this book, to be free of errors. Readers are advised to keep in mind that statements, data, illustrations, procedural details or other items may inadvertently be inaccurate.

Library of Congress Card No.: applied for
British Library Cataloging-in-Publication Data:
A catalogue record for this book is available from the British Library.
Bibliographic information published by Die Deutsche Bibliothek
Die Deutsche Bibliothek lists this publication in the Deutsche Nationalbibliografie; detailed bibliographic data is available in the Internet at http://dnb.ddb.de.

© 2006 WILEY-VCH Verlag GmbH & Co. KGaA, Weinheim

All rights reserved (including those of translation into other languages). No part of this book may be reproduced in any form – by photoprinting, microfilm, or any other means – nor transmitted or translated into a machine language without written permission from the publishers. Registered names, trademarks, etc. used in this book, even when not specifically marked as such, are not to be considered unprotected by law.

Typesetting TypoDesign Hecker GmbH, Leimen
Printing and Bookbinding   Markono Print Media Pte Ltd, Singapore

Printed in Singapore
Printed on acid-free paper

ISBN-13: 978-3-527-31052-4
ISBN-10: 3-527-31052-5

Table of Contents

Preface IX

Symbols XV

1	**Principles of X-ray Diffraction** 1	
1.1	The Basic Phenomenon 1	
1.2	The $\theta/2\theta$ Scan 11	
1.3	Intensity of Bragg Reflections 14	
1.3.1	Atomic Form Factors 17	
1.3.2	Structure Factor 19	
1.3.3	Multiplicity 24	
1.3.4	Geometry Factor 25	
1.3.5	Preferred Orientation (Texture) 25	
1.3.6	Polarization Factor 26	
1.3.7	Absorption Factor 26	
1.3.8	Integration of the Interference Function 29	
1.4	Applications 37	
	Exercises 39	
	References 41	
2	**Identification of Chemical Phases** 43	
2.1	Histogram-Based Techniques 43	
2.2	Linear Attenuation Coefficient μ 55	
2.3	Determination and Interpretation of the μt Product 60	
2.4	Analysis of Phase Mixtures 66	
2.5	Amorphous Thin Films 70	
2.6	Accurate Determination of Lattice Parameter 74	
2.7	Applications 80	
	Exercises 81	
	References 83	

Thin Film Analysis by X-Ray Scattering. M. Birkholz
Copyright © 2006 WILEY-VCH Verlag GmbH & Co. KGaA, Weinheim
ISBN: 3-527-31052-5

3	**Line Profile Analysis** 85
3.1	Model Functions and Peak Parameters 86
3.2	Instrumental Line Profile 97
3.3	Deconvolution by Fourier Techniques 101
3.4	Reflection Broadening by Small Crystallite Size Only 107
3.4.1	Scherrer Equation 108
3.4.2	Column Height Distribution 111
3.4.3	Crystallite Shapes Other Than Cubes 112
3.4.4	Determination of the Column Height Distribution Function 115
3.4.5	Determination of the Crystallite Size Distribution Function 118
3.5	Concomitant Occurrence of Size and Strain Broadening 120
3.5.1	Analysis According to Williamson and Hall 122
3.5.2	Method of Warren and Averbach 126
3.5.3	Single-Line Analysis 129
3.5.4	Techniques of Whole-Pattern Fitting 130
3.6	Applications 134
	Exercises 136
	References 138

4	**Grazing Incidence Configurations** 143
4.1	Grazing Incidence X-ray Diffraction (GIXRD) 148
4.2	Penetration Depth and Information Depth 155
4.3	Depth-Dependent Properties 158
4.4	Refractive Index for X-rays 160
4.5	Total External Reflection and Critical Angle 161
4.6	X-ray Reflectivity (XRR) 165
4.6.1	Reflectivity of a Substrate 166
4.6.2	Reflectivity of a Single Layer 168
4.6.3	Reflectivity of Multilayers and Superlattices 171
4.7	Grazing Incidence Diffraction (GID) 175
4.8	Applications 177
	Exercises 179
	References 181

5	**Texture and Preferred Orientation** 183
5.1	Texture Factors 188
5.2	Pole Figures 191
5.3	Measurement of Pole Figures 195
5.4	Directions, Orientations and Inverse Pole Figures 200
5.5	Fiber Textures or Layer Textures 204
5.5.1	Harmonic Method 204
5.5.2	Whole Pattern Techniques 207
5.5.3	Rocking Curves (ω Scans) 211
5.6	Biaxial and Fully General Textures 216
5.6.1	Azimuthal Scans (ϕ Scans) 218
5.6.2	General Orientation Distribution 220

5.6.3	Determination of Fully General Texture 225
5.7	Depth Dependence of Thin-Film Textures 228
5.8	Applications 230
	Exercises 234
	References 235

6	**Residual Stress Analysis** 239
	Mario Birkholz and Christoph Genzel
6.1	Ceiiinnossstttuv 241
6.2	Fundamental Equation of XSA 246
6.3	Measurement of d_ψ Distributions 249
6.4	Diffraction Elastic Constants (DECs) s_1 and $1/2s_2$ 258
6.5	Grain Interaction Models 261
6.6	The Effect of Texture 265
6.7	Classification of Stresses 268
6.7.1	Classification by Dimension 268
6.7.2	Residual Stresses in Multiphase Materials 269
6.7.3	Origin of Residual Stresses: Extrinsic and Intrinsic Stresses 271
6.8	Effect of Residual Stress Gradients 273
6.8.1	General Considerations 273
6.8.2	The Biaxial Stress State 274
6.9	Detection of Residual Stress Gradients in Thin Films 276
6.9.1	Basic Relations 276
6.9.2	X-ray Penetration Depth for the General Case of Asymmetric Diffraction 278
6.9.3	Special Methods for X-ray Stress Gradient Analysis 281
6.9.4	Grazing-Incidence Diffraction (GID) 282
6.9.5	The Scattering Vector Method 284
6.9.6	Realization of *H* Mode on a Four-Circle Diffractometer 286
6.10	Applications 289
	Exercises 291
	References 291

7	**High-Resolution X-ray Diffraction** 297
	Mario Birkholz and Paul F. Fewster
7.1	Strain, Strain Relaxation and Composition in Epitaxial Layers 303
7.2	High-Resolution Rocking Curves 306
7.3	Mosaicity and Extinction 314
7.4	Dynamical Theory of Ewald and Extensions 319
7.5	High-Resolution Rocking Curves and Profiles from Layer Structures 324
7.6	Reciprocal Space Mapping 332
7.7	Diffuse Scattering 337
7.8	Extensions to High-Resolution Diffraction 338
	Exercises 339
	References 340

Preface

Mario Birkholz

Thin films have become an important branch of materials science and technology over the last few decades. A thin film is considered in this book as having a thickness between about 1 nm and some 10 μm. Their first application was probably in the field of decorative coatings, but in the last century many other applications in microelectronics, optics, data storage, sensorics, protection and other purposes have had a large impact on the development of thin films and related deposition techniques. Figure 1 displays a variety of thin-film applications from the areas mentioned. Depending on the intended application, thin films are made of metals, inorganic compounds, organic compounds or from biological molecules. The task of the thin-film developer is easily described by stating that the deposition process has to be optimized such that the arrangement of atoms enables the film to fulfill the intended functionality.

Since structure and function are intimately related properties in any material, the characterization of structural properties is a very relevant issue in thin-film development. This book is concerned with the structural analysis of thin films by x-ray scattering procedures. There exist various other characterization techniques like electron microscopy, scanning tunneling methods, ion beam scattering, magnetic resonance, optical spectroscopy and others by which important structure properties may be elucidated. Here, however, the focus is on x-ray scattering. The suitability of this technique for thin-film analysis is mainly motivated by two reasons:
1. The wavelengths of x-rays are of the order of atomic distances in condensed matter, which especially qualifies their use as structural probes.
2. X-ray scattering techniques are nondestructive and leave the investigated sample or – more importantly – the produced device intact.

Electron microscopy might be considered of comparative importance for the characterization of structure and morphology. This technique is a complementary one to x-ray scattering, since it probes a rather confined volume of the sample, whereas x-ray scattering yields information from a much larger volume. Therefore, some micrographs from electron microscopy will appear in the text, but the reader is referred to the special literature for an introduction to the subject.

Thin Film Analysis by X-Ray Scattering. M. Birkholz
Copyright © 2006 WILEY-VCH Verlag GmbH & Co. KGaA, Weinheim
ISBN: 3-527-31052-5

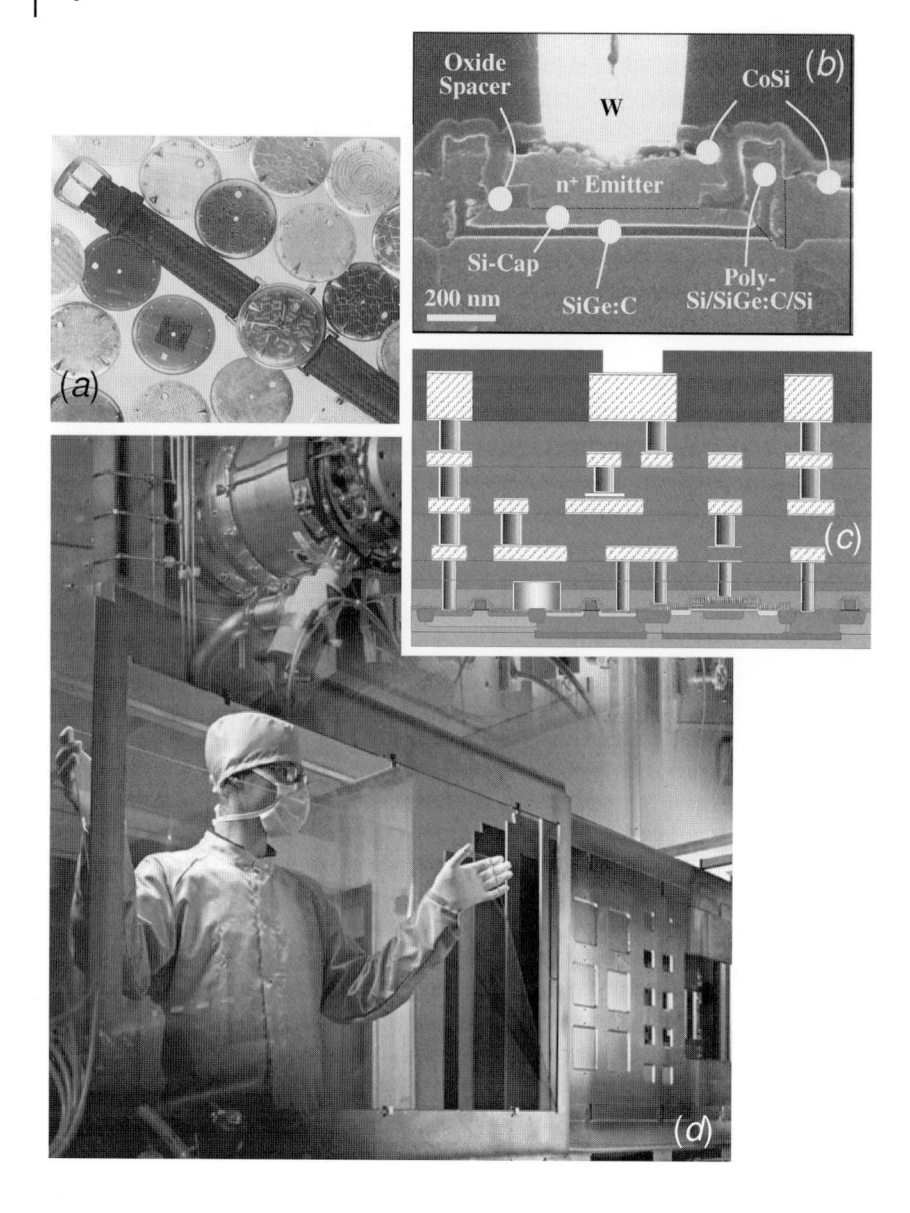

Various examples from thin film applications: (a) decorative coatings on metal dials, (b) scanning electron micrograph of cross-section from SiGe:C heterojunction bipolar transistor, (c) schematic of SiGe:C BiCMOS architecture with four metal layers, (d) processing of window glass for optical coating (Fotograf: Rainer Maier, BFF, Wittmar), (e) identity card with optical data storage made from bacterial purple membrane containing the protein bacteriorhodopsin and example of stored pixel patterns, (f) cemented carbide cutting insert with c-BN protective coating and (g) tip of a diamond-coated abrasive pencil of 60 μm diameter (figures kindly provided by (a, d, f and g) Fraunhofer IST, Braunschweig [1], (b, c) IHP, Frankfurt (Oder) [2], (e) Prof. N. Hampp, University of Marburg [3]).

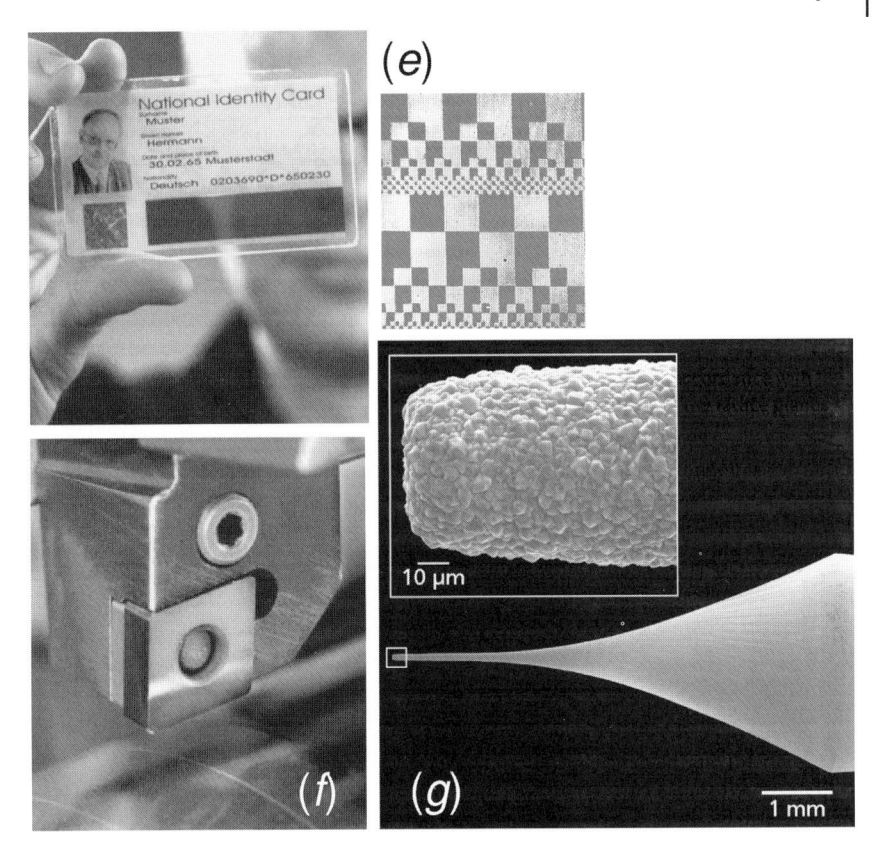

This book is intended to give overviews of the relevant x-ray scattering techniques for thin-film work and to equip scientists and engineers with the basic understanding to apply them. It has to be emphasized that for each x-ray technique presented in one of the following chapters there are authoritative and comprehensive textbooks available; these are listed at the end of each chapter and the reader is referred to them for further consultation. It seems, however, that there exists a gap between the highly developed and complex structural sciences on the one hand and the daily needs of materials scientists on the other. Many of the conclusive, effective and powerful techniques that have been developed for structural investigations appear to be not as extensively used in thin-film technology as they would deserve. It is the aim of this book to bridge this gap by introducing the concepts of x-ray techniques that appear most interesting to elucidate the close relations between structure, function and growth of thin films.

Chapter 1 introduces the basic phenomenon of x-ray diffraction by a crystalline lattice. In Chapter 2 methods for the identification of chemical phases are presented. Chapter 3 is related to the line profile analysis of diffraction peaks with respect to film microstructure. Measurement geometries characterized by a grazing incident x-ray beam are introduced in Chapter 4. The preferred orientation of crystallites and residual stresses in thin films are dealt with in Chapters 5 and 6, respectively. Up to this point mostly polycrystalline films will be considered and use will be made of the kinematic theory only. Epitaxial thin films are in the focus of Chapter 7, where high-resolution x-ray diffraction is outlined and the first grounding of dynamical theory is introduced. The majority of the material presented is based on the physical phenomenon of diffraction, but some parts – as for instance the presentation of reflectometry in chapter four – are related to the more general phenomenon of x-ray scattering. This is the reason for the title of the book.

It is recommended to start reading with Chapters 1 and 2, which might be helpful even for those readers to whom the basics are already known in order to become familiar with the conventions and notation used. After this introductory training the reader may consult any other chapter presenting the method that might be expected of relevance for his or her actual work. The emphasis of the book is on x-ray scattering with laboratory setups in contrast to synchrotron radiation beam lines. However, many of the measurement concepts presented are equally realized at synchrotron facilities and may also be applied in experiments with the much higher intensity available at synchrotron sources.

Two concepts or quantities meander through the following chapters like a thread through the tows of the former British royal navy ("roter Faden" [4]). The first of these quantities is the scattering vector that is abbreviated by Q here. The scattering vector is met in almost every chapter since interatomic distances are probed by diffraction only along the direction of Q. Diffraction or scattering experiments may be considered as intensity mappings under complex rotations of the sample with respect to the scattering vector. These reorientations are dealt with by the use of three reference frames $\{l_i\}$, $\{s_i\}$ and $\{c_i\}$, one for each frame of the laboratory, the sample and the crystallographic unit cell. The different frames are sometimes confusing for the newcomer to the field. It is recommended that when one seems of having lost the "roten Faden" it might be taken up again by answering the question "what are the coordinates of the scattering vector Q within the respective reference frame?"

The second recurring quantity is that of the x-ray attenuation coefficient μ. In condensed matter x-rays are attenuated on a length scale of some 10 to some 100 μm. These penetration depths are accordingly often larger than the film thickness t and special methods have been developed to restrict the probing beam to the sample volume. In almost all of the forthcoming chapters we have to derive how the μt product affects the measured scattering intensities. It may even be stated that the μt product can be regarded as the central physical quantity in thin-film analysis by x-ray scattering.

The chapters end with an application section, where studies and works related to the issue of the chapter are presented. The selection of these examples and those mentioned in the main body of the text is probably highly selective and reflects the interest and working areas of the author(s). Since each chapter covers a large field of research activities it was hardly possible to overview fully the many interesting x-ray scattering investigations that have been carried out in the appropriate areas. A collection of exercises is given at the end of each chapter, by the solution of which the reader may verify the understanding of the text. Solutions to the exercises can be found on the internet [5].

Two further issues are considered in parallel with the main text. These are related, firstly, to the instrumentation in x-ray scattering experiments and, secondly, to the structure of selected material classes. In the instrumentation boxes, the instrumentation required by the experiments described in the chapter is detailed. The structure boxes present crystallographic structures, structural parameters and selected physical properties of relevant materials. The material systems have been chosen in accordance with their relevance to illustrate the x-ray scattering technique in the chapter. The selected material systems are in order of increasing chapter number metals, semiconductors, nanocomposites, optical thin films, dielectric and superconducting materials, hard coatings and finally semiconductors for micro- and optoelectronics. Other combinations would have equally been possible.

Two remarks have to be made on notation: (a) SI units have been used throughout the text and (b) it has been endeavored to use a consistent notation throughout the text. However, this turned out a difficult task, since every chapter covers a highly developed subfield of x-ray scattering with its own nomenclature. It could thus not be completely avoided to make use of the same symbol with different meaning in different chapters. These cases will explicitly be pointed out. In case of doubt, the appropriate meaning of a symbol can be identified by consulting the symbol list.

In some cases, the names of inventors or scientific pioneers are mentioned. It should be borne in mind, however, that scientific achievements always rely on the communication among different researchers exchanging their ideas and imaginations. This statement is illustrated by the famous discussion between P.P. Ewald and M. Laue that laid the basis for the first x-ray diffraction experiments by Friedrich, Knipping and Laue [6]. Scientific progress has always been based on teamwork, even if the protagonists did not know each other personally. This fact is explicitly stated here, since we cannot be sure in every case that all the researchers that should be credited were adequately indicated when one or more of them are mentioned. Since x-ray diffraction is about 100 years old, it may be possible that future research in the history of science will reveal personal contributions of which we were not fully aware at the time of writing. The reader interested in the early history of x-ray diffraction is referred to a paper collection published for the International Union of Crystallography [7].

This book project would not have become reality without the help of some friends and colleagues. Firstly, I would like to thank Paul Fewster and Christoph Genzel for their co-authorship of Chapters 6 and 7. They both agreed on being co-authors to chapters they would be much more qualified of writing themselves and let me assume the role of the first author in order to maintain consistency with the rest of the book. Their great expertise in the respective fields helped enormously to formulate these two state-of-the-art chapters. I enjoyed the work with both of them very much. I am indebted to Daniel Chateigner, Carl Krill, Paolo Scardi, Thomas Schröder, Antonella Tagliente, Mark Vaudin, Thomas Wieder, Don Williamson, Joachim Woitok and Peter Zaumseil, who carefully red draft versions of single chapters, gave valuable recommendations and pointed me to some examples from the work of their own and others. Moreover, I like to thank those colleagues, who permitted the reprint of figures from their publications, which are (in addition to those already mentioned) J. Almer, U. Balachandran, M. Deutsch, J. Driscoll, J. Gubicza, N. Hampp, K. Helming, Y. Iijima, D. Knoll, S. Kondruweit, L. Lutterotti, M. A. Meyer, C. Michaelsen, M. Morales, A. Navarro-Quezada, S. Peist, A. J. Perry, M. Li, R. P. Reade, Á. Révész, J. Ricote, A. Saerens, K. Saito, K. Schiffmann, S. Sun, H. Suo, T. Ungár, J. J. Wells and E. Zschech. Their work on thin film analysis will probably be very helpful to the reader by illustrating the theoretical derivations with real-world examples. Thanks are also to Jürgen Altmann and Wulff Pfeiffer for their support of this project, to the companies AMD, Bruker AXS, IfG, PANalytical, Seifert for providing pictures from their thin film or x-ray equipment products, to Cyrille Boudias and Daniel Monceau, who draw the crystallographic figures in the structure boxes by virtue of their software package CaRIne Crystallography [8] and to the Wiley-VCH Verlag, who followed all my suggestions in preparing an editing the manuscript. Most of all, I would like to thank my wife, Johanna, for her patience and continuous support of this project.

References

Monographs, conference proceedings and internet sites of general relevance

E. Prince (Ed.): *International Tables for Crystallography – Vol. C*, 3rd ed., Kluwer Academic, Dordrecht, 2004.
M. Ohring: *The Materials Science of Thin Films*, Academic Press, San Diego, 2nd ed., 2002.
International Union of Crystallography: www.iucr.ac.uk.
Proceedings of the annual Denver X-Ray Conferences (DXC) published by the International Center for Diffraction Data (ICDD)
Advances in X-ray Analysis, Vol. 48, Proceedings of the 2004 Conference, 2005, etc.

Special papers

[1] Institut für Schicht- und Oberflächentechnik, *Jahresbericht 2003*, Braunschweig, 2004.
[2] IHP, *Jahresbericht 2003*, Frankfurt (Oder), 2004, and D. Knoll, personal communication.
[3] N. Hampp, *Bakteriorhodopsin als photochromes Sicherheitspigment und biologischer Datenspeicher* (vdi, Düsseldorf, 2000).
[4] J. W. Goethe, *Wahlverwandtschaften*, Weimar, 1809.
[5] Exercise solutions: www.thinfilm-at.com.
[6] G. Hildebrandt, *75 Jahre Röntgenstrahlinterferenzen in Kristallen*, Phys. Bl. **43** (1987) 430.
[7] J. M. Bijvoet, W. G. Burgers, G. Hägg (eds.), *Early Papers on Diffraction of X-rays by Crystals*, (Published for the International Union of Crystallography by A. Oosthoek's Uitgeversmaatschappij N. V., Utrecht, 1969).
[8] C. Boudias & D. Monceau, CaRIne Crystallography 4.1, User's Guide (2004), and http://pro.wanadoo.fr/carine.crystallography/.

Symbols

Symbols consistently used throughout the book are listed in the sequence of appearance in the chapters.

Chapter 1

2θ	scattering angle
α, β, γ	angles between edges of crystallographic unit cell
δ	divergence of x-ray beam
ε_0	permittivity of a vacuum
ϕ_h	phase of diffracted plane wave
λ	x-ray wavelength
μ	linear x-ray attenuation or absorption coefficient
μ_m	mass attenuation coefficient
μt	product of attenuation coefficient and film thickness
σ, π	indices to denote perpendicular and parallel polarized beam components
ρ	mass density
ρ_e	electron density
$\tau_{1/e}$	$1/e$ penetration depth
θ	Bragg angle
$\theta_B, 2\theta_B$	Bragg angle and scattering angle of a Bragg reflection
$\{c_i\}$	crystal reference frame of individual crystallites (superscript C)
$\{l_i\}$	laboratory reference frame (superscript L)
$\{s_i\}$	specimen reference frame (superscript S)
a, b, c	lattice parameters or cell edges of crystallographic unit cell
a_{ij}^{XY}	transformation matrices from reference frame X to Y (with $X, Y =$ C, L, S)
b_i	unit cell vectors of reciprocal lattice
A	absorption factor in XRS experiment
$A(2\theta), A_{\theta/2\theta}$	absorption factor for symmetric $\theta/2\theta$ experiment
B	isotropic temperature damping factor

Thin Film Analysis by X-Ray Scattering. M. Birkholz

Symbols

c	velocity of light in vacuum
C	polarization factor
$\overline{C^2}$	average polarization factor
Cu Kα	x-ray radiation from copper source (Kα line)
e	electron charge
d	interplanar spacing between crystallographic lattice planes
E	energy of radiation
$\mathbf{E_0}, \mathbf{E}$	electrical field vector of incoming and scattered beam
f	atomic form factor
f'	real part of anomalous scattering factor
f''	imaginary part of anomalous scattering factor
$\mathbf{F, F^*}$	structure factor and its complex conjugate
G	geometry factor
$\mathbf{h}$	abbreviated form of Miller indices hkl, also used as subscript
hkl	Miller indices of Bragg reflection from (hkl) lattice planes
$(hkl), \{hkl\}$	Miller indices of crystallographic lattice plane
$[hkl], \langle hkl \rangle$	Miller indices of crystallographic direction
i	imaginary unit
$\mathfrak{I}$	interference function
I	x-ray intensity
I_0	intensity of the incident beam
I_B	maximum intensity of x-ray reflection at Bragg angle θ_B
$I_{\mathbf{h}}$	integrated intensity of a Bragg reflection from (hkl) lattice planes
k	geometry factor for XRS experiment
$k(2\theta), k_{\theta/2\theta}$	geometry factor for symmetric $\theta/2\theta$ scan
$\mathbf{K_0}$	wave vector of incoming x-ray beam outside the sample
$\mathbf{K}$	wave vector of scattered x-ray beam outside the sample
ℓ	path length of x-ray in the sample
L	Lorentz factor
Lp	Lorentz polarization factor
$m_{hkl}, m_{\mathbf{h}}$	multiplicity of a Bragg reflection hkl in a powder pattern
m_0	rest mass of an electron
n	order of reflection in Bragg equation
N	number of unit cells in a crystallite
n_1, n_2, n_3	integers to denote the position of a unit cell in a crystallite
N_1, N_2, N_3	number of unit cells of a single crystallite along its orthogonal directions
N_L	Loschmidt's number
P	power of scattered x-ray beam
$\mathbf{Q}$	scattering vector
r	atomic distance or bond length
$\mathbf{r}_{n_1 n_2 n_3}$	distance vector from the origin of a crystallite to unit cell of number (n_1, n_2, n_3)
r_e	classical radius of the electron
R	goniometer radius

R_{FC}	radius of the focusing circle
SCF	scaling factor for x-ray intensity received by the detector
S	surface
s_3	substrate normal vector
t	thickness of thin film
t'	time
T	temperature
$T_{\mathbf{h}}$	texture factor of $\mathbf{h}$th reflection
$\overline{u^2}$	mean-square oscillation amplitude due to thermal vibrations of the atom
V	irradiated sample volume
V_{uc}	unit cell volume
x_n, y_n, z_n	fractional coordinates of nth atom in the unit cell
z	thin film depth coordinate
Z	number of electrons of an atom

Chapter 2

α_1, α_2	subcomponents of α x-ray emission line
δ_a	x-ray beam divergence along the θ and 2θ axis (axial divergence)
δ_p	x-ray beam divergence in the scattering plane (equatorial divergence)
δ_{ds}	opening angle of divergence slit
δ_{rec}	opening angle of receiving slit
δ_{Sol}	opening angle of Soller slits
γ	angle between scattering vector and bond vector
ϕ	azimuth, i.e. angle of in-plane sample rotation
σ	cross-section for x-ray scattering
θ_0	centroid of Bragg reflection
$(\Delta 2\theta)_i$	instrumental shifts of centroid from Bragg peak position
a_0, a_1	coefficients in the linear expansion of unit cell edge a
b	height misalignment of sample surface
B_a, B_p	extension of x-ray foot step on the sample along axial and in-plane directions
j	numeration of layers in a multilayer stack
J	number of layers in a multilayer stack
M	atomic mass
S	subscript to denote quantities related to a substrate
T_{dep}	deposition temperature
T_{hom}	homologous temperature
T_{mel}	melting temperature
TF	subscript to denote quantities related to a thin film
w_j	weight of the jth phase in a thin-film sample
W	total weight of a thin film
x	stoichiometry coordinate of a binary compound $A_{1-x}B_x$

Chapter 3

β	integral breadth of Bragg reflection
$\beta_{2\theta}$	integral breadth on scattering angle scale
β_Q	integral breadth on the scale of momentum transfer
γ	variance of crystallite sizes in lognormal distribution function
Γ_{hkl}	ratio of the permutation invariants of fourth order polynomials of Miller indices
ε	lattice strain
ε_{rms}	root-mean square strain
$\tilde{\varepsilon}$	weighted average strain
λ_1, λ_2	wavelengths of subcomponents of α x-ray emission line
θ_S, θ_E	start and end value of scattering angle range
ρ_d	dislocation density
η	weight parameter of pV profile function indicating the Cauchy fraction
$2w$	full width at half maximum
$A_n, A(n)$	coefficients of Fourier expansion
A^S, A^{size}	coefficients of Fourier expansion accounting for size broadening
A^D, A^{dis}	Fourier expansion coefficients accounting for broadening by lattice distortions
b	Burger's vector to describe geometrical structure of dislocation
$B(2\theta)$	background function in whole pattern fitting
C	Cauchy profile function
C_{hkl}	dislocation contrast factor
$\bar{C}_{hkl}$	average dislocation contrast factor
d_0	position of the centroid of a Bragg peak
D	crystallite size parameter
D_0	crystallite size median in lognormal distribution function
D_{cub}	length of a cube-shaped crystallite
D_{sph}	diameter of a spherical crystallite
$\langle D^n \rangle$	nth statistical moment of crystallite size distribution
$\langle D \rangle_A$	area-weighted average crystallite size
$\langle D \rangle_V$	volume-weighted average crystallite size
f	sample line profile free from instrumental effects
$\mathcal{F}(y(x))$	Fourier transform of function $y(x)$
g	instrumental line profile
$g(D)$	crystallite size distribution function
$g_{LN}(D)$	logarithmic normal distribution function of crystallite size
G	Gauss profile function
$G_n, G(n)$	coefficients of Fourier expansion of standard peak
h	convolution of instrumental and sample-broadened line profile
$H_n, H(n)$	coefficients of Fourier expansion of measured sample peak
I_0	maximum peak intensity

k	parameter in Voigt profile function
K_S	Scherrer constant
K_D	scaling factor in Williamson–Hall relation and derivatives
L	correlation length
$\langle L^n \rangle$	nth statistical moment of column height distribution
m	parameter in Pearson profile function
M	dislocation configuration parameter
N	number of data points subjected to a Fourier transformation
N_3	number of unit cells in a unit cell column
pV	pseudo Voigt profile function
P	Pearson (VII) profile function
$p(L)$	column height distribution function of the sample
$P(L)$	column height distribution function of one crystallite
R	radius of spherical crystallite
R_c	core radius of dislocation
R_e	effective outer cut-off radius of dislocations
R_w	weighted R factor
R_{21}	intensity ratio of α_2 over α_1 intensity
V	Voigt profile function
$\mathbf{u}(\mathbf{r})$	displacement field or distortion field
$y(2\theta)$	model function of intensity in whole pattern fitting

Chapter 4

α, α_i	incidence angle of incident x-ray with respect to sample surface
α_f	exit angle of diffracted x-ray with respect to sample surface
α_c, θ_c	critical angle for total external reflection
α_t	angle of transmitted x-ray with respect to surface plane
β	negative imaginary part of x-ray refractive index
δ	negative deviation of real part of x-ray refractive index from unity
$\Delta(2\theta)$	shift of Bragg peak position due to refraction
k_α	geometry factor for GIXRD experiment
θ_m	maximum of Kiessig fringes in reflectivity curve
σ	root mean square roughness of film surface or interface
ρ_a	density of atoms
$\tau_{1/e}$	penetration depth
τ_{63}	penetration depth due to scattered intensity
$\bar{\tau}$	average information depth in thin film of finite thickness
$\bar{\tau}_\alpha$	average information depth in GIXRD
A_α	absorption factor for GIXRD experiment
B_+, B_-	real and negative imaginary part of α_t
E_t	amplitude of transmitted electrical field
I_R	intensity of the reflected beam
$\mathbf{k}$	wave vector of x-rays within the sample

$k_{z,j}$	z component of wave vector in the *j*th layer of a multilayer stack
$\mathscr{L}$	Laplace transform
m	order of Kiessig fringe
M	transfer matrix in XRR
n	real part of index of refraction
$\hat{n}$	complex index of refraction
$P(z)$	depth-dependent property
$\bar{P}$	property averaged over depth
$\bar{P}_\alpha$	depth-averaged property in GIXRD
$\mathbf{R}_{j,j+1}$	refraction matrix in XRR
$r_{j,j+1}$	reflection coefficient between the *j*th and (*j*+1)th layer in a multilayer stack
t_{per}	superlattice period
$\mathbf{T}_{j,j+1}$	translation matrix in XRR
$z(x, y)$	height function of sample surface
$\bar{z}$	average sample height

Chapter 5

$\Delta\psi$	tilt in crystallite orientation distribution, i.e. FWHM of $I_h(\psi)$
$\Delta\phi$	twist in crystallite orientation distribution, i.e. FWHM of $I_h(\phi)$
$\varphi_1, \Phi, \varphi_2$	Euler angles for the S → C transformation
ω	sample rotation angle on Θ axis
k_ψ	geometry factor for pole figure measurement
k_ω	geometry factor in Ω mode
$\psi_{1/2}$	FWHM of volume share of fiber fraction
ψ	tilt angle or polar angle
$\bar{\tau}_\psi$	average information depth in ψ scan
A_ψ	absorption factor in Ψ mode
A_ω	absorption factor in Ω mode
C_l^μ	coefficients in linear expansion of inverse pole figure
C_l^{mn}	coefficients in linear expansion of harmonic ODF
$C_l^{\mu\nu}$	linear coefficients in symmetry-adapted ODF expansion
d	direction vector
$f(g)$	orientation distribution function
$f(\varphi_1, \Phi, \varphi_2)$	orientation distribution function
$F_l(\mathbf{h})$	weight factors in $I_h(\psi)$ expansion
$F_e^\nu(\mathbf{h})$	pole figure coefficients of **h**th reflection
g	orientation of a crystallite with respect to the sample reference frame
G	volume fraction of the textured phase (March model)
$I_h(\psi)$	intensity distribution of Bragg reflection **h** for varying tilt angles
$I_h(\phi)$	course of intensity of Bragg reflection **h** for varying azimuths
$I_{rc}(\omega)$	intensity distribution in a rocking curve scan
J	texture index

$k_l^\mu(\Phi, \beta)$	symmetrized spherical harmonics
l_{max}	order of maximum expansion of orientation distribution function
$M(l)$	number of ODF expansion coefficients of order l
$\mathbf{n_h}$	normal on $\mathbf{h}$ lattice planes
$P_l(x)$	Legendre function
$\bar{P}_l^m(x)$	normalized associated Legendre polynomial
$P_l^{mn}(x)$	generalized associated Legendre polynomial
r	ratio of t_i over t_f (March model)
$R(\beta, \Phi)$	inverse pole figure
t_i, t_f	initial and final thickness of rolled metal sheet
$T_l^{mn}(x)$	generalized harmonics
$T_l^{\mu\nu}$	symmetry-adopted generalized harmonics
$V_f(\psi)$	volume share of fiber texture between tilt angle 0 and ψ

Chapter 6

α, β	incidence and exit angle of x-rays
α	coefficient of thermal expansion
$\Delta\alpha$	difference of coefficients of thermal expansion between layer and substrate
ε_{ij}	strain tensor
$\varepsilon_{ij}^L, \varepsilon_{ij}^S$	strain tensor in the laboratory and specimen reference frame
$k_{\alpha\beta}$	geometry factor for general incidence and exit angle
k_η	geometry factor in the scattering vector technique
ν	Poisson ratio
η	angle of rotation around the scattering vector
H	rotation axis parallel to the scattering vector
σ_{ij}	stress tensor
$\sigma_\|$	residual stress in the film plane
$\sigma_\|(z)$	stress gradient for in-plane residual stress
$\sigma_{ij}^L, \sigma_{ij}^S$	stress tensor in the laboratory and specimen reference frame
$\sigma^I, \sigma^{II}, \sigma^{III}$	residual stresses of first, second and third kind
ψ^*	stress-free tilt angle
$\tau_{\psi\eta}(\theta)$	penetration depth in scattering vector mode
$\tau_{1/e}(\alpha)$	$1/e$ penetration depth for incidence angle α under inclusion of absorption
$A_{\alpha\beta}$	absorption factor for general incidence and exit angle
A_η	absorption factor in the scattering vector mode
c_{ijkl}, C_{ijkl}	elastic stiffness tensor of single crystal and polycrystalline aggregate
d_0	stress-free interplanar spacing
$d_\perp$	interplanar spacing of surface-parallel lattice planes
$d_{\phi\psi}$	interplanar spacing as determined for azimuth ϕ and tilt angle ψ
$d(\psi), d_\psi$	distribution of d values as a function of tilt angle
E	Young's modulus

F_{ij}		anisotropic stress factors
R		bending radius of layer–substrate composite due to residual stress
s_{ijkl}, S_{ijkl}		elastic compliances tensor of single crystal and polycrystalline aggregate
s_0		anisotropy constant
$s_1, 1/2s_2$		diffraction elastic constants DEC
t_{ijkl}		elastic polarizability tensor

Chapter 7

Δ		standard deviation in orientation distribution of mosaic blocks and crystallites
Δ_φ		standard deviation of in-plane orientation distribution (twist)
Δ_ψ		standard deviation of out-of-plane orientation distribution (tilt)
$\Delta\lambda$		bandwidth of x-ray radiation
γ		asymmetry parameter
γ_o		sin of the inclination angle to the film surface
γ_h		sin of the angle of emergence to the film surface
κ		elastic stiffness-dependent parameter
$\tilde{\mu}$		extinction coefficient
$\nu 0_L, \nu 1_L$		linear expansion coefficient of Poisson's ratio
η		deviation parameter
χ		crystal polarizability
τ_{ex}		extinction length
Δa		absolute lattice mismatch
$\Delta d_\perp$		lattice spacing difference between substrate and layer peak in symmetric scan
a_L, a_S		unit cell edges of cubic layer (L) and substrate (S) material
$a_\parallel, a_\perp$		in-plane and out-of-plane unit cell edges of tetragonally distorted layers
$a^0{}_L, a^1{}_L$		linear coefficients of unit cell edge in Vegard's rule
$f(p)$		primary extinction correction
$\mathbf{D}$		dielectric displacement vector
$\mathbf{k}_0$		wave vector of primary x-ray beam inside the sample
$\mathbf{k}_h$		wave vector of diffracted x-ray beam inside the sample
m		relative lattice mismatch
$\mathbf{q}$		scattering vector inside the sample
Q_x, Q_z		in-plane and out-of-plane component of scattering vector
R		relaxation parameter
t_c		critical thickness
W		tilt distribution function of mosaic blocks
x		composition variable
X_0, X_h		excitation errors (Anregungsfehler)

1
Principles of X-ray Diffraction

Diffraction effects are observed when electromagnetic radiation impinges on periodic structures with geometrical variations on the length scale of the wavelength of the radiation. The interatomic distances in crystals and molecules amount to 0.15–0.4 nm which correspond in the electromagnetic spectrum with the wavelength of x-rays having photon energies between 3 and 8 keV. Accordingly, phenomena like constructive and destructive interference should become observable when crystalline and molecular structures are exposed to x-rays.

In the following sections, firstly, the geometrical constraints that have to be obeyed for x-ray interference to be observed are introduced. Secondly, the results are exemplified by introducing the $\theta/2\theta$ scan, which is a major x-ray scattering technique in thin-film analysis. Thirdly, the $\theta/2\theta$ diffraction pattern is used to outline the factors that determine the intensity of x-ray reflections. We will thereby rely on numerous analogies to classical optics and frequently use will be made of the fact that the scattering of radiation has to proceed coherently, i.e. the phase information has to be sustained for an interference to be observed.

In addition, the three coordinate systems as related to the crystal $\{c_i\}$, to the sample or specimen $\{s_i\}$ and to the laboratory $\{l_i\}$ that have to be considered in diffraction are introduced. Two instrumental sections (Instrumental Boxes 1 and 2) related to the $\theta/2\theta$ diffractometer and the generation of x-rays by x-ray tubes supplement the chapter. One-elemental metals and thin films composed of them will serve as the material systems for which the derived principles are demonstrated. A brief presentation of one-elemental structures is given in Structure Box 1.

1.1
The Basic Phenomenon

Before the geometrical constraints for x-ray interference are derived the interactions between x-rays and matter have to be considered. There are three different types of interaction in the relevant energy range. In the first, electrons may be liberated from their bound atomic states in the process of photoionization. Since energy and momentum are transferred from the incoming radiation to the excited electron, photoionization falls into the group of inelastic scattering processes. In

Thin Film Analysis by X-Ray Scattering. M. Birkholz
Copyright © 2006 WILEY-VCH Verlag GmbH & Co. KGaA, Weinheim
ISBN: 3-527-31052-5

addition, there exists a second kind of inelastic scattering that the incoming x-ray beams may undergo, which is termed Compton scattering. Also in this process energy is transferred to an electron, which proceeds, however, without releasing the electron from the atom. Finally, x-rays may be scattered elastically by electrons, which is named Thomson scattering. In this latter process the electron oscillates like a Hertz dipole at the frequency of the incoming beam and becomes a source of dipole radiation. The wavelength λ of x-rays is conserved for Thomson scattering in contrast to the two inelastic scattering processes mentioned above. It is the Thomson component in the scattering of x-rays that is made use of in structural investigations by x-ray diffraction.

Figure 1.1 illustrates the process of elastic scattering for a single free electron of charge e, mass m and at position R_0. The incoming beam is accounted for by a plane wave $E_0\exp(-iK_0R_0)$, where E_0 is the electrical field vector and K_0 the wave vector. The dependence of the field on time will be neglected throughout. The wave vectors K_0 and K describe the direction of the incoming and exiting beam and both are of magnitude $2\pi/\lambda$. They play an important role in the geometry of the scattering process and the plane defined by them is denoted as the scattering plane. The angle between K and the prolonged direction of K_0 is the scattering angle that will be abbreviated by 2θ as is general use in x-ray diffraction. We may also define it by the two wave vectors according to

$$2\theta = \arccos\frac{\langle K, K_0 \rangle}{KK_0} \quad (1.1)$$

The formula is explicitly given here, because the definition of angles by two adjoining vectors will be made use of frequently.

The oscillating charge e will emit radiation of the same wavelength λ as the primary beam. In fact, a phase shift of 180° occurs with the scattering, but since this shift equally arises for every scattered wave it has no effect on the interference pattern in which we are interested and will be neglected. If the amplitude of the scattered wave $E(R)$ is considered at a distance R we may write according to Hertz and Thomson

$$E(R) = E_0 \frac{1}{4\pi\varepsilon_0 R} \frac{e^2}{mc^2} \sin\angle(E_0, R)\exp(-iKR) \quad (1.2)$$

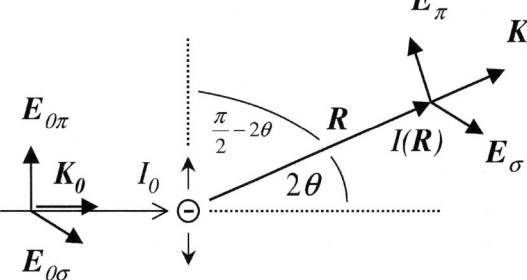

Figure 1.1 Scattering of x-rays by a single electron.

where ε_0 and c are the vacuum permittivity and velocity of light. The field vector $\mathbf{E}$ and wave vector $\mathbf{K}$ are oriented perpendicular to each other as is usual for electromagnetic waves. The sin term is of significance when the state of polarization is considered for which two extreme cases may arise. In one case, the exciting field $\mathbf{E}_0$ is confined to the scattering plane and in the second case it is normally oriented. In classical optics these two cases are named π polarization and σ polarization. The field vectors in both cases will be denoted by $\mathbf{E}_\pi$ and $\mathbf{E}_\sigma$. The angle between $\mathbf{E}_\sigma$ and $\mathbf{R}$ is always 90° and the sin term will equal unity. For the case of π polarization, however, it may be expressed by virtue of the scattering angle according to $\sin\angle(\mathbf{E}_0, \mathbf{R}) = |\cos 2\theta|$. If the character C abbreviates the sin term it may be written

$$C = \begin{cases} 1 & \sigma\text{-polarization} \\ |\cos 2\theta| & \pi\text{-polarization} \end{cases} \quad (1.3)$$

Since the intensity is obtained from the sum of the square of both field vectors the expression

$$\left(\frac{1}{4\pi\varepsilon_0 R}\right)^2 \left(\frac{e^2}{mc^2}\right)^2 \left(E_\sigma^2 + E_\pi^2 \cos^2 2\theta\right) \quad (1.4)$$

is obtained. In a nonpolarized beam both polarization states will have the same probability of occurring,

$$\overline{E_\sigma^2} = \overline{E_\pi^2} = I_0/2$$

and it is finally arrived at the intensity of the scattered beam at distance R

$$I(\mathbf{R}) = I_0 \frac{r_e^2}{R^2} \frac{1 + \cos^2 2\theta}{2} \quad (1.5)$$

Here, use has been made of the notion of the classical radius of the electron, $r_e = e^2/(4\pi\varepsilon_0 mc^2)$, that amounts to 2.82×10^{-15} m. The intensity of the scattering is seen to scale with the inverse of R^2 as might have been expected. It can also be seen that $I(\mathbf{R})$ scales with the ratio of squares of r_e over R. Since distances R of the order of 10^{-1} m are realized in typical laboratory setups the probability of observing the scattering by a single electron tends to zero. The situation substantially improves if the number of scattering objects is of the same order of magnitude as Loschmidt's number N_L – as usually is the case in experiments.

It also becomes evident from this equation as to why the scattering from atomic nuclei has not been considered in the derivation. In fact, the equation would also hold for the scattering from atomic nuclei, but it can be seen from Eq. (1.4) that the nuclei component will only yield a less than 10^{-6} smaller intensity compared to an electron. The difference is simply due to the mass difference, which is at least larger by a factor of 1836 for any atomic species. The scattering of x-rays by nuclei may, therefore, confidently be neglected. From the viewpoint of x-ray scattering an atom can thus be modeled by the number of Z electrons, which it contains according to its rank in the periodic table. In terms of the Thompson scattering model Zr_e may be written in Eq. (1.3) instead of r_e in order to describe the scattering from an atom, since the primary beam is then equally scattered by all electrons. In addition, it will be assumed temporarily that all electrons are confined to the origin of the atom. The consequences that follow from a refinement of the model by assuming a spatially extended charge distribution will be postponed to a later section. Hence, we have a first quantitative description for the x-ray elastic scattering from an atom.

In the next step consideration is given to what the scattering will look like if it occurs for a whole group of atoms that are arranged in a periodically ordered array like a crystal lattice. Figure 1.2 visualizes such an experiment where the crystal is irradiated with monochromatic x-rays of wavelength λ. In the special case considered here, each atom is surrounded by six neighbor atoms at distance a and the angle between two atomic bonds is always 90° or multiples of it. Atomic positions can then be described by the lattice vector $r_{n_1 n_2 n_3} = n_1 a c_1 + n_2 a c_2 + n_3 a c_3$ with c_1, c_2 and c_3 being the unit vectors of the three orthogonal directions in space. The c_i axes are the unit vectors of the crystal coordinate system $\{c_i\}$, which is assigned to the crystal. For some properties of the crystal this coordinate system will turn out to be extremely useful and the notion will be used throughout the book. The shape of the crystal is assumed to be that of a parallelepiped as is accounted for by the inequalities $0 \leq n_i = N_i - 1$ for $i = 1, 2, 3$. Each node of adjacent cubes is thus occupied by an atom. Such a structure is called simple cubic in crystallography. Only a single element crystallizes in this structure, which is polonium exhibiting an interatomic distance of $a = 0.3359$ nm. Although this metal has only very few applications, the case shall be considered here in detail, because of its clarity and simplicity.

It will now be calculated at which points in space interferences of x-rays might be observed that arise due to the scattering at the crystal lattice. The task is to quantify the strength of the scattered fields at a point R when elastic scattering occurs according to Eq. (1.5) at all atoms. The reference point of R is chosen such that it starts at the origin of the crystal lattice r_{000}. This means that we relate the phase difference in the summation of all scattered fields to their phase at r_{000}. This choice is arbitrary and any other lattice point might have been equally selected.

The wave vector of the primary beam K_0 is assumed to be parallel to the [100] direction of the crystal. The scattering plane defined by K_0 and K may coincide with one of the (010) planes. The wavefronts of the incoming plane waves which are the planes of constant phase are then oriented parallel to (100) planes. An atom on the position $r_{n_1 n_2 n_3}$ would then cause a scattering intensity to be measured at R of the strength

$$E_0 \exp(-iK_0 r_{n_1 n_2 n_3}) \frac{Zr_e}{|R - r_{n_1 n_2 n_3}|} \sin \angle (E_0, R - r_{n_1 n_2 n_3}) \exp(-iK(R - r_{n_1 n_2 n_3})) \quad (1.6)$$

This expression differs from Eq. (1.2) essentially by the fact that $R - r_{n_1 n_2 n_3}$ occurs instead of R, and for $n_1 = n_2 = n_3 = 0$ it becomes equal to Eq. (1.2). The solution of our task would simply consist in a summation over all fields scattered by the number of $N_1 \times N_2 \times N_3$ atoms comprising the crystal. However, the physics of the solution will become more transparent when an important approximation is made.

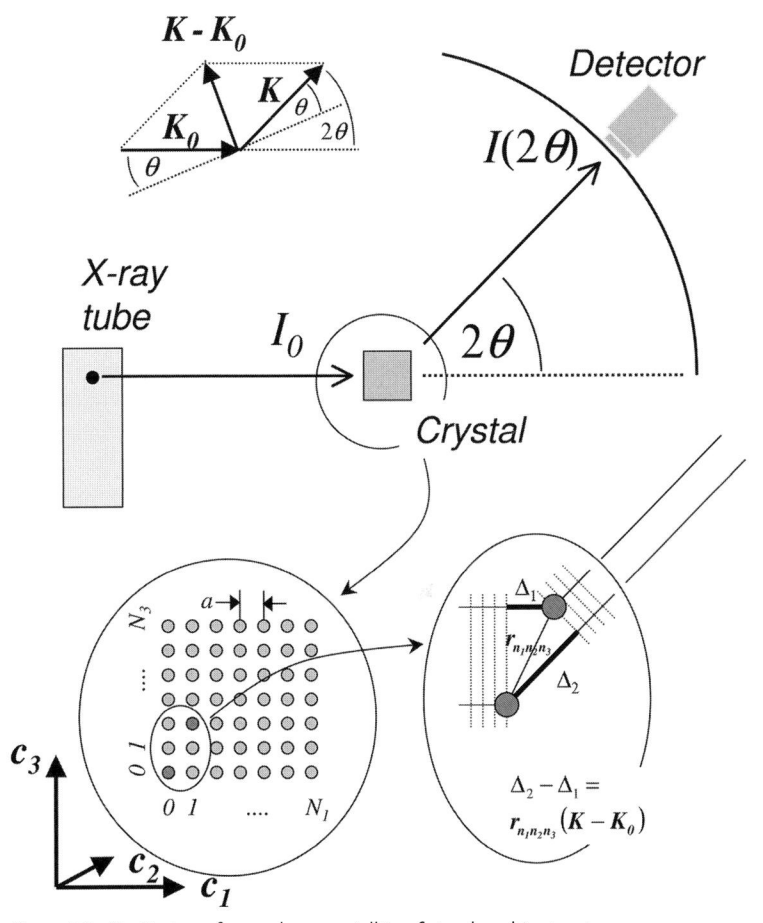

Figure 1.2 Scattering of x-rays by a crystallite of simple cubic structure.

It will be assumed that the interatomic distances r_{n1n2n3} (~10^{-10} m) are much smaller than the distances to the point of the intensity measurement $R - r_{n1n2n3}$ (~10^{-1} m). The denominator in Eq. (1.6) and in the sin term $\boldsymbol{R} - \boldsymbol{r}_{n_1n_2n_3}$ may then be replaced by $\boldsymbol{R}$ without introducing a large error. This substitution, however, is not allowed in the exponent of the last factor, since the interatomic distances are of the order of the wavelength and every phase shift according $Kr_{n_1n_2n_3} = 2\pi r_{n1n2n3}/\lambda$ has to be fully taken into account in the summation procedure. If these rules are applied the sin term may be replaced by the polarization factor C and the sum over all scattered fields reads

$$E_0 \frac{Zr_e}{R} C \exp(-iKR) \sum_{n_1n_2n_3} \exp\left(-i(\boldsymbol{K} - \boldsymbol{K}_0)\boldsymbol{r}_{n_1n_2n_3}\right) \qquad (1.7)$$

All terms independent of the lattice vector $r_{n_1n_2n_3}$ could be placed in front of the summation symbol. The approximation of which we have made use of is named Fraunhofer diffraction, which is always a useful approach when the distances between scattering objects are much smaller than the distance to the measurement point. In contrast to this approach stands the so-called Fresnel diffraction, for which interference phenomena are investigated very close to the scattering objects. The case of Fresnel diffraction will not be of interest here.

We have achieved a significant progress in solving our task by applying the Fraunhofer approximation and arriving at Eq. (1.7). It can be seen that the scattered field scales with two factors, where the first has the appearance of a spherical wave while the second is a sum over exponentials of vector products of wave vectors and lattice vectors. In order to improve our understanding of the summation over so many scattering centers the geometry is shown in the lower part of Fig. 1.2. A closer look at the figure reveals that the phase shift for two waves (a) scattered at r_{000} and (b) scattered at $r_{n_1n_2n_3}$ comprises two components due to $K_0 r_{n_1n_2n_3}$ and to $K r_{n_1n_2n_3}$. The strength of the total scattered field of Eq. (1.7) thus sensitively depends on the spatial orientation of the wave vectors K_0 and K with respect to the crystal reference frame $\{c_i\}$.

Because a single phase shift depends on the vector product between the lattice vector and the wave vector difference $K - K_0$ the latter quantity is recognized as a physical quantity of its own significance and is named the scattering vector

$$Q = K - K_0 \tag{1.8}$$

The scattering vector has the dimensionality of an inverse length, while its direction points along the bisection of incoming and scattered beam. The geometry is demonstrated in Fig. 1.3 and a closer inspection tells that the relation $|Q| = 4\pi \sin\theta/\lambda$ holds for the scattering vector magnitude. This relation will be made use of extensively throughout the book and the reader should be fully aware of its derivation from Fig. 1.3. It should be realized that $|Q|$ depends on both (a) the geometry of the scattering process via θ and (b) the wavelength λ of the probing x-ray beam. The physical meaning of Q in a mechanical analogy is that of a momentum transfer. By analogy with the kinetic theory of gases the x-ray photon

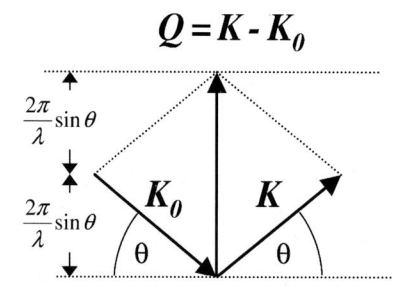

Figure 1.3 Geometry of scattering vector construction.

is compared to a gas molecule that strikes the wall and is repelled. The direction of momentum transfer follows from the difference vector between the particle's momentum before and after the event, $p - p_0$, while the strength of transferred momentum derives from $|p - p_0|$. In the case considered here the mechanical momentum p just has to be replaced by the wave vector K of the x-ray photon. This analogy explains why the scattering vector Q is also named the vector of momentum transfer. It has to be emphasized that the scattering vector Q is a physical quantity fully under the control of the experimentalist. The orientation of the incident beam (K_0) and the position of the detector (K) decide the direction in which the momentum transfer (Q) of x-rays proceeds. And the choice of wavelength determines the amplitude of momentum transfer to which the sample is subjected. From these considerations it is possible to understand the collection of a diffraction pattern as a way of scanning the sample's structure by scattering vector variation.

If the summation factor of Eq. (1.7) is expanded into three individual terms and the geometry of the simple cubic lattice is used it is found that the field amplitude of the scattered beam is proportional to

$$\sum_{n_1=0}^{N_1-1} \sum_{n_2=0}^{N_2-1} \sum_{n_3=0}^{N_3-1} \exp\left(-iQ\left[n_1 a c_1 + n_2 a c_2 + n_3 a c_3\right]\right) \tag{1.9}$$

where the scattering vector Q has already been inserted instead of $K - K_0$. This expression can be converted by evaluating each of the three terms by the formula of the geometric sum. In order to arrive at the intensity the resultant product has to be multiplied by the complex conjugate and we obtain the so-called interference function

$$\Im(Q) = \frac{\sin^2(N_1 a Q c_1 / 2)}{\sin^2(a Q c_1 / 2)} \cdot \frac{\sin^2(N_2 a Q c_2 / 2)}{\sin^2(a Q c_2 / 2)} \cdot \frac{\sin^2(N_3 a Q c_3 / 2)}{\sin^2(a Q c_3 / 2)} \tag{1.10}$$

that describes the distribution of scattered intensity in the space around the crystallite. For large values of N_1, N_2 and N_3 the three factors in $\Im(Q)$ only differ from zero if the arguments in the $\sin^2$ function of the denominator become integral multiples of π. Let us name these integers h, k and l in the following. The necessary condition to realize the highest intensity at R accordingly is

$$\Im(Q) \to \max \Leftrightarrow \begin{aligned} a Q c_1 &= 2\pi h \\ a Q c_2 &= 2\pi k \\ a Q c_3 &= 2\pi l \end{aligned} \tag{1.11}$$

Here, the integers h, k, l may adopt any value between $-\infty$ and $+\infty$. The meaning of these integers compares to that of a diffraction order as known in optics from diffraction gratings. The hkl triple specifies which order one is dealing with when the primary beam coincides with zero order 000. However, the situation with a crystalline lattice is more complex, because a crystal represents a three-dimensional grating and three integral numbers instead of only one indicate the order of a diffracted beam. The set of Eqs. (1.11) are the Laue conditions for the special case of cubic crystals that were derived by M. von Laue to describe the relation between lattice vectors $r_{n_1 n_2 n_3}$ and scattering vector Q for crystals of arbitrary symmetry at the position of constructive interference.

The severe condition that is posed by Eq. (1.11) to observe any measurable intensity is illustrated in Fig. 1.4. The plot shows the course of the function $\sin^2 Nx/\sin^2 x$, for $N = 15$, which is the one-dimensional analogue of Eq. (1.10). It can be seen that the function is close to zero for almost any value of x except for $x = \pi h$, with h being an integer. At these positions the $\sin^2 Nx/\sin^2 x$ function sharply peaks and only at these points and in their vicinity can measurable intensity be observed. The sharpness of the peak rises with increasing N and a moderate value of N has been chosen to make the satellite peaks visible. It should be noted that in the case of diffraction by a crystal the three equations of Eq. (1.11) have to be obeyed simultaneously to raise $I(R)$ to measurable values. As a further property of interest it has to be mentioned that $\sin^2 Nx/x^2$ may equally be used instead of $\sin^2 Nx/\sin^2 x$ for $N \gg x$. This property will enable some analytical manipulations of the interference function, which would otherwise be possible only on a numerical basis.

In order to gain further insight into the significance of the condition for observable intensity, we will investigate the Laue conditions with respect to the magnitude of the scattering vector. The magnitude of Q at $I(R) \to$ max can be obtained from the three conditional Eqs. (1.11) by multiplying by the inverse cell parameter $1/a$, adding the squares and taking the square root. This yields as condition for maximum intensity

$$I(R) \to \max \Leftrightarrow \frac{|Q|}{2\pi} = \frac{\sqrt{h^2 + k^2 + l^2}}{a} \tag{1.12}$$

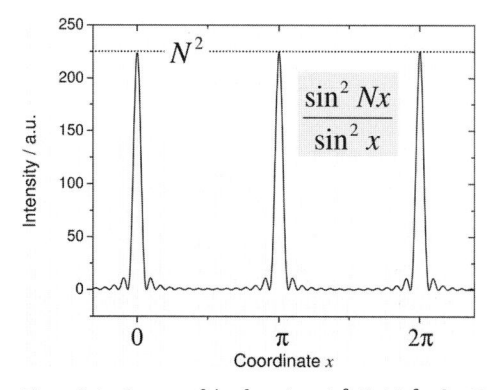

Figure 1.4 Course of the function $\sin^2 Nx/\sin^2 x$ for $N = 15$.

which can be rewritten by inserting the magnitude of the scattering vector, $|Q| = 4\pi \sin\theta/\lambda$, known from geometrical considerations

$$I(R) \to \max \Leftrightarrow 2\frac{a}{\sqrt{h^2 + k^2 + l^2}} \sin\theta = \lambda \tag{1.13}$$

This is an interesting result that may be read with a different interpretation of the hkl integer triple. The high degree of order and periodicity in a crystal can be envisioned by selecting sets of crystallographic lattice planes that are occupied by the atoms comprising the crystal. The planes are all parallel to each other and intersect the axes of the crystallographic unit cell. Any set of lattice planes can be indexed by an integer triple hkl with the meaning that a/h, a/k and a/l now specify the points of intersection of the lattice planes with the unit cell edges. This system of geometrical ordering of atoms on crystallographic planes is well known to be indicated by the so-called Miller indices hkl. As an example, the lattice planes with Miller indices (110) and (111) are displayed in Fig. 1.5 for the simple cubic lattice.

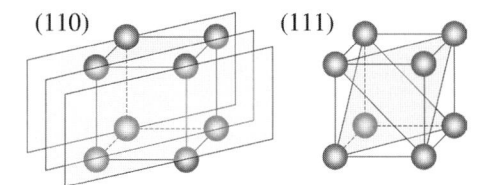

Figure 1.5 Lattice planes with Miller indices (110) and (111) in a simple cubic lattice.

The distance between two adjacent planes is given by the interplanar spacing d_{hkl} with the indices specifying the Miller indices of the appropriate lattice planes. For cubic lattices it is found by simple geometric consideration that the interplanar spacing depends on the unit cell parameter a and the Miller indices according to

$$d_{hkl} = \frac{a}{\sqrt{h^2 + k^2 + l^2}} \tag{1.15}$$

Keeping this meaning of integer triples in mind, Eq. (1.13) tells us that to observe maximum intensity in the diffraction pattern of a simple cubic crystal the equation

$$2d_{hkl} \sin\theta_B = \lambda \tag{1.15}$$

has to be obeyed. The equation is called Bragg equation and was applied by W.H. Bragg and W.L. Bragg in 1913 to describe the position of x-ray scattering peaks in angular space. The constraint $I(R) \to \max$ has now been omitted, since it is implicitly included in using θ_B instead of θ which stands for the position of the maximum. In honor of the discoverers of this equation the peak maximum position has been named the Bragg angle θ_B and the interference peak measured in the reflection mode is termed the Bragg reflection.

The Laue conditions and the Bragg equation are equivalent in that they both describe the relation between the lattice vectors and the scattering vector for an x-ray reflection to occur. Besides deriving it from the Laue condition, the Bragg equation may be obtained geometrically, which is visualized in Fig. 1.6. A set of crystallographic lattice planes with distances d_{hkl} is irradiated by plane wave x-rays impinging on the lattice planes at an angle θ. The relative phase shift of the wave depends on the configuration of atoms as is seen for the two darker atoms in the top plane and one plane beneath. The phase shift comprises of two shares, Δ_1 and Δ_2, the sum of which equals $2d\sin\theta$ for any arbitrary angle θ. Constructive interference for the reflected wave, however, can only be achieved when the phase shift $2d\sin\theta$ is a multiple of the wavelength. Therefore, Bragg's equation is often written in the more popular form $2d\sin\theta_B = n\lambda$, where the integer n has the meaning of a reflection order. Because we are dealing with three-dimensional lattices that act as diffraction gratings, the form given in Eq. (1.14) is preferred. It should be emphasized that the Bragg equation (Eq. (1.14)) is valid for any lattice structure, not only the simple cubic one. The generalization is easily performed by just inserting the interplanar spacing d_{hkl} of the crystal lattice under investigation. Table 1.1 gives the relation of d_{hkl} and the unit cell parameters for different crystal classes.

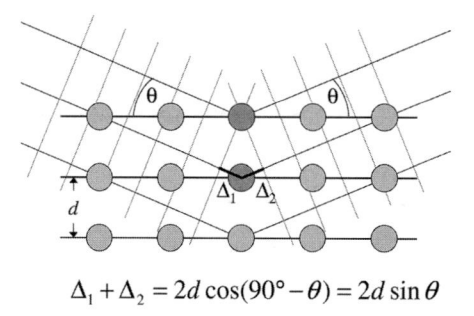

$$\Delta_1 + \Delta_2 = 2d\cos(90° - \theta) = 2d\sin\theta$$

Figure 1.6 Visualization of the Bragg equation. Maximum scattered intensity is only observed when the phase shifts add to a multiple of the incident wavelength λ.

Having arrived at this point it can be stated that we have identified the positions in space where constructive interference for the scattering of x-rays at a crystal lattice may be observed. It has been shown that measurable intensities only occur for certain orientations of the vector of momentum transfer Q with respect to the crystal coordinate system $\{c_i\}$. Various assumptions were made that were rather crude when the course of the intensity of Bragg reflections is of interest. It has been assumed, for instance, that the atom's electrons are confined to the center of mass of the atom. In addition, thermal vibrations, absorption by the specimen, etc., were neglected. More realistic models will replace these assumptions in the following. However, before doing so it should be checked how our first derivations compare with the measurement of a thin metal film and how diffraction patterns may be measured.

Table 1.1 Interplanar spacings d_{hkl} for different crystal systems and their dependency on Miller indices hkl. Parameters a, b and c give the lengths of the crystallographic unit cell, while α, β and γ specify the angles between them.

Crystal system	Constraints	$\dfrac{1}{d_{hkl}^2} =$
Cubic	$a = b = c$ $\alpha = \beta = \gamma = 90°$	$\dfrac{h^2 + k^2 + l^2}{a^2}$
Tetragonal	$a = b$ $\alpha = \beta = \gamma = 90°$	$\dfrac{h^2 + k^2}{a^2} + \dfrac{l^2}{c^2}$
Orthorhombic	$\alpha = \beta = \gamma = 90°$	$\dfrac{h^2}{a^2} + \dfrac{k^2}{b^2} + \dfrac{l^2}{c^2}$
Hexagonal	$a = b$ $\alpha = \beta = 90°$ $\gamma = 120°$	$\dfrac{4}{3}\dfrac{h^2 + hk + k^2}{a^2} + \dfrac{l^2}{c^2}$
Trigonal/ Rhombohedral	$a = b = c$ $\alpha = \beta = \gamma$	$\dfrac{(h^2 + k^2 + l^2)\sin^2\alpha + 2(hk + hl + kl)(\cos^2\alpha - \cos\alpha)}{a^2(1 - 3\cos^2\alpha + 2\cos^3\alpha)}$
Monoclinic	$\alpha = \gamma = 90°$	$\dfrac{h^2}{a^2 \sin^2\beta} + \dfrac{k^2}{b^2} + \dfrac{l^2}{c^2 \sin^2\beta} - \dfrac{2hl\cos\beta}{ac\sin^2\beta}$
Triclinic	None	Exercise 4

1.2
The θ/2θ Scan

An often-used instrument for measuring the Bragg reflection of a thin film is the θ/2θ diffractometer. Let us introduce its operation principle by considering the results obtained with the question in mind as to how x-ray scattering experiments are preferably facilitated. What we are interested in is the measurement of Bragg reflections, i.e. their position, shape, intensity, etc., in order to derive microstructural information from them. The intensity variation that is associated with the reflection is included in the interference function like the one given in Eq. (1.10), while the scattered intensity depends on the distance from the sample to the detection system **R**. We therefore should configure the instrument such that we can scan the space around the sample by keeping the sample–detector distance **R** constant. This measure ensures that any intensity variation observed is due to the interference function and is not caused by a dependency on **R**. The detector should accordingly move on a sphere of constant radius **R** with the sample in the center of it. In addition, the sphere reduces to a hemisphere above the sample, since we are only interested in the surface layer and data collection will be performed in reflection mode. The geometry is shown in Fig. 1.7.

Because the scattering of x-rays depends sensitively on the orientation of the crystal with respect to the scattering vector, we carefully have to define the various coordinate systems with which we are dealing. A sample reference frame $\{s_i\}$ is introduced for this purpose that is oriented with s_1 and s_2 in the plane of the thin film, while s_3 is equivalent to the surface normal.

1 Principles of X-ray Diffraction

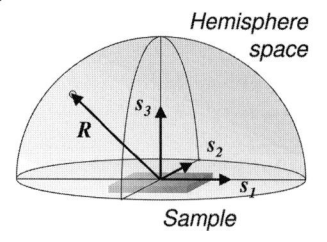

Figure 1.7 Sample reference frame $\{s_i\}$ and hemisphere above it.

The working principle of a $\theta/2\theta$ scan is visualized in Fig. 1.8 in the hemisphere of the sample reference frame. The sample is positioned in the center of the instrument and the probing x-ray beam is directed to the sample surface at an angle θ. At the same angle the detector monitors the scattered radiation. The sample coordinate vectors s_1 and s_3 lie in the scattering plane defined by K_0 and K. During the scan the angle of the incoming and exiting beam are continuously varied, but they remain equal throughout the whole scan: $\theta_{in} = \theta_{out}$. Note that the angle convention is different from the one used in optics: in x-ray diffraction the angles of incoming and exiting beam are always specified with respect to the surface plane, while they are related to the surface normal in optics. The $\theta/2\theta$ scan can also be understood as a variation of the exit angle when this is determined with respect to the extended incoming beam and this angle is 2θ for all points in such a scan. This is the reason for naming the measurement procedure a $\theta/2\theta$ scan. The quantity measured throughout the scan is the intensity scattered into the detector. The results are typically presented as a function of $I(2\theta)$ type.

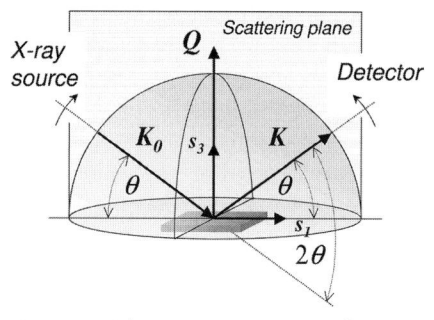

Figure 1.8 Schematic representation of a $\theta/2\theta$ scan from the viewpoint of the sample reference frame $\{s_i\}$.

These $\theta/2\theta$ scans are extensively used for the investigation of polycrystalline samples. The measurement of polycrystals is somewhat easier than that of single crystals due to the fact that, among other reasons, the scattered intensity for constant scattering angle is distributed on a circle rather than focused to a few points in space. Interestingly, in a $\theta/2\theta$ scan the scattering vector Q is always parallel to the substrate normal s_3. This fact is evident from Fig. 1.8 and the graphical definition of Q in Fig. 1.3. Due to this geometrical constraint only those lattice planes hkl that are oriented parallel to the surface plane can contribute to a Bragg reflection. The selective perception of certain subsets of crystallites in a $\theta/2\theta$ scan is visualized in Fig. 1.9. If various reflections hkl are measured they all stem from distinct subsets of crystallites – except they are of harmonic order, i.e. $h'k'l' = n(hkl)$.

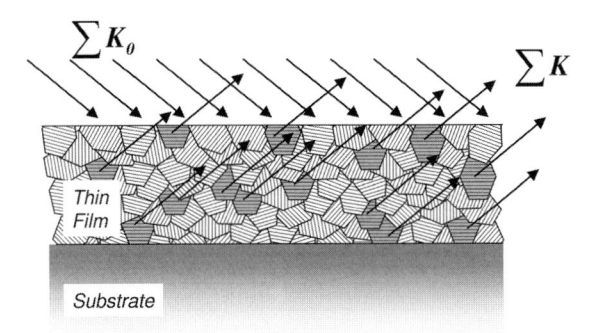

Figure 1.9 Selection principle for exclusive measurement of surface-parallel lattice planes in a $\theta/2\theta$ scan.

In order to demonstrate the principles developed so far, the simulation of a $\theta/2\theta$ scan of a 500 nm thin Al film is shown in Fig. 1.10. The simulation was calculated for the characteristic radiation of a copper x-ray tube having $\lambda(\text{Cu K}\alpha) = 0.154$ nm (see Instrumental Box 1 for further information). Various interesting features are realized from this plot, which displays eight Bragg reflections in the scattering angle range from 25° to 125°. The reflections may be assigned to their Miller indices when use is made of the Bragg equation and the unit cell parameter of the Al lattice, $a = 0.4049$ nm. For this purpose the d values of the $2\theta_B$ reflex positions have been calculated according to the Bragg equation $d = \lambda/(2\sin\theta_B)$ and checked for the solution of $(a/d)^2 = h^2 + k^2 + l^2$. It is seen that various reflections like 111 and 200 are observed, but other peaks like 100, 110, etc., are missing. This phenomenon has to be understood in the sense of destructive interference, which is caused by the structure of the Al lattice, which is distinct from the simple cubic lattice. It has to be noted that a splitting of peaks into an α_1 peak and an α_2 peak cannot be observed, although the feature was included in the simulation. The absence is explained from the broadness of the Bragg peaks causing a severe overlap between both peaks such that they remain unresolved. Broad reflections are caused by small grain sizes and crystal lattice faults that are often observed in thin polycrystalline films and are discussed in more detail in Chapter 3. Moreover, the diffraction pattern exhibits a pronounced decrease of scattered intensity with increasing scattering angle. Therefore, the diffraction pattern is also shown in the inset with a $\sqrt{I}$ ordinate in order to emphasize the smaller peaks. The square-root intensity plot is an often-used presentation mode. It is concluded that the basic features of Section 1.1

are in accordance with the simulated measurement of a thin Al film, but some aspects remain to be clarified.

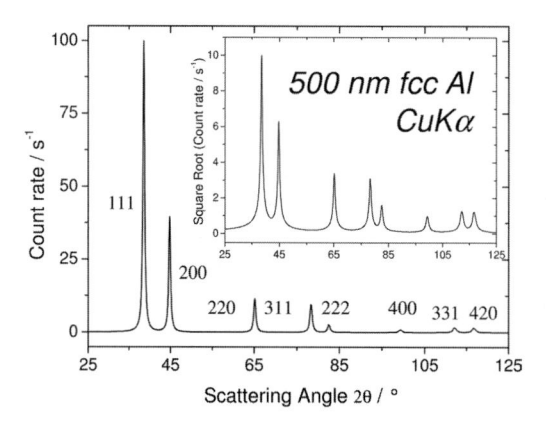

Figure 1.10 Simulation of a $\theta/2\theta$ scan of a 500 nm thin Al film measured with Cu Kα radiation. The inset shows the same pattern with a $\sqrt{I}$ ordinate.

1.3
Intensity of Bragg Reflections

The necessary refinement of the expression for the intensity of a Bragg reflection is now developed. For this purpose the finding will be used that was made by deriving the Bragg equation and the Laue conditions. It has been realized that the amplitude of the total scattered field from a charge distribution in the Fraunhofer approximation is characterized by a phase factor $\exp(-iQr_{n_1n_2n_3})$ comprising the scattering vector Q and the distance $r_{n_1n_2n_3}$ between all pairs of point charges. This result may be generalized by subjecting the sum in Eq. (1.9) to a continuous limit. Instead of writing a discrete distance vector $r_{n_1n_2n_3}$ the continuous variable r is used and it is argued that the scattered field depends as

$$\int \rho_e(r)\exp(-iQr)dr \tag{1.16}$$

on the electronic charge distribution $\rho_e(r)$ of the scattering object. The integration has to be performed over the volume dr to which the scattering electrons are confined. Because ρ_e has the dimensionality of an inverse volume the integration yields a dimensionless quantity, which is in accordance with our starting point. This new expression can now be applied to the scattering objects in which we are interested, i.e. atoms and crystallographic unit cells, to check whether the provisional intensity function is improved.

1.3 Intensity of Bragg Reflections

Instrumental Box 1:

θ/2θ Diffractometer

The basic measurement geometry of by far the most frequently used x-ray diffraction instrument is depicted in Fig. i1.1. The sample should preferably exhibit a plane or flattened surface. The angle of both the incoming and the exiting beam is θ with respect to the specimen surface. A vast number of organic and inorganic powder samples have been measured with these instruments from which the naming of powder diffractometer is understood. Its measurement geometry may also be applied to the investigation of thin films, especially if the layer is polycrystalline and has been deposited on a flat substrate, as is often the case.

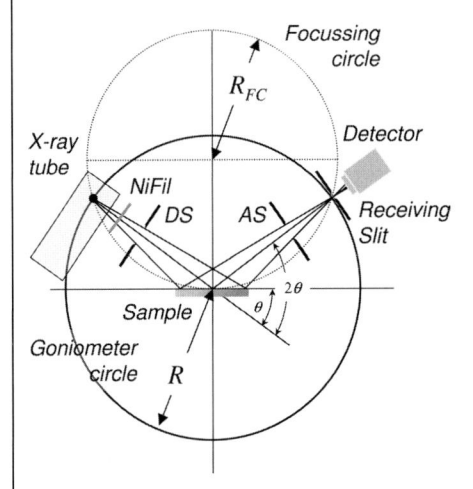

Figure i1.1 Schematic representation of θ/2θ diffraction in Bragg–Brentano geometry.

The diffraction pattern is collected by varying the incidence angle of the incoming x-ray beam by θ and the scattering angle by 2θ while measuring the scattered intensity $I(2\theta)$ as a function of the latter. Two angles have thus to be varied during a θ/2θ scan and various types of powder diffractometers are in use. For one set of instruments the x-ray source remains fixed while the sample is rotated around θ and the detector moves by 2θ. For other systems the sample is fixed while both the x-ray source and the detector rotate by θ simultaneously, but clockwise and anticlockwise, respectively. The rotations are performed by a so-called goniometer, which is the central part of a diffractometer. A goniometer of a powder diffractometer comprises at least two circles or – equally – two axes of rotation. Typically the sample is mounted on the rotational axis, while the detector and/or x-ray source move along the periphery, but both axes of rotation coincide. In most laboratory θ/2θ diffractometers the goniometer radius, which is the sample-to-detector distance, is in the range 150–450 mm. Highly precise goniometers with 0.001°

precision and even lower on both the θ and the 2θ circles are commercially available. The collected diffraction pattern $I(2\theta)$ consists of two sets of data: a vector of $2\theta_i$ positions and a second vector with the appropriate intensities I_i. The step size $\Delta 2\theta_i$ between two adjacent $2\theta_i$ should be chosen in accordance with the intended purpose of the data. For chemical phase analysis (Chapter 2) the full width of half the maximum of the tallest Bragg peak in the pattern should be covered by at least 5 to 7 measurement points. However, for a microstructural analysis (Chapter 3) in excess of 10 points should be measured on the same scale. The appropriate value of $\Delta 2\theta_i$ will also depend on the slit configuration of the diffractometer. The preset integration time of the detector per step in $2\theta_i$ should allow the integral intensity of the smallest peak of interest to exceed the noise fluctuations $\sigma(I)$ by a factor of 3 or 5, etc., according to the required level of statistical significance.

The control of the x-rays beam bundle suffers from the constraint that lenses and other refractive elements are not as easily available as those used for visible light. For this reason the beam conditioning in $\theta/2\theta$ diffractometers is mostly performed by slits and apertures and may be termed shadow-casting optics. In addition, powder diffractometers have to deal with the divergent beam characteristic that is emitted by an x-ray tube. Most systems operate in the so-called Bragg–Brentano or parafocusing mode. In this configuration a focusing circle is defined as positioned tangentially to the sample surface (see Fig. i1.1). The focusing condition in the Bragg–Brentano geometry is obeyed when the x-ray source and detector are positioned on the goniometer circle where it intersects the focusing circle. True focusing would indeed occur only for a sample that is bent to the radius of the focusing circle R_{FC}. Since R_{FC} differs for various scattering angles 2θ, true focusing cannot be obtained in a $\theta/2\theta$ scan and the arrangement is thus termed parafocusing geometry.

In a $\theta/2\theta$ scan the scattering vector Q is always parallel to the substrate normal. It is, however, evident from the above considerations and from Fig. i1.1 that this is strictly valid only for the central beam, while slight deviations from the parallel orientation occur for the divergent parts of the beam. If the most divergent rays deviate by $\pm\delta$ from the central beam their scattering vector is tilted by δ from the sample normal – at least for those scattering events that are received by the detector. In many configurations of diffractometer optics it suffices to consider only the central beam.

The analysis and interpretation of x-ray diffraction measurements necessitates distinguishing three different reference frames that are assigned to the laboratory, the sample and the crystallites and symbolized by $\{l_i\}$, $\{s_i\}$ and $\{c_i\}$, respectively. The unit vectors in each system are denoted by l_i, s_i or c_i, with i ranging from 1 to 3 for the three orthogonal directions. Transformations between these coordinate systems are frequently used, for which unitary transformation matrices a_{ij} are defined with superscripts LS, SC, CL, etc., indicating the initial and the final reference frame. The relations are visualized in Fig. i1.2.

1.3 Intensity of Bragg Reflections

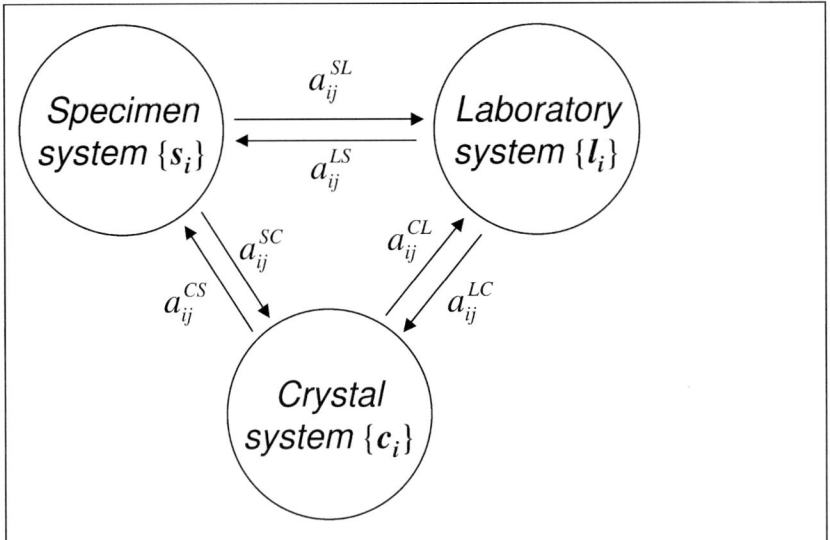

Figure i1.2 The three reference frames used in x-ray diffraction and the appropriate transformation matrices between them.

1.3.1
Atomic Form Factors

Formula (1.16) can be applied to atoms by inserting the square of electronic wavefunctions for the charge density $\rho_e(r)$. Before the results of this procedure are presented let us first investigate what might be expected from basic physical considerations. For this purpose the electrons may temporarily be imagined in the atomic model of Bohr to move in circular orbits around the nuclei. If the scattering from any two arbitrary electrons from this atom could be obtained it is evident that the scattering may occur for many different distance vectors r being associated with a large variation of phase shifts $-iQr$. The orbital smearing of the electron density will thus lead to a cessation of coherency and a reduction in the coherently scattered intensity. This reduction will be stronger the larger Q becomes, because it is the scalar product Qr that determines the phase shift.

The ansatz is made that the scattering of an atom depends on the shape of the electron density function or on its form, and we thus define an atomic form factor f by

$$f = \int_{at} \rho_e(r)\exp(-iQr)dr \qquad (1.17)$$

In the limit of $Q = 0$ the integration just runs over the charge distribution and yields the number of electrons of the atom Z. For $Q \neq 0$ the form factors are reasonably presented as a function of $|Q|$ or $\sin\theta/\lambda$. Atomic form factors have been calculated with various quantum mechanical methods of increasing sophistication. A compilation of values for all chemical elements and some important ions is given in Ref. [1]. Moreover, very often an approximation of f in the form of the model function

$$f = \sum_{j=1}^{4} a_j \exp\left(-b_j \sin^2\theta / \lambda^2\right) + c_j \qquad (1.18)$$

is used. By this approach a precision of 10^{-6} is achieved for the form factors and only nine coefficients have to be given for any atom or ion to model the whole $\sin\theta/\lambda$ range. The coefficients a_j, b_j and c_j are also tabulated in Ref. [1].

It is concluded that point charges Ze have to be substituted by fe in all the foregoing expressions in order to deal correctly with the extension of atomic charge distributions. For some metallic atoms the atomic form factors as calculated by Eq. (1.18) are displayed in Fig. 1.11. For low scattering angles they can be seen to reach values close to the atomic number Z, but a steep decrease with increasing $\sin\theta/\lambda$ is clearly seen for all of them. It should be noted that the intensity scales with the square of the atomic form factor and that an even stronger decrease will occur for f^2. For the example of Nb the form factor for the fivefold ion Nb^{5+} is also given. It can be seen that a difference between atoms and their ions is only significant for f values at low $\sin\theta/\lambda$, which is a general tendency for all atoms and ions, not just for Nb.

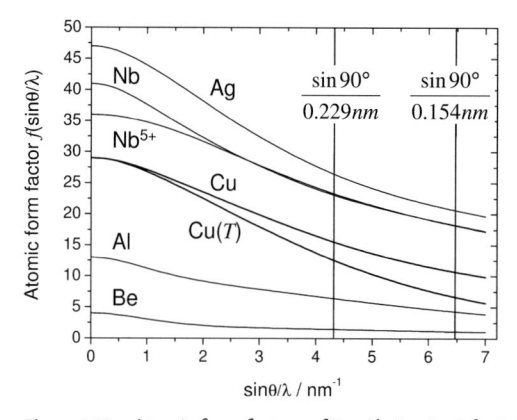

Figure 1.11 Atomic form factors of Be, Al, Cu, Cu@fccCuRT, Nb, Nb^{5+} and Ag.

For some investigations the inelastic scattering of x-rays cannot be neglected and the concept of the atomic form factors will then have to be extended by including real and imaginary anomalous scattering factors, f' and f'', that have to be added to the atomic form factors f given above. In most cases, anomalous scattering factors f' and f'' are small when compared with f. Numerical values for f' and f'' are given in the Ref. [1].

1.3.2
Structure Factor

The crystallographic unit cell is the smallest unit by which the periodic order in the crystal is repeated. In the simple cubic lattice that has been considered to derive the Bragg and Laue equations there is only one atom per unit cell. The scattered intensity was found to scale with the square of the charge of this atom – or the form factor as should be said now – and the interference function, see Eq. (1.10). For more complex structures the integration has to be extended over the total charge distribution of the unit cell (uc) rather than over a single atom. This quantity is denoted as the structure factor $\boldsymbol{F}$ that is given by

$$\boldsymbol{F} = \int_{uc} \rho_e(\boldsymbol{r}) \exp(-i\boldsymbol{Qr}) d\boldsymbol{r} \tag{1.19}$$

We will symbolize it consistently by a bold letter, since it is a complex quantity. The expression for the structure factor may be simplified by recalling that the unit cell comprises N atoms, numbered by n from 1 to N. It is thus possible to decompose the structure factor into single shares due to the individual atoms (at)

$$\boldsymbol{F} = \sum_{n=1}^{N} \int_{at} \rho_e(\boldsymbol{r}) \exp\left(-i\boldsymbol{Q}(\boldsymbol{r} - \boldsymbol{r}_n)\right) d\boldsymbol{r} \tag{1.20}$$

and the integration just has to be performed over the charge distributions of individual atoms. These values are known: they are given by the atomic form factors f_i of the nth atom. Accordingly, the structure factor can be written

$$\boldsymbol{F} = \sum_{n=1}^{N} f_n \exp(i\boldsymbol{Qr}_n) \tag{1.21}$$

The product of the scattering factor with the positions $\boldsymbol{r}_n$ of the N various atoms in the unit cell thus has to be evaluated. The latter are specified by their fractional coordinates (x_n, y_n, z_n) that read for the cubic cell $\boldsymbol{r}_n = x_n a\boldsymbol{c}_1 + y_n a\boldsymbol{c}_2 + z_n a\boldsymbol{c}_3$. We know that reflection intensity may only be observed when the Laue conditions are simultaneously obeyed which may be applied to simplify the phase factor by

$$\boldsymbol{Q}(x_n a\boldsymbol{c}_1 + y_n a\boldsymbol{c}_2 + z_n a\boldsymbol{c}_3) = hx + ky + lz \tag{1.22}$$

Only if this equation is obeyed does measurable intensity from interfering x-rays enter into the detector and the scattering of the crystal scales with

$$\boldsymbol{F}(hkl) = \sum_{n=1}^{N} f_n \exp\left[2\pi i(hx_n + ky_n + lz_n)\right] \tag{1.23}$$

The structure factor thus depends on the Miller indices of the reflection under consideration, the positions of the atoms in the unit cell and the atomic scattering factor. In monoatomic lattices the form factor is the same for all atoms and can be placed in front of the sum. For the simple cubic structure $N = 1$ and $x = y = z = 0$ and thus $\boldsymbol{F} = f$ for all hkl and reflections are observed for each order; however, for more complicated structures the full structure factor has to be investigated. Although the derivative has only been given for the cubic lattice it has to be emphasized that the expression for the structure factor, Eq. (1.23), is valid for crystals of arbitrary symmetry.

The majority of one-elemental metals are found in either the face-centered cubic (fcc), the body-centered cubic (bcc) or the hexagonal close-packed (hcp) structure. The relative arrangement of atoms in theses lattices is presented in Structure Box 1. Aluminum, for instance, crystallizes in the fcc structure. In this case the Bragg equation might be obeyed for certain lattice planes hkl, but for some combinations of hkl the phase shift in the x-rays scattered by neighboring atoms may amount to π or odd multiples of it. The scattered beams then interfere destructively and the reflections for these lattice planes are not extincted. In the fcc structure, for instance, destructive interference occurs for hkl = 100, 110, etc.

The extinction conditions can be derived for any crystal lattice by performing the same summation procedure that has been performed for the simple cubic lattice in the first section and it is an instructive exercise to do so (Exercise 7). One will then arrive at conditions comparable to Eq. (1.11) which predict under which orientation of Q towards $\{s_i\}$ reflections might be observed. A simpler approach instead is the calculation of the structure factor. Inserting the fractional coordinates of all four atoms of the fcc structure in Eq. (1.23) yields the result

$$F_{fcc}(hkl) = f\begin{bmatrix} 1+\exp(i\pi(h+k))+ \\ \exp(i\pi(h+l))+\exp(i\pi(k+l)) \end{bmatrix} = \begin{cases} 4f & \text{all } hkl \text{ even/odd} \\ 0 & hkl \text{ mixed} \end{cases} \quad (1.24)$$

The expression is seen to vanish for certain hkl and the lower equation is thus denoted as an extinction condition. It means that Bragg reflections are only observed for the fcc lattice if all Miller indices are either even or odd. For mixed triples destructive inference occurs and these reflections are systematically absent. It is evident from the $\theta/2\theta$ scan in Fig. 1.10 that this pattern in fact is in accordance with the extinction conditions of the fcc structure. Mathematically speaking, the Bragg equation is a necessary but not a sufficient condition for x-ray reflections to arise.

The structure factor of the bcc lattice can be obtained in the same way and results in

$$F_{bcc}(hkl) = f\left[1+\exp(i\pi(h+k+l))\right] = \begin{cases} 2f & h+k+l = 2n \\ 0 & \text{otherwise} \end{cases} \quad (1.25)$$

The extinction condition now derives from a sum over Miller indices and reads that the sum must yield an even number for the reflection to occur. Only if this condition is obeyed is the interference nondestructive and can be detected at the position predicted by Bragg's equation (Eq. (1.15)). The structure factor for the hcp structure can be derived as an exercise from the atom coordinates in the unit cell.

The structure factor F_h from a reflection h is of central importance in x-ray diffraction, because it relates the position of the atoms in the unit cell to the intensity of a reflection. Here, the Miller index triple hkl has been abbreviated by the subscript h which will be used very often in the following. The intensity scales with the product of F and its complex conjugate F^*. As can be seen from the examples of the

Structure Box 1:

Elementary Metals

The simple cubic structure that is used in this chapter to derive the basic formulas of x-ray diffraction only rarely occurs in nature. It is instead observed that one-elemental crystal lattices often take the face-centered cubic (fcc) or the body-centered cubic (bcc) structure. Figure s1.1 displays both of them. The interatomic distances are fully specified by the unit cell edge a. Each atom is surrounded by eight neighbor atoms in the bcc structure or twelve in the fcc structure. Besides the cubic structures, various metals are found to crystallize in the hexagonal close-packed (hcp) structure, also shown in Fig. s1.1. This structure has two degrees of freedom, namely the interatomic distance a in the base plane and the distance between two of the planes, $c/2$. Accordingly, two distinct interatomic distances r_1 and r_2 occur between next neighbors in the hcp structure (Exercise 1.12). An inspection of the periodic system reveals that the majority of elements assume one of these three basic structures under thermodynamic standard conditions (298 K, 101.6 kPa).

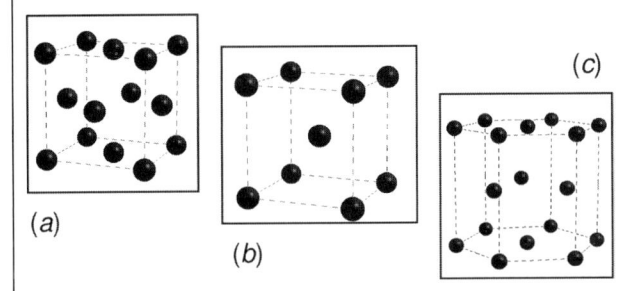

Figure s1.1 Crystallographic unit cells of the most frequently occurring structures of one-elemental metals: (a) face-centered cubic, (b) body-centered cubic and (c) hexagonal close-packed structure.

In both the fcc and hcp structures the atoms are arranged according to the model of close-packed spheres. It is thereby assumed that the atoms can be modeled by rigid spheres that all exhibit the same radius r_{at}. The value of r_{at} is chosen such that atomic neighbors are in contact via their surfaces. According to this scheme the atomic radius can be calculated and is found to be $r_{at,fcc} = a\sqrt{2}/4$ in the fcc structure. In the hcp structure the condition can only be obeyed when $r_1 = r_2$ holds, which leads to $r_{at,hcp} = a$ or equivalently $c/a = \sqrt{8/3} = 1.633$. The c/a parameter in general serves to define the ideal hcp structure, which is a close-packed one. In one-elemental metals with the hexagonal Mg structure the c/a ratio is always found to be very close to this ideal value (see Table s1.1). Although these structures deviate slightly from the ideal hcp structure, they are often considered hcp structured anyway and the value of c/a is specified additionally. If the volume of atomic spheres in both the fcc and the hcp structure is calculated and normalized with respect to the unit cell volume V_{uc}, a value of 0.74 results for the volume

Table s1.1 The three most common crystal structures of one-elemental metals. Unit cell edges under standard conditions are given.

Structure	Strukturbericht designation	Space group	Atomic positions		Examples with lattice parameters a and c (nm)
Face-centered cubic (fcc)	Copper structure (A1)	$F\bar{m}3m$ (225)	000 $\frac{1}{2}0\frac{1}{2}$	$0\frac{1}{2}\frac{1}{2}$ $\frac{1}{2}\frac{1}{2}0$	Al: 0.4049; Ni: 0.3524 Cu: 0.3615; Ag: 0.4086
Body-centered cubic (bcc)	Tungsten structure (A2)	$I\bar{m}3m$ (229)	000	$\frac{1}{2}\frac{1}{2}\frac{1}{2}$	Cr: 0.2884; Fe: 0.2866 Nb: 0.3307; W: 0.3165
Hexagonal close-packed (hcp)	Magnesium structure (A3)	$P6_3/mmc$ (194)	000	$\frac{2}{3}\frac{1}{3}\frac{1}{2}$	Mg: 0.3209, 0.5210 Ti: 0.2950, 0.4879 Zn: 0.2665, 0.4947 Zr: 0.3231, 0.5147

ratio. This is the largest value of spatial filling that might be achieved by the packing of spheres all having the same diameter.

It should be noted that crystallographic lattice planes in the hcp and also in other hexagonal structures are indexed by four Miller indices $(hkil)$, where always $i = -(h + k)$ holds. This indexing results from the usage of three unit vectors in the basal plane of hexagonal unit cells. In a widely used abbreviation a period is simply inserted for the third index: $(hk.l)$. One immediately realizes from the occurrence of both types of Miller index symbol that a hexagonal structure is being considered.

There exists an interesting relation between the close-packed fcc and hcp structures. The relation becomes evident when all atoms in the fcc lattice are decomposed into atomic (111) planes and compared with the (00.1) planes in the hcp structure (see Fig. s1.2). The coordination within the plane is the same, i.e. each sphere is surrounded by six neighbors to yield the highest packaging density of spheres within the plane. Looking from above on the plane stacking reveals that there exist three distinct positions where atoms might become situated, which are named A, B and C. In each plane atoms are positioned at A, B or C. It turns out that the stacking of planes may be accounted for by the sequences ...ABCABC... in the fcc structure, but by ...ABABAB... in the hcp structure. Therefore, both structures just differ by the vertical stacking sequence of fully occupied atom planes.

Figure s1.2 Stacking of close-packed planes in (a) fcc and (b) hcp structures.

fcc and bcc lattice the magnitude of $\boldsymbol{F}$ maximally equals the number of atoms in the unit cell multiplied by their atomic form factor. This situation is rarely observed for more complicated structures, because the scattering of the different groups of atoms often causes a partial destructive interference. This fact is demonstrated from the structure factors of technologically relevant compounds that are found in the various structure boxes of subsequent chapters.

The structure factor has the mathematical form of a discrete Fourier transform. The reverse transformation from the intensity of observed reflections would thus allow the determination of the atomic positions in the unit cell. However, the intensity scales with the product of the structure factor and its complex conjugate, $F_h F_h^*$, which is associated with a severe loss of information. If the structure factor is plotted in the Euler plane of complex numbers it may be characterized by its magnitude $|F_h|$ and its phase ϕ_h. In this picture the information loss can be envisaged as a loss of phase information, which is the well-known phase problem in the structure determination by x-ray diffraction.

Regarding the effect of thermal vibrations the same arguments apply as given above to justify the reduction in coherency by the spatial extension of electronic charge distribution. It is well known that the atoms in a solid oscillate at their equilibrium positions r_n. Temperature vibrations entail a reduction of phase coherence in the scattered beam and thus reduce the measured intensity. The phenomenon can quantitatively be accounted for by the mean quadratic deviation $\overline{u^2}$ of the atom from its average position r_n. The atomic form factors f have then to be replaced by the temperature-dependent expression

$$f_T = f \exp\left(-8\pi \overline{u^2} \sin^2\theta / \lambda^2\right) \tag{1.26}$$

Again, it can be seen that the scattering amplitude is exponentially damped with increasing scattering angle and that the damping coefficient scales with the square of momentum transfer $4\pi\sin\theta/\lambda$. The $8\pi\overline{u^2}$ factor is often abbreviated by the symbol B in the literature. Typically the average displacements of atoms $\sqrt{\overline{u^2}}$ at room temperature are in the range between 0.005 and 0.03 nm, which translates into a few percent to more than 10% of the bond length. In the fcc structured Cu lattice, for instance, $\sqrt{\overline{u^2}}$ amounts to about 6% of $d_{\text{Cu-Cu}}$. The effect of the temperature vibrations can be seen from Fig. 1.11, where in addition to the zero-temperature f also the atomic form factor of Cu in the Cu lattice is shown. It is evident that the scattering strength may be significantly reduced by thermal vibration, which holds in particular for high scattering angles.

These results are applied to the structure factor simply by replacing the form factor with its temperature-dependent value

$$F_{hkl}(T) = \sum_{n=1}^{N} f_n \exp\left(-B_n(T)\sin^2\theta/\lambda^2\right)\exp\left(2\pi i(hx_n + ky_n + lz_n)\right) \tag{1.27}$$

Because the thermal vibration amplitudes increase with increasing temperature the damping of Bragg reflections will also increase. This causes the Bragg reflection to sink into a background of diffuse scattered intensity when the temperature

1.3.3
Multiplicity

The multiplicity specifies the number of equivalent lattice planes that may all cause reflections at the same θ_B position. The phenomenon is visualized in Fig. 1.12, for the laboratory in reference frame $\{l_i\}$. In this coordinate system the position of the incoming beam is set constant with its direction pointing along the $\{l_i\}$ system unit vector l_1. While K_0 is fixed, K moves on a circle during a $\theta/2\theta$ scan as does the substrate normal s_3. The figure displays the position during the scan when θ is at the Bragg angle of Al (111). In the case where the sample comprises a single Al crystal of (111) orientation three further reflections would equally be excited at the intersection of the $2\theta_{111}$ cone with the $\{l_i\}$ sphere. The reflections would be caused by the equally probable scattering of the incoming x-ray beam at lattice planes ($\bar{1}11$), ($1\bar{1}1$) and ($11\bar{1}$) that all exhibit the same interplanar spacing d_{111} to obey the Bragg equation. In the case of a polycrystalline sample being measured, however, the intensity would look totally different. Because of the random orientation of crystallites the intensity of all equivalent (111) planes would be equally distributed on a cone of opening angle 4θ rather than being concentrated in a few singular spots. The intensity would be smeared out over a ring shown as a grey line in Fig. 1.12. There are $m_{111} = 8$ equivalent (111) planes, but only $m_{200} = 6$ for (200) and it is evident that the multiplicity m_h will enter the expression of a Bragg reflection intensity as a scaling factor.

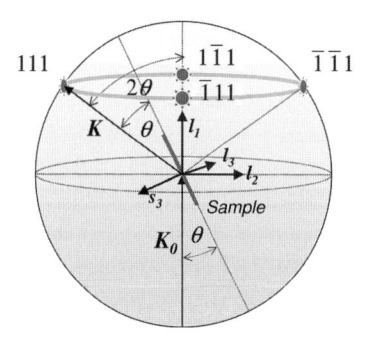

Figure 1.12 Scattering in the laboratory reference frame $\{l_i\}$ for a 111 reflection from an Al single crystal of [111] orientation and a polycrystalline Al powder sample.

1.3.4
Geometry Factor

The spreading of the Bragg peak over a circular segment of the $\{l_i\}$ sphere as discussed above introduces a further θ dependency into the diffraction pattern of a $\theta/2\theta$ scan. The effect is visualized in Fig. 1.13 where the set of all diffracted intensity for scattering angle 2θ is symbolized by a cone of opening angle 4θ. The circumferences of the intensity rings scale with $\sin 2\theta$ causing a dilution of intensity by $1/\sin 2\theta$. There also arise a variety of scattering vectors Q that lie on a cone. The scattered intensity will scale with their density, which is $\sin(\pi/2 - \theta) = \cos\theta$. The geometry factor is the product of both density functions and it is finally obtained as $G = \cos\theta/\sin 2\theta = 1/(2\sin\theta)$.

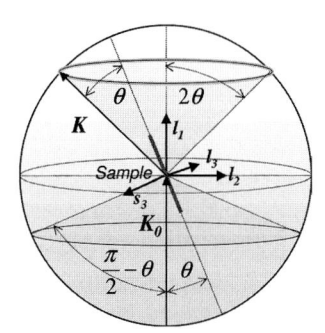

Figure 1.13 Scattering in the laboratory reference frame $\{l_i\}$ to derive the geometry factor G.

1.3.5
Preferred Orientation (Texture)

For a powder sample it may generally be assumed that all grain orientations occur with the same probability, i.e. that the distribution function of grain orientations is isotropic. It is a characteristic structural feature of thin polycrystalline films that certain crystallographic lattice planes can occur with a greater probability than others. This phenomenon is termed preferred orientation or texture. It is evident from Fig. 1.12 that a texture might have a significant influence on the diffraction pattern, where density-enhanced lattice planes will be associated with an increase of the corresponding Bragg reflection intensity I_h. The intensity then has to be scaled with the density of crystallite orientations that are indicated by the texture factors T_h. For a random orientation $T_h = 1$ holds for all of them. The measurement of texture and the determination of orientation distribution functions are outlined in detail in Chapter 5.

1.3.6
Polarization Factor

The x-ray radiation emitted from a laboratory x-ray tube is of random polarization. Therefore, the scattering by a polycrystalline sample has to be decomposed into a σ component and a π component. These considerations have already been outlined for the scattering by a single electron and they equally apply to the case considered here. In the case that $I_\sigma = I_\pi = I_0/2$ is valid on the average, the polarization factor takes the form

$$\overline{C^2} = \frac{1+\cos^2 2\theta}{2} \qquad (1.28)$$

and it is by this factor that the intensity received by the detector has to be scaled.

The geometry factor may be different for measurement configurations other than the $\theta/2\theta$ scan. The experimentalist should check this point carefully if integral intensities have to be analyzed quantitatively. The dependency of the geometry factor G and the polarization factor $\overline{C^2}$ are both shown in Fig. 1.14 as a function of scattering angle 2θ. Also the Lorentz factor L is shown that will be derived later and the product $G\overline{C^2}L$ of all three factors. The product function is seen to exhibit a pronounced minimum close to $2\theta = 120°$. Up to this point $G\overline{C^2}L$ continuously decreases, but recovers for high scattering angles close to $2\theta_{max} = 180°$.

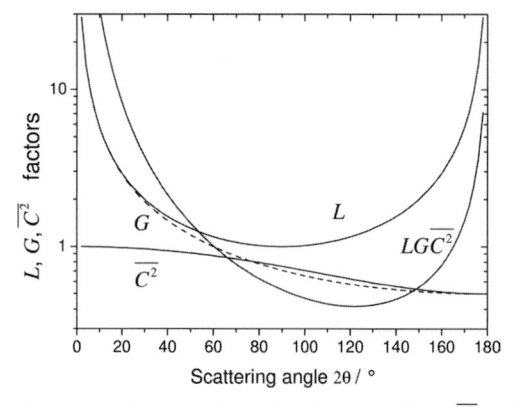

Figure 1.14 Geometry factor G, polarization factor $\overline{C^2}$ and Lorentz factor L as a function of 2θ. Also the product of the three factors is shown. Note the logarithmic ordinate scale.

1.3.7
Absorption Factor

During their transit through matter x-rays suffer from an attenuation of intensity caused by their absorption. The Lambert–Beer law, well known from optics, can describe the absorption effect. The intensity I_0 that enters into the sample will be exponentially damped to an amount $I_0\exp(-2\mu\ell)$ after a path of 2ℓ. The parameter μ

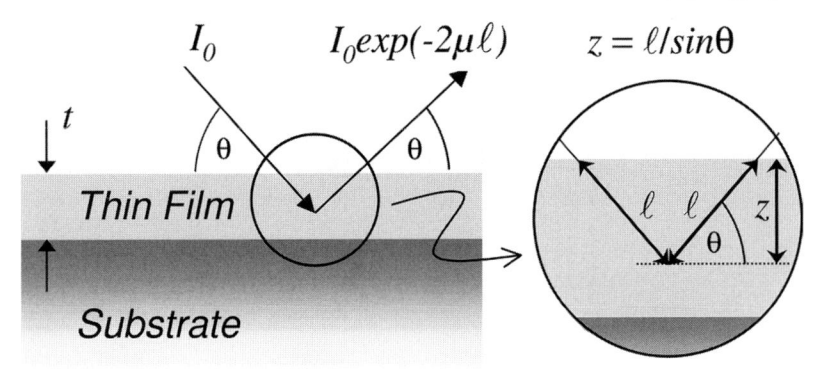

Figure 1.15 Schematic representation of the absorption effect for a thin-film sample in a $\theta/2\theta$ scan.

is named the linear attenuation coefficient and depends on the wavelength of the radiation used, the chemical composition of the sample and its density. The inverse of μ would give a penetration depth for normal incidence $\tau_{1/e} = 1/\mu$ that specifies the path length for which the intensity I_0 drops to $1/e$ of its initial value. The dimensions of the attenuation coefficient are m^{-1} or µm^{-1}. Often, the value of the mass absorption coefficient μ_m is listed in various tables that can be converted into $\mu = \rho\, \mu_m$ by multiplication with the mass density ρ. For many substances attenuation coefficients of the order of 10^5 to 10^7 m^{-1} are obtained for Cu Kα or comparable wavelengths. This corresponds to penetration depths $\tau_{1/e}$ of 0.1 to 10 µm and thus is in the range of a typical layer thickness. It can be concluded that absorption effects might significantly affect Bragg reflections of thin films.

The dominant effect the absorption factor has on a diffraction pattern is the variation of the scattered intensity. Its derivation is shown in Fig. 1.15 for the case of a $\theta/2\theta$ scan. For any x-ray beam that has traveled through a sample to become scattered into the detector the primary intensity has been reduced by the factor $\exp(-2\mu\ell)$. The reduction of intensity of the total x-ray beam is the sum over all possible paths of the beam within the limits of 0 to ℓ_{max}

$$\int_0^{\ell_{max}} \exp(-2\mu\ell)d\ell \qquad (1.29)$$

The path 2ℓ that is traversed by the x-ray beam may be expressed by the depth variable z for which $\ell = z/\sin\theta$ holds. Then ℓ is substituted by $z/\sin\theta$, $d\ell$ by $dz/\sin\theta$ and the integration is performed from 0 up to the thickness t of the film. Here, $z = 0$ accounts for the surface of the film and $z = t$ for the film–substrate interface. The solution of the integral yields

$$\frac{1}{2\mu}\left\{1 - \exp\left(-\frac{2\mu t}{\sin\theta}\right)\right\} \qquad (1.30)$$

In the limit of an infinitely thick sample, $t \to \infty$ which is equivalent to $t \gg 1/\mu$, the result $1/(2\mu)$ is obtained. In the following the absorption factor is denoted by the ratio of the absorption for a sample of finite thickness with respect to an infinitely thick sample

$$A = \int_0^t \Big/ \int_0^\infty \quad (1.31)$$

The application of this procedure results in the absorption factor for the $\theta/2\theta$ configuration

$$A_{\theta 2\theta} = \left(1 - \exp\left(-\frac{2\mu t}{\sin\theta}\right)\right) \quad (1.32)$$

The subscript $\theta/2\theta$ has been added in order to indicate the measurement geometry. We will become acquainted with various A factors in the following chapters for different diffractometer configurations. The A factor is also termed the thickness factor and it is seen to cause the measured intensity to cease as a function of increasing scattering angle 2θ. In Fig. 1.16 the $A_{\theta 2\theta}$ factor is displayed for thin Al and Nb films of 500 nm and 1 μm thickness measured with Cu Kα radiation, where use has been made of the material parameters

Al: μ_m = 486.7 m² kg⁻¹, ρ = 2700 kg m⁻³ ⇒ μ = 1.31 × 10⁶ m⁻¹

Nb: μ_m = 1492 m² kg⁻¹, ρ = 8550 kg m⁻³ ⇒ μ = 1.276 × 10⁷ m⁻¹

The general decrease of $A_{\theta 2\theta}$ can easy be understood from the fact that the x-ray beam enters the sample under increasingly steeper angles and accordingly "sees" less sample volume for increasing 2θ. It is seen from the plot that $A_{\theta 2\theta}$ approaches unity the thicker the film or the larger is μ. Accordingly, the absorption factor depends on the product of both, which is called the μt product in the following. It often happens in thin-film work that t and $1/\mu$ are of the same order of magnitude and $A_{\theta 2\theta}$ has to be considered for a proper interpretation of the pattern. The μt product turns out to be a decisive quantity in the x-ray investigations of thin films.

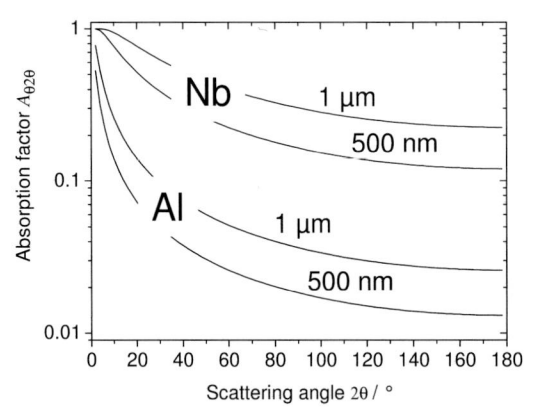

Figure 1.16 Absorption factor A for 500 nm and 1 μm thin Al and Nb films as function of 2θ.

1.3.8
Integration of the Interference Function

The central question of this section is now tackled regarding the integral intensity of a Bragg peak in a $\theta/2\theta$ scan. What we are interested in is the energy E that is received by the surface dS of the detector in the time element dt'

$$dE = I(\mathbf{R})dSdt' \tag{1.33}$$

The geometry of the quantities involved is shown in Fig. 1.17. If all the factors outlined above, i.e. square of the structure factor $|F_h|^2$, multiplicity m, texture factor T_h, geometry factor G, polarization factor $\overline{C^2}$ and absorption factor $A_{\theta 2\theta}$, are included into the intensity formula, the integration has to be done for

$$I(\mathbf{R})dSdt' = I_0 \frac{r_e^2}{R^2}|F_h|^2 m_h T_h \overline{GC^2} \frac{A_{\theta 2\theta}(t)}{2\mu} \prod_{i=1}^{3} \frac{\sin^2(N_1 a Q c_i / 2)}{(aQc_i/2)^2} dSdt' \tag{1.34}$$

This expression is rather complex, but its main problem is related to the fact that the variables t' and S over which the integration has to be performed are not realized at first sight in the integrand. Therefore, the differential term has to be transformed to quantities over which the integration might be performed. For this purpose we aim at transforming dS and dt' to quantities that allow one to integrate over (aQc_i) terms of the interference function. It is now assumed that the intensity of one single crystallite shall be estimated, while the corrections for the polycrystallinity of the sample shall be considered by multiplying by the factors given above. Without restricting the general validity of the derivation the crystallite is assumed to be (001) oriented and the reflection under consideration is of 00l type (see Fig. 1.17).

An important constraint in any x-ray scattering experiment is that the probing beam and the detected beam unavoidably exhibit a finite divergence δ. A divergence-free beam is a theoretical idealization that does not exist. Consequently, if the x-ray source and the detector are both at Bragg position θ_B one could still trace incoming and exiting beams that are off the diffraction position. The off-angle beams

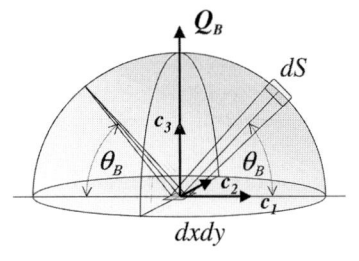

Figure 1.17 Geometrical quantities involved in the integration of the interference function.

exhibit diffraction angles of $\theta_B + \Delta\theta$. This also means in terms of the scattering vector that besides the ideal vector Q_B other Q vectors with $Q = Q_B + \Delta Q$ would be operative at the Bragg position.

It is outlined in Instrumental Box 2 that a $\theta/2\theta$ scan may be carried out by rotating the sample by θ and the detector by 2θ. The detector thus rotates with a circular velocity of $d(\Delta\theta)/dt' = \dot{\theta}$ with respect to the sample which allows the first substitution to be made, namely $dt' = d(\Delta\theta)/\dot{\theta}$ and we obtain

$$dE = I(R)\frac{1}{\dot{\theta}}dSd(\Delta\theta) \tag{1.35}$$

The surface dS is proportional to the square of the sample–detector distance R. A small area $dxdy$ is illuminated on the sample surface by the incoming x-ray beam. This infinitesimally small area corresponds by

$$dS = R^2 \sin\theta dxdy \tag{1.36}$$

with the receiving area of the detector. The $\sin\theta$ term is simply caused by the tilt of the sample area when it is portrayed on the detector window. The illuminated area of the sample surface $dxdy$ depends on the divergence of the beam or on how strong ΔQ deviates from Q_B. The infinitesimal area element can thus be expressed by the scalar product of Q with the crystal lattice vectors c_1 and c_2 giving

$$dxdy = \left(\frac{\lambda}{4\pi\sin\theta}\right)^2 d(Qc_1)d(Qc_2) \tag{1.37}$$

The prefactor has to be introduced to normalize with respect to the length of Q. The surface detector element can thus be expressed by the sample surface element according to

$$dS = \frac{R^2\lambda^2}{4a^2 4\pi^2 \sin\theta} d(aQc_1)d(aQc_2) \tag{1.38}$$

It should be noted that Qc_1 equals ΔQc_1, because the product of Q_B with c_1 – and also c_2 – yields zero under the assumed geometry. The integration over $d(aQc_1)d(aQc_2)$ is thus an integration in the vicinity of the Bragg reflection.

The transformation from $d(\Delta\theta)$ to $d(aQc_3)$ can be obtained by the derivative of the latter quantity with respect to the first. For this purpose the decomposition $\theta = \theta_B + \Delta\theta$ is applied

$$\frac{d(Qc_3)}{d(\Delta\theta)} = \frac{d}{d(\Delta\theta)}\frac{4\pi\sin(\theta_B + \Delta\theta)}{\lambda} = \frac{4\pi\cos\theta}{\lambda} \tag{1.39}$$

Finally, the differentials are adopted to the arguments in the denominator and one arrives at

$$\frac{dS}{R^2}d(\Delta\theta) = \frac{\lambda^3}{4a^3(2\pi)^3 \sin 2\theta}\prod_{i=1}^{3}d(aQc_i) \tag{1.40}$$

1.3 Intensity of Bragg Reflections

Instrumental Box 2:

Generation of X-rays by X-ray Tubes

X-rays are generated when electrons with kinetic energies in the keV range and above impinge on matter. The emission spectrum comprises a continuous part, called Bremsstrahlung, and some discrete lines indicative of the chemical elements of the target material. In laboratory x-ray tubes electrons are accelerated onto an anode plate made from a specific metal of high purity. The cross-section of such a tube is shown in Fig. i1.1. The electrons are emitted from the cathode filament and accelerated towards the anode plate. The anode is typically fabricated from copper, chromium, molybdenum or another metal as listed in Table i1.1. The electron current between filament and anode may be adjusted by tuning the filament current in the range of some 10 mA. When impinging upon the anode the electrons are decelerated by their interaction with the target plate atoms leading to the emission of x-rays. The acceleration voltage (in kV) must be greater than the energy of the characteristic radiation required by the experiment (in keV). Energy E and wavelength λ of the radiation are related by

$$E \text{ [keV]} = \frac{1.2398}{\lambda \text{ [nm]}} \tag{i2.1}$$

As a rule of a thumb line energies may be converted to wavelengths by relating them to the set point 11.1 keV ≈ 0.111 nm. This relation is precise to 0.6% and may easier be remembered than Eq. (i2.1). Often the anode is made of copper to make use of the characteristic Cu Kα line. The Cu Kα wavelength is 0.154 nm which is adequate to 8.04 keV. Other x-ray tubes may be used if the diffraction pattern has to be contracted or expanded or if the excitation of x-ray fluorescence from the sample is to be avoided. A prominent example for the latter effect is given by Fe-containing samples that cause a strong background when Cu Kα radiation is applied.

Table i2.1 Metals used as anodes of x-ray tubes with characteristic wavelengths λ, typical Kβ filter materials and their appropriate K-edge wavelengths λ_K.[a]

Target element	Z	λ (nm)				Kβ filter	λ_K (nm)
		Kα_1	Kα_2	K$\bar{\alpha}$	Kβ_1		
Cr	24	0.228975	0.229365	0.22909	0.20849	V	0.2269
Fe	26	0.193631	0.194002	0.19373	0.17567	Mn	0.1896
Co	27	0.178900	0.179289	0.17905	0.16208	Fe	0.1744
Ni	28	0.165794	0.166178	0.16591	0.15002	Co	0.1608
Cu	29	0.154059	0.154441	0.15418	0.139225	Ni	0.1488
Mo	42	0.709317	0.713607	0.7107	0.63230	Zr	0.6889
Ag	47	0.559422	0.563813	0.5608	0.49708	Rh	0.5339

a From various tables of the *International Tables for Crystallography*, Vol. C, Section 4.2.2.

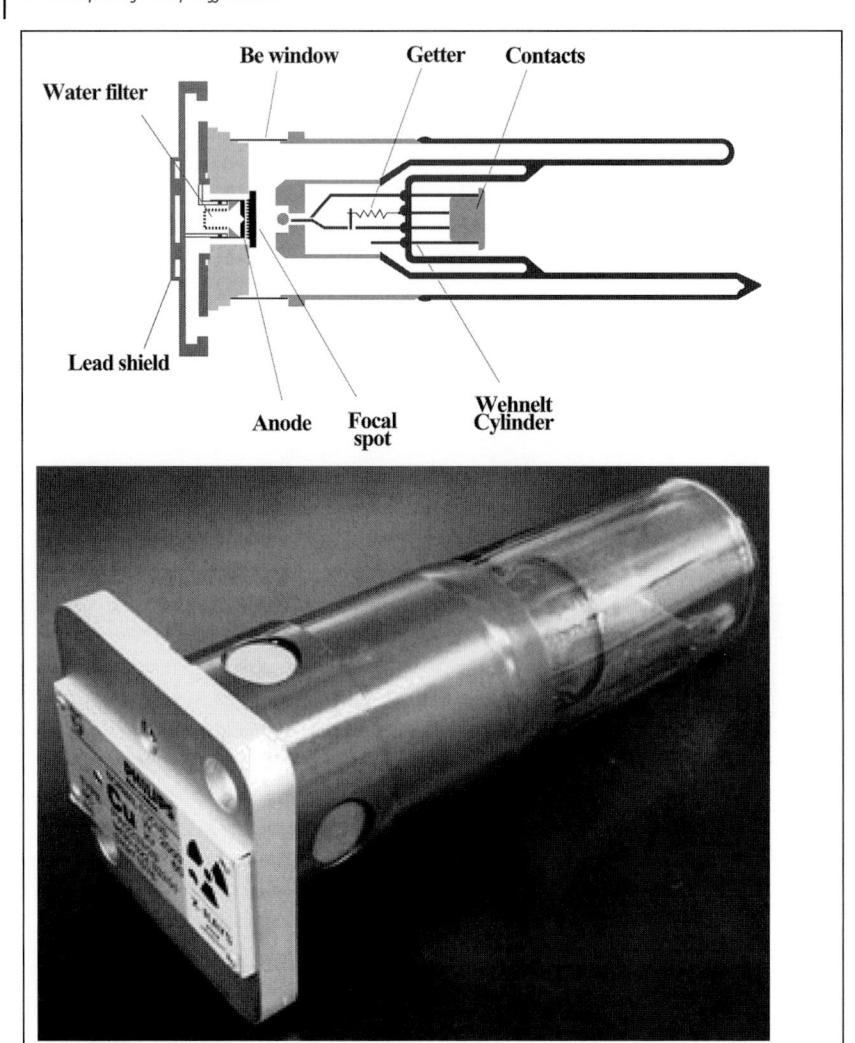

Figure i2.1 Schematics and photographs of laboratory x-ray tubes. This page: conventional x-ray tube type. Next page: ceramic x-ray tube for less laborious change from line to point focus mode and vice versa. (Kindly provided by PANalytical.)

The process of impact ionization and relaxation of an x-ray emitting atom is shown schematically in Fig. i2.2. In this sequence of processes, bound electrons are first released and subsequently an electron from a higher energy level relaxes into the emptied state. The relaxation is associated with the emission of radiation the energy of which corresponds to the energy difference between the final and initial state. This energy thus sensitively depends upon the chemical nature of the emitting atom. According to the levels involved the relaxation of the emitted radiation is denoted by $K\alpha_1$, $K\beta_2$, $L\alpha$, etc. In most x-ray scattering experiments only the characteristic emission lines are of interest.

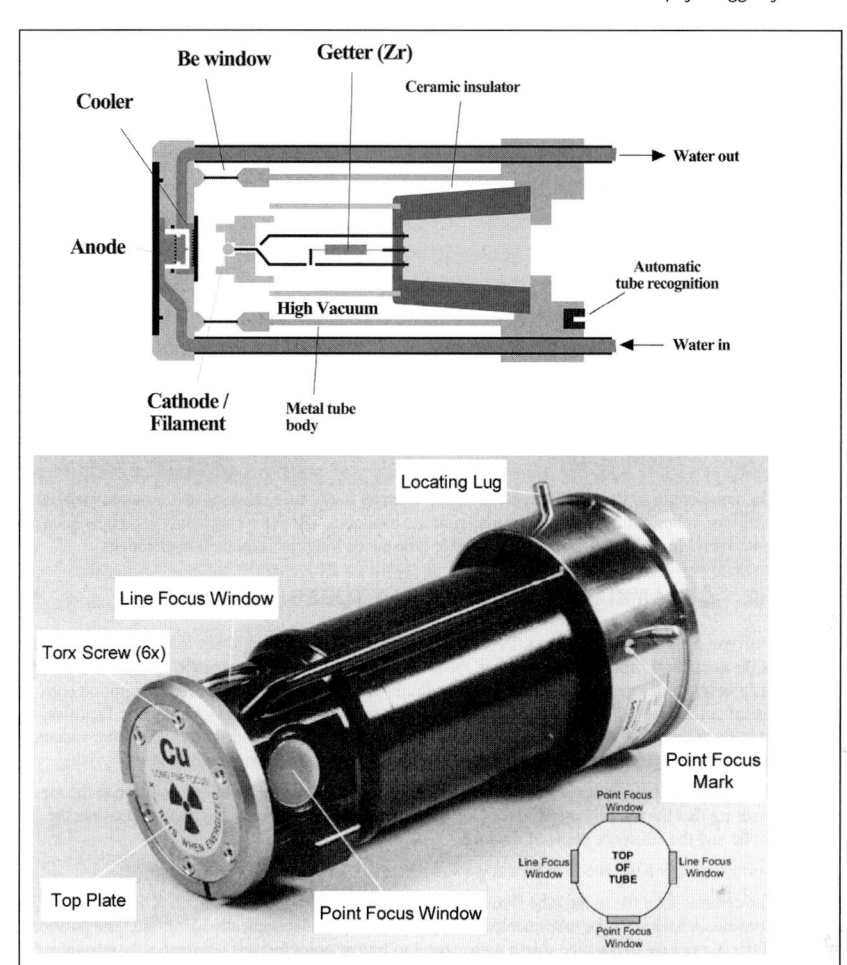

Figure i2.1 continued

In the majority of cases investigations are performed with Kα radiation, because of its higher intensity compared to Kβ. A closer look at the Kα radiation reveals its doublet nature. The doublet resolution increases for Bragg peaks with increasing scattering angle 2θ. Moreover, the occurrence of the splitting strongly depends on the microstructure of the sample, and in many polycrystalline thin-film investigations the α_1–α_2 splitting may not be observed. In such cases the weighted average of both lines is used denoted by K$\bar{\alpha}$ or simply Kα. The Kβ line may severely disturb the interpretation of the diffraction pattern and various techniques are applied for its suppression. Very common are edge filters that are introduced into the incident beam path and make use of the fact that the photoionization of deep levels requires a certain minimum energy. In the case of copper, $E(\text{Cu K}\beta_1) = 8.9$ keV and thus a thin Ni foil can be used as an edge filter that significantly

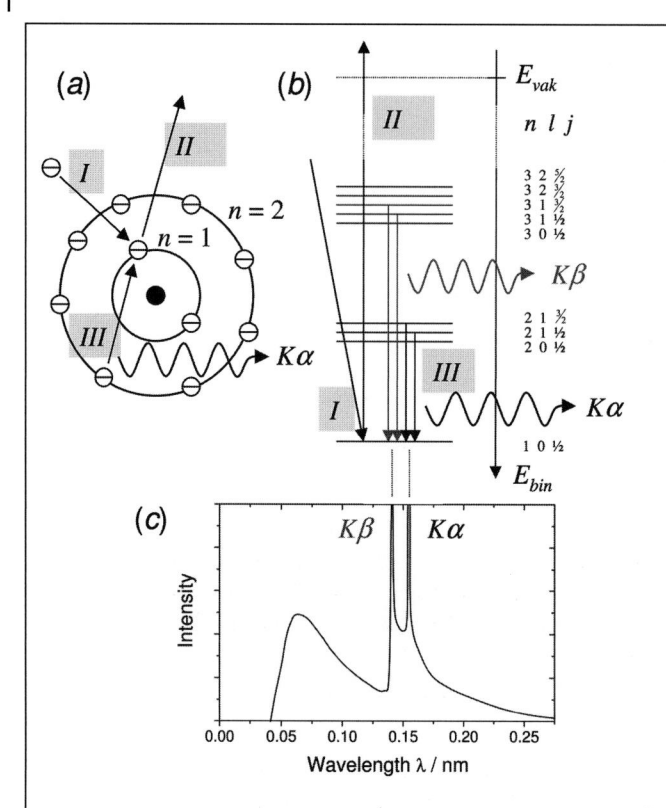

Figure i2.2 (a) Emission of x-rays by an atom through sequential processes of electron impact (I), ionization (II) and simultaneous electronic relaxation and emission of radiation (III). (b) Elementary processes visualized in the term scheme, where only transitions that obey the selection rules $\Delta l = \pm 1$ and $\Delta j = 0, \pm 1$ are shown. (c) Schematic x-ray spectrum emitted by an anode. It is composed of a continuous Bremsstrahlung background and discrete single lines.

suppresses the Kβ line (see Exercise 1.14). Other foil materials for other tubes are specified in Table i2.1.

The filament and anode operate in a high-vacuum environment of the sealed tube. The decoupling of radiation is facilitated via one of the four Be windows (300 μm thickness) at each side of the tube. The dimensions of the focal-spot size on the anode plate range between (0.4 × 10) mm and (2 × 12) mm. The anode focus is "seen" by the sample under a take-off angle between 2° and 6°. The focus either appears as a line or a spot according to which window is used for the decoupling. The x-ray tube may thus be operated in line focus mode or point focus mode depending on the experimental necessities. Remember that the point focus is not a point focus, but a line focus viewed from the side (see Fig. i.1.3). Due to the high power dissipation at the focal spot the back of the anode plate is water cooled.

1.3 Intensity of Bragg Reflections

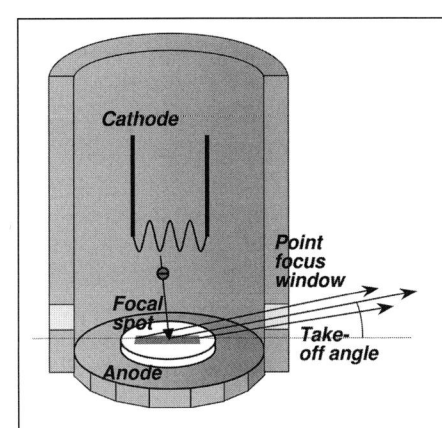

Figure i2.3 Schematic of x-ray tube head to visualize point focus operation.

In addition to the sealed x-ray tubes discussed above, so-called rotating anode x-ray generating systems are also in use. These make use of an improved dissipation of the heat introduced by impinging electron current on the anode by rotating it. Higher emission rates of x-rays by a factor of about 3–5 may be obtained. Orders of magnitude higher radiation intensities are available at synchrotron beamlines.

The factor in front of the differential product is called the Lorentz factor L. Frequently, only the term $L = 1/\sin 2\theta$ is denoted as the Lorentz factor. Its effect on the course of intensity in conjunction with the geometry factor G and the polarization factor $\overline{C}^2$ is depicted in Fig. 1.14.

Inserting the transformation Eq. (1.40) into the differential form for the energy received by the detector

$$I(R)dSd(\Delta\theta) = I_0 \frac{r_e^2}{R^2 \overline{\theta}}|F_h|^2 m_h T_h \overline{GC^2} \frac{A_{\theta 2\theta}(t)}{2\mu} \prod_{i=1}^{3} \frac{\sin^2(N_i a Q c_i / 2)}{(a Q c_i / 2)^2} dSd(\Delta\theta) \qquad (1.41)$$

the integration may be carried out and is found to yield

$$\iiint \prod_{i=1}^{3} \frac{\sin^2(N_i a Q c_i / 2)}{(a Q c_i / 2)^2} d(a Q c_i / 2) = N_1 N_2 N_3 \qquad (1.42)$$

The three integers N_i give the length of the crystallite in units of cell edges and are related to the volume of the sample V and the unit cell V_{uc} by

$$N_1 N_2 N_3 = \frac{V}{a^3} = \frac{V}{V_{uc}} \qquad (1.43)$$

If the substitution of V_{uc} for a^3 is also performed for the Lorentz factor and the other results from above are applied we finally end up with the expression for the integral intensity of a Bragg reflection in a $\theta/2\theta$ scan

$$I_h = I_0 \frac{r_e^2}{\theta} \lambda^3 m_h T_h \frac{1+\cos^2\theta}{\sin\theta \sin 2\theta} \frac{|F_h|^2}{V_{uc}^2} V \frac{A_{\theta 2\theta}(t)}{2\mu} \qquad (1.44)$$

Various points concerning the integral intensity Eq. (1.44) are of interest. Firstly, it is realized that the intensity is proportional to $|F_h|^2/V_{uc}^2$. This ratio has the character of a squared electron density and underlines again the statement made above that x-ray diffraction patterns scale with the square electron density ρ_e^2 of the sample. Secondly, there are various factors active that may cause a severe reduction of intensity with increasing scattering angle 2θ. The ratio of trigonometric functions occurring in Eq. (1.44) has already been plotted in Fig. 1.14 and can be seen to reduce the integral intensity of a reflection by an order of magnitude, depending on where it occurs. Also the structure factor ceases strongly as a function of 2θ and even more so for the square of it. A special feature in thin-film analysis is the damping caused by the absorption factor $A_{\theta 2\theta}(t)$ which also increases with increasing scattering angle. The different factors make comprehensible the features of the thin Al film diffraction pattern shown in Fig. 1.10.

In the majority of investigations the integral intensity is specified by virtue of a scaling factor SCF that lumps together all instrumental settings like scan velocity $\dot\theta$, slit width, etc., and allows the comparison of integral intensities from one diffraction pattern on a relative scale

$$I_h = \text{SCF} \cdot m_h \cdot T_h \cdot Lp \cdot |F_h|^2 \cdot A_{\theta 2\theta}(t) \qquad (1.45)$$

The convention is also often used to abbreviate the product of Lorentz, geometry and polarization factor by $Lp = G\overline{C}^2 L$ and denoting it as the Lorentz-polarization factor.

In practice, it may be a reliable rule of a thumb in thin-film work to extend the $\theta/2\theta$ scan to $2\theta_{max} = 60°$ to $80°$ when Cu Kα radiation is used while higher 2θ angles are associated with too low a signal-to-noise ratio. The use of Cu Kα radiation with $\lambda = 0.154$ nm is widespread for the investigation of inorganic layers because it portrays interplanar spacings d from 0.355 to 0.154 nm into a 2θ range of 25–60°. However, the rule should only be taken as a recommendation and not as dogma. In any case, the analyst should check the range of scattering angles useful for the material and the analytical question under investigation. As a first step it might be helpful to calculate the strength of the μt product and hence the course of $A_{\theta 2\theta}(t)$ if the thin film thickness and composition – or reasonable estimates – are available.

An important point regarding the integral intensity of an x-ray reflection has to be mentioned. If the measurement is not made with a polycrystalline thin film, but with a single crystalline material a modification of Eq. (1.44) might become necessary. In highly perfect single crystals it may happen that the incoming beam is diffracted twice or even multiple times, which can effectively cause an attenuation of the reflection much stronger than accounted for by the absorption factor given above. The calculation of the diffracted intensity then has to take into account all wave fields that propagate in the crystal. The description of these phenomena is facilitated within the framework of the dynamical theory of x-ray diffraction, while the considerations outlined above are denoted as kinematical theory. For many applications the results of the kinematical theory as condensed in Eq. (1.44) are sufficient to explain the diffraction pattern of thin films – at least if polycrystalline layers are concerned. Solely in high-resolution work with epitaxial layers the dynamical theory sometimes has to be applied as is outlined in Chapter 8.

1.4
Applications

Metals have found applications in thin-film technology due to their luster and high electrical conductivity. Both properties are caused by the high mobility of electronic charge carriers in the crystal lattice. Decorative and anticorrosion coatings of chromium, zinc and derivative alloys for armatures, metal work, kitchen fittings, automotive parts, etc., are mostly deposited by electroplating. The technique of gilding of jewelry, porcelain, relics, etc., by gold leaf of only micrometers in thickness has been continuously developed for more than 4000 years. Typical modern applications of thin metal films are Al layers of less than 100 nm thickness used as barrier coatings on food packaging and as reflective coatings in light housings for automobile headlights to mention only a few examples.

Laterally structured thin metal films also play an enormous role in the semiconductor industry as electrical contacts between micrometer- and even submicrometer-sized devices. For this purpose, metals are deposited by electroplating, chemical or physical vapor deposition and subsequently structured by lithographic procedures to yield conducting plain contacts and metal level connecting vias. Al plays a vital role in these applications, but a severe problem that has to be addressed for metallic interconnects is electromigration limiting the mean time to failure of integrated circuits. Due to its higher electrical conductivity, Cu has recently partly replaced Al, as depicted in Fig. 1.18 where the highly complex three-dimensional stacking of Cu layers is shown [2]. Cu films deposited by electroplating were found to exhibit a remarkable microstructural evolution after plating [3]. Among the microstructural properties of interest in Cu interconnects are the texture and grain size [4] that can both affect the electromigration resistance. It should be mentioned, however, that the introduction of Cu in the Si industry has been avoided for a long time, because Cu is an extremely fast diffuser in Si and may severely degrade integrated circuit performance by acting as a deep-level recombination center. The diffusion of Cu into the semiconductor has to be safely prohibited by introducing anti-diffusion barriers, the functionality of which has been effectively investigated using x-ray diffraction procedures enabling their optimization [5].

Thin metal films are also of interest because of their mechanical properties and various basic investigations have been performed with x-ray diffraction techniques due to their model character for thin-film systems in general. Refractory Nb films, for instance, are qualified by the high yield strength of Nb for engineering applications and were found to exhibit pronounced depth dependencies during growth

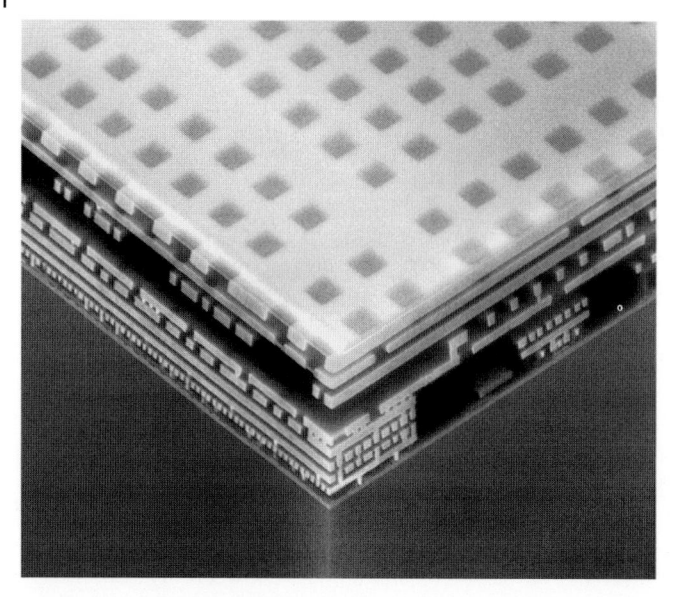

Figure 1.18 Thin film metallic interconnect structure as used in a modern microelectronic chip. The multilayer system exhibits a total thickness in the μm range (figure kindly provided by AMD, Dresden [2]).

[6]. The same phenomenon was identified for titanium films that are applied as adhesion layers for hard coating applications [7]. Increasing attention has been paid in recent years to the interesting plastic properties of nanocrystalline metals. For instance, plastic deformations in electroplated nanocrystalline Ni films could be revealed by *in situ* investigations to be reversible upon unloading instead of producing a residual dislocation network [8]. X-ray scattering investigations of thin metal films are used extensively to investigate grain sizes, crystalline lattice faults, texture and residual stress. The various x-ray scattering techniques to elucidate these microstructural features are outlined in the following chapters.

Exercises

1.1 The determination of the classical radius of the electron proceeds by assuming that the total electric field energy $\int \mathbf{D} \cdot \mathbf{E} dV$ is confined to the space surrounding the spherical particle of radius r_e. Show that $r_e = e^2/(4\pi\varepsilon_0 mc^2)$ is obtained when the total field energy is identified with the energy of the rest mass mc^2.

1.2 Consider a preferably plane transparent thin film on a reflecting substrate (like a Si wafer). Examine it in daylight and in a room illuminated by neon light. Which optical phenomena can be observed in both cases? Why is the optical length of coherence longer in one case than in the other? How does the coherence length depends on the frequency spectrum of the radiation?

1.3 What is the momentum of an 8 keV x-ray photon? Compare this to the average momentum of an Ar gas atom confined to a vessel under thermodynamic standard conditions.

1.4 Give the general expression for $1/d_{hkl}^2$ (Table 1.1) for the triclinic crystal by starting from $1/d_{hkl}^2 = (h\mathbf{b}_1 + k\mathbf{b}_2 + l\mathbf{b}_3)^2$, with the reciprocal lattice vectors $\mathbf{b}_i$ depending like $\mathbf{b}_i = (\mathbf{a}_j \times \mathbf{a}_k)/(\mathbf{a}_i (\mathbf{a}_j \times \mathbf{a}_k))$ on the unit cell vectors $\mathbf{a}_j$.

1.5 Show that the summation over all scattering centers according to Eq. (1.9) indeed yields Eq. (1.10). Make use of the geometrical sum.

1.6 Derive the atomic form factor of He by modeling it with the spherical charge distribution of two electrons $\rho(r) = (2e/\pi)(2/a_0^3)\exp(-4r/a_0)$ and performing the integration (1.17). Find a numerical approximation in the form of eq. (1.18) through a non-linear regression. What is the precision of the fit in the $\sin\theta/\lambda$ range from 0 to 10 nm^{-1} when the numerical value for Bohr's radius $a_0 = 0.0529$ nm is inserted?

1.7 Derive the structure factor F_h and extinction conditions of the three lattice types bcc, fcc, hcp from the general expression (1.23). Show that the evaluation of an expression comparable to Eq. (1.9) for the bcc lattice yields an equivalent result.

1.8 What are the multiplicities m_h of Bragg peaks of the thin Al film shown in Fig. 1.10?

1.9 Suppose a Bragg reflection in a $\theta/2\theta$ pattern to be observed at $2\theta_1$ when Cu Kα_1 monochromatized radiation is applied. What would be the appropriate position of $2\theta_2$ for the Cu Kα_2 line? In absolute values for $2\theta_1 = 32°$?

1.10 A diffractometer is run without suppression of Kβ radiation. Where do the β peaks occur in the diffraction pattern compared to the α peaks? General: are the interference lines of higher energy radiation shifted towards lower or higher scattering angles 2θ?

1.11 Calculate the c/a ratios for the hcp-structured metals given in Table s1.1. How do they deviate from the "ideal" value of 1.633 and what is the meaning for the two different interatomic distances d_1 and d_2 in the hcp structure?

1.12 Determine the distance d of an atom in the fcc and bcc structure to its nearest neighbors as a function of the lattice parameter a. Do the same for the hcp structure for interatomic distances within the basal plane and in the out-of-plane direction. Assume that atoms may be modeled by spheres and calculate the volume ratio of space occupied by spherical atoms V_{at} and the complete unit cell volume V_{uc} for all three structure types.

1.13 Calculate the linear attenuation coefficient μ and the penetration depth $\tau_{1/e}$ for Be, Zn, Fe, Ag and Au for irradiation with Cu Kα radiation (8.04 keV). Determine the density and mass attenuation coefficient of the materials from tables like Table 2.1 or from the internet.

1.14 The mass absorption coefficients for Cu Kα and Cu Kβ_1 radiation in Ni are 48.8 and 279 cm^2 g^{-1} while the mass density of Ni is 8.9 g cm^3. Calculate the intensity ratios $I(K\alpha)/I(K\beta_1)$ and $I(K\alpha)/I_0(K\alpha)$ after Cu K radiation with an initial ratio $I_0(K\alpha)/I_0(K\beta_1)$ of 3:1 has been transmitted through a Ni foil of (a) 8, (b) 15 and (c) 20 µm thickness. Which of these foils would you consider as the optimum Kβ filter?

1.15 In cubic substances two group of lattice planes may scatter simultaneously for the same scattering angle 2θ, because $d_{hkl} = \sqrt{h^2+k^2+l^2}/a$ is not a one-to-one unique function of Miller indices hkl. Which are the first double reflections to occur in a $\theta/2\theta$ scan? Give the first three of them for fcc and bcc structures.

1.16 An x-ray copper tube is operated with 40 kV and 40 mA and is assumed to have efficiency for x-ray generation of 1%. X-rays are emitted with cos^2 characteristic and the take-off angle is 6°. The divergence slit is 1°. By how many x-ray photons is the sample irradiated per second? How many Cu Kα photons? Give order of magnitude estimations.

1.17 What is the varied measurement quantity in all kinds of spectrometers and why is it inadequate to denote a diffractometer as a spectrometer? The same arguments are valid in distinguishing a diffraction pattern from a spectrum.

1.18 Derive the dependence of the radius of the focusing circle R_{FC} on the diffraction angle θ and goniometer radius R.

1.19 The element Polonium has only found few technical applications, since its isotopes are all unstable with the most long-living ones decaying under emission of α particles. The most relevant application of Po is as initiator in nuclear fission bombs. Devices based on uranium, for instance, operate by the gun principle and comprise two undercritical U masses that are shot into each other for ignition. The chain reaction of fission processes is reported to be accelerated by covering one of the masses with Po and the other by Be. The α particles emitted from Po are then captured by Be nuclei leading to higher Be isotopes that act as neutron emitters accelerating the U fission reaction. About 2×10^4 nuclear weapons with the explosive power of more than 10^6 Hiroshima bombs are still stored in the arsenals of the atomic powers although they pledged in the Non-Proliferation Treaty of 1970 "to pursue negotiations in good faith on effective measures relating to cessation of the nuclear arms race at an early date and to nuclear disarmament, and on a Treaty on general and complete disarmament under strict and effective international control" (Article VI). Various organizations exists, in which scientists are acting for disarmament and the abolition of nuclear weapons like the International Network of Scientists and Engineers for Global Responsibility (www.inesglobal.com), the Pugwash movement (pugwash.org), the Union of Concerned Scientists (ucsusa.org) and others. Visit the internet sites of these organizations and evaluate what scientists can do to assure the humane and useful application of their work.

References

Monographs relevant to this chapter

B. E. Warren, *X-Ray Diffraction,* Addison-Wesley, Reading, MA, 1969.
B. D. Cullity, *Elements of X-Ray Diffraction,* Addison-Wesley, Reading, MA, 1967.

Special papers

[1] E. N. Maslen, A. G. Fox, M. A. O'Keefe, X-ray Scattering, in: E. Prince (Ed), *International Tables for Crystallography,* Vol. C (Kluwer Academic, Dordrecht, 2004), p. 554.
[2] M. A. Meyer & E. Zschech, AMD, Dresden (2004).
[3] J. M. E. Harper, C. Cabral, P. C. Andricacos, L. Gignac, I. C. Noyan, K. P. Rodbell, C. K. Hu, Mechanism for microstructure evolution in electroplated copper thin films near room temperature, *J. Appl. Phys.* **86** (1999) 2516.
[4] C. Lingk, M. E. Gross, W. L. Brown, X-ray diffraction pole figure evidence for (111) sidewall texture of electroplated Cu in submicron damascene trenches, *Appl. Phys. Lett.* **74** (1999) 682.
[5] Y. S. Gong, J.-C. Lin, C. Lee, Interdiffusion and reactions in the Cu/TiN/Si thin film system, *Appl. Surf. Sci.* **92** (1995) 335.
[6] B. Okolo, P. Lamparter, U. Welzel, E. J. Mittemeijer, Stress, texture, and microstructure in niobium thin films sputter deposited onto amorphous substrates, *J. Appl. Phys.* **95** (2004) 466.
[7] M. Birkholz, C. Genzel, T. Jung, X-ray diffraction study on residual stress and preferred orientation in thin titanium films subjected to a high ion flux during deposition, *J. Appl. Phys.* **96** (2004) 7202.
[8] Z. Budrovic, H. Van Swygenhoven, P. M. Derlet, S. Van Petegem, B. Schmitt, Plastic Deformation with Reversible Peak Broadening in Nanocrystalline Nickel, *Science* **304** (2004) 273.

2
Identification of Chemical Phases

A typical task in thin-film analysis is the identification of chemical phases. In many thin-film projects a materials system is to be developed and the prepared layers have to be checked as to whether an intended chemical phase has successfully been prepared. Information on phase composition is also needed for the deposition process in order to optimize the process conditions. X-ray diffraction yields different diffraction patterns for different chemical phases and this chemical phase selectivity makes it well suited for such an analytical task. It might be conjectured that the majority of x-ray diffraction investigations of thin solid films are devoted to this type of analysis.

This chapter outlines how a symmetric $\theta/2\theta$ scan has to be analyzed to identify the chemical phases present in a sample. The x-ray attenuation coefficient is considered in more detail because of its close relation to the chemical nature of the thin film. In addition, the attenuation coefficient and the absorption coefficient that derive from it can be used to quantify the layer thickness when a substrate reflection or harmonic reflections occur in the pattern. Also the question of phase mixtures and amorphous phases is addressed. The composition of alloy compounds or stoichiometry deviations may be tackled by Vegard's rule, and, therefore, the obstacles to determining accurate lattice parameters are considered. Since these investigations are all based on a symmetric $\theta/2\theta$ scan, further information on beam conditioning in the $\theta/2\theta$ diffractometer is given in Instrumental Box 3.

The illustrative examples of this chapter are related to polycrystalline and amorphous semiconductors composed from either group IV elements or II–VI compounds. Structure Boxes 2 and 3 are thus devoted to the diamond structure and the sphalerite and wurtzite structure.

2.1
Histogram-Based Techniques

Crystallographic structures of all chemical elements and an enormous number of organic and inorganic solids have been investigated by x-ray diffraction procedures for about 100 years. Such a huge number of crystallographic structures have already been solved that nowadays the thin-film analyst will rarely be confronted with

Instrumental Box 3:

Divergent Beam Optics

The shaping of the probing beam and the focusing on the sample is more difficult to perform in an x-ray experiment than in conventional optics. The difficulties lie in the small deviation of the x-ray refractive index n from unity or, equivalently, in the small difference between n and the vacuum refractive index n_0 (see Chapter 4). Beam-shaping elements thus have to rely on physical principles other than refraction and became available for laboratory diffractometers only recently. Therefore, the manipulation of the x-ray beam in a $\theta/2\theta$ diffractometer is essentially performed by virtue of "shadowing optics". For this purpose, undesired beam parts are simply cut off by apertures, slits and masks and are thus not received by the detector (see Fig. i3.1 for a typical configuration). The fundamental quantity in beam conditioning is the divergence δ, i.e. the degree of spatial extension the beam undergoes after traveling over a certain distance from the point of emission. The divergence can be decomposed into two orthogonal components, which are the divergence δ_p parallel to the scattering plane and the axial divergence δ_a in the direction of the θ and 2θ axes (equatorial and axial divergence in earlier work, e.g. Ref. [1]).

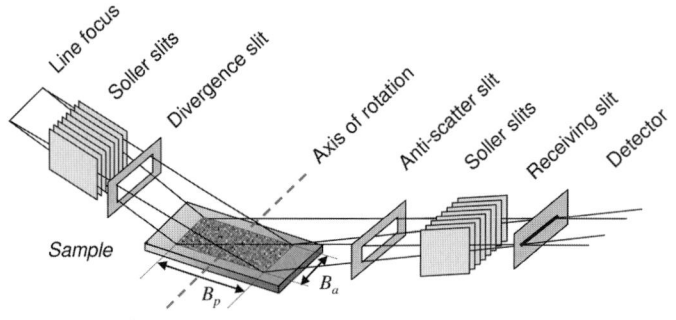

Figure i3.1 Schematic representation of beam conditioning in a $\theta/2\theta$ diffractometer.

Considered first are the technical means that affect the scattering plane divergence δ_p. The radiation leaves the tube via the tube window and illuminates the sample under an angle of incidence θ for the central beam. The initial divergence depends on the shape of the focus on the anode and the geometry of the window. The initial divergence is typically reduced to values of 0.1–2° by a primary aperture that is called divergence slit (DS). The divergence slit aperture δ_{ds} determines the extension of the focal spot on the sample in the direction of the scattering plane. The incidence angles of the most outlying beams are $\theta - \delta_{ds}/2$ and $\theta + \delta_{ds}/2$. The length of the illuminated sample surface B_p depends on the opening angle of the divergence slit and incidence angle θ as

$$B_p = R \frac{\sin\theta \sin\delta_{ds}}{\sin^2\theta - \sin^2(\delta_{ds}/2)} \tag{i3.1}$$

where R is the radius of the goniometer. The illuminated sample length B_p is shown as a function of scattering angle 2θ for various divergence slits δ_{ds} and two diffractometer radii R in Fig. i3.2. It can be seen that B_p easily exhibits values of more than 10 and 20 mm for certain δ_{ds} which might exceed the sample length in some cases. This point ultimately has to be considered when different Bragg peaks in a pattern are compared quantitatively. The size of the illuminated area on the sample holder can also be determined experimentally for various slit settings. For this purpose plates or disks with fluorescing coatings are placed instead of the specimen on the holder that convert the impinging x-rays into visible light. It is very useful to perform such a test prior to a measurement, because the x-ray footprint must be confined to the sample surface for the whole range of scattering angles. Modern diffractometer systems often have programmable divergence slits that enable a constant sample area to be illuminated throughout the whole scan.

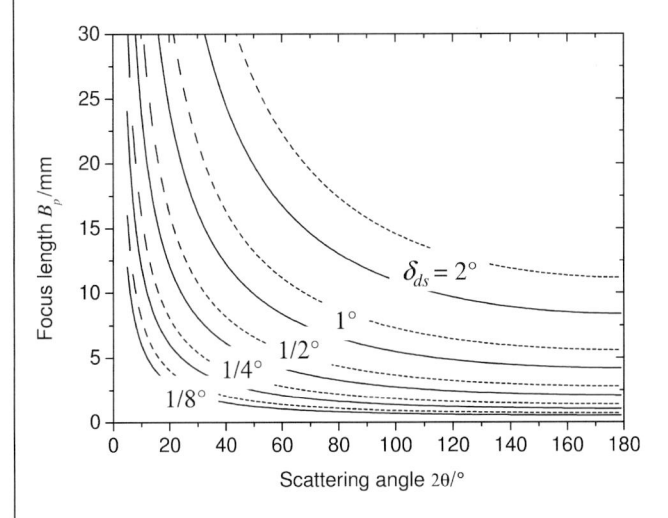

Figure i3.2 Illuminated sample length B_p in the scattering plane as a function of scattering angle 2θ and for various divergence slit opening angles δ_{ds} and two different diffractometer radii R of 240 mm (solid curves) and 320 mm (dashed curves).

On the secondary side, the decisive aperture for the divergence component δ_p is the detector slit. The distance of the detector slit from the θ and 2θ axis should perfectly compare with the distance towards the tube focus which is the diffractometer radius R. The opening angle of the detector slit δ_{det} determines the resolution of the measurement in 2θ, i.e. affects the peak width. Whether the $\alpha_1-\alpha_2$ splitting of Bragg peaks may be observed in the pattern thus sensitively depends on the choice of δ_{det} (in addition, the peak width is affected by the focus size and the sample's microstructure).

We now discuss the tailoring of the beam divergence δ_a in the direction perpendicular to the scattering plane. If the x-ray tube is operated in the line focus mode the initial value of δ_a scales with the length of the line focus. In order to reduce δ_a sets of parallel plates are introduced into the beam path that are all aligned parallel to the scattering plane. According to the work of W. Soller these slit sets are named Soller slits. They are comprised of a number of thin metal plates that reduce the normal beam divergence to typical values between $\delta_{Sol} = 1°$ to $5°$. Soller slits are often installed on both the primary and secondary side.

Also in the axial direction the extension of the x-ray focus on the sample has to be checked prior to a $\theta/2\theta$ scan. If the focal spot size exceeds the width of the specimen the detector would also receive diffraction information from the sample holder. This should be avoided by using a beam mask that is installed on the primary side and cuts off the axial beam flanks. For samples with a width in the few millimeters range retrofitting to the point focus mode of the x-ray tube might be a reliable alternative, since the restriction of beam width by masks is achieved at the expense of intensity.

On its path from the x-ray tube to the detector the beam might not only be scattered by the sample, but also by air and aperture edges. This scattering can cause an unacceptably high background or unexplainable peaks in the diffraction pattern. In order to avoid these artifacts various anti-scatter devices are used the design of which depend on the measurement configuration. In symmetrical $\theta/2\theta$ scans anti-scatter slits (AS) are used that may be positioned on either the primary or secondary side depending on the type of diffractometer.

Which slit setting should be chosen for thin-film work? To answer this question one should be aware of the fact that in most cases the inverse attenuation coefficient $1/\mu$ is larger than the film thickness and, therefore, the measurement has to be performed under maximum intensity conditions to arrive at a reliable counting statistics. Maximum intensity for a thin-film specimen is obtained by illuminating as much sample area as possible, i.e. choosing the largest divergence slit. The choice should thus be made in accordance with Fig. i3.2 to avoid the illumination of the sample holder. In addition, the detector slit should be set in accordance with the desired resolution. The analyst is recommended to perform some tests prior to the full 2θ scan with different slit settings to optimize between the conflicting demands on intensity and resolution. Also a possible thickness inhomogeneity has to be considered when slits and masks are chosen, because too great a t variation may be prohibitive for comparing different peak intensities I_h on a quantitative basis. This results from the fact that the illuminated sample area varies by varying the scattering angle 2θ.

The manipulation of divergences δ_p and δ_a affects the beam's geometrical shape. There are also beam conditioners in use that influence the spectral composition. The simplest one is the β filter that is implemented as a thin metal foil and is installed in most diffractometers very close to the tube window. In addition to Kβ-filtering foils secondary monochromators are also used for the reduction of the Kβ line. They belong to the group of crystal monochromators and perform the beam shaping by selecting only those parts that fit the appropriate Bragg angle to which the crystal is set. For a long time crystal monochromators were the only available dispersive elements for beam shaping in x-ray diffraction and they are discussed in more detail in Instrumental Box 8 in Chapter 7. The monochromators used in powder diffraction serve to suppress Kβ radiation and fluorescence background which is achieved by installing the crystal plate such that only the Kα share of the beam obeys the Bragg condition. Highly oriented pyrolytic graphite (HOPG) is widely used in secondary monochromators that are installed behind the receiving slit (see Fig. i3.3). The graphite monochromator acts as a second parafocusing diffractometer with radius R_2 and is preferably realized as a bent plate with a bending radius of $R_2/2$. In graphite monochromators the incidence angle of the beam is set to the Bragg angle of the graphite 00.2 reflection having $\theta_{SM} = 13.191°$ for Cu Kα radiation. Under these geometrical constraints only the Kα component is scattered into the detector. Evidently, the setting of the monochromator is specific to the tube anode material and cannot be simply readjusted if the tube is changed to another wavelength. When a monochromator crystal is used the polarization factor has to be modified to an expression that also comprises θ_{SM} (Exercise 11).

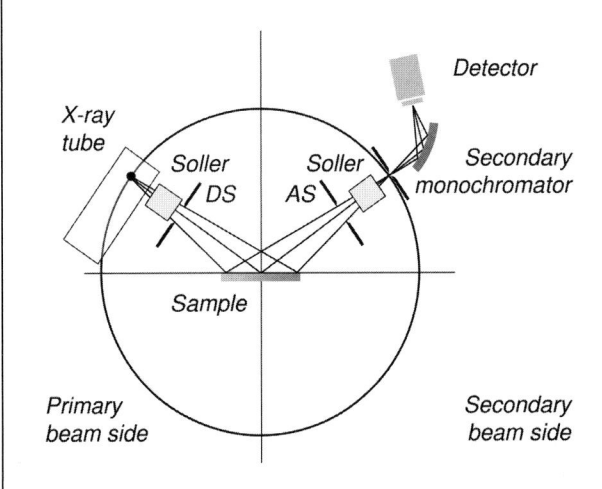

Figure i3.3 Secondary monochromator configuration.

an experimental pattern of unknown provenance. Large databases are available containing some hundreds of thousands of diffraction patterns that were either experimentally collected or theoretically calculated. The essential information that is usefully stored from a $\theta/2\theta$ diffraction pattern of a pure substance is the set of scattering angles $2\theta_i$ where Bragg reflections are observed and their corresponding integral intensities I_i. This gives a two-component vector $(2\theta_i, I_i)$ that looks like a stick pattern and is named a histogram when plotted in a $(2\theta, I)$ coordinate frame. Powder diffraction histograms of many substances are stored in large databases in combination with their chemical composition. The pattern is usually normalized by setting the most intense Bragg reflection to an intensity of 100%, relative to which all other peak intensities are scaled. For many patterns also the hkl triples are specified for each reflection together with information on structure, measurement

Structure Box 2:

The Diamond Structure

The structure of the diamond lattice may be derived from the face-centered cubic (fcc) lattice in which many one-elemental metals crystallize. The diamond structure is composed of two fcc lattices with the second one shifted by ¼ ¼ ¼ with respect to the first. The unit cell thus contains eight atoms of the same element. The structure is shown in Fig. s2.1 and it can be seen that each atom is surround by four neighbor atoms in a tetrahedral coordination. The distance between nearest neighbors amounts to $\sqrt{3}/4$ in fractions of the unit cell edge a. The point symmetry of each lattice point is T_d.

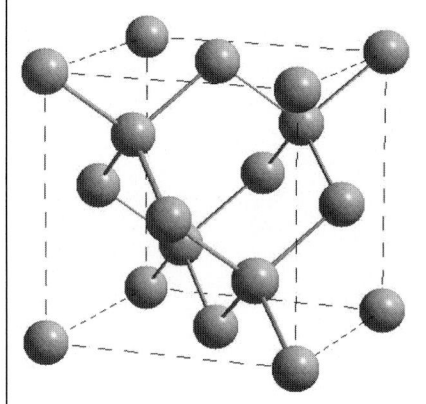

Figure s2.1 Unit cell of the diamond structure with local atomic coordination.

The elements C, Si, Ge and Sn are found in nature to crystallize in the diamond structure (see Table s2.1) as do some of their alloys. However, there exists a tendency to polymorphism among these elements and they are thus also found in other crystal structures, like carbon in the graphite modification. The chemical bonds are strongly directed in the diamond structure and concentrated to the vicinity of interatomic axes. In consequence, the electronic $E(k)$ relations are characterized by the occurrence of energy gaps with missing states for certain electronic energies. These band gaps are 5.2 eV for diamond, 1.1 eV for Si and 0.67 eV for Ge under standard conditions. Therefore, diamond is an isolator, while the electronic conductivity can be tailored through impurity doping in the semiconductors Si and Ge.

The structure factor of the diamond structure yields in analogy with the fcc structure that h, k, l either have to be (a) all even or (b) all odd for constructive interference to be observed. Mixed hkl triples cause destructive interference. However, a distinction has to be introduced for case (a) and (b), because they apply to give

Table s2.1 Structure and symmetries in the diamond structure.

Strukturbericht designation	Space group (number)	Atomic positions		Examples (cell parameter a, nm)
Diamond structure (A4)	Fd3m (227)	000	0½½	C (0.3567)
		½0½	½½0	Si (0.5430)
		¼¼¼	¾¾¼	Ge (0.5658)
		¾¼¾	¼¾¾	α-Sn (0.6489)

$$F(hkl) = \begin{cases} \text{(a)} \ f\left(4 + \left(\cos\frac{3\pi h}{2} + \cos\frac{\pi h}{2}\right)\left(\cos\frac{3\pi k}{2} + \cos\frac{\pi k}{2}\right)\cos\frac{\pi l}{2}\right) \\ \text{(b)} \ f\left(4 - i\left(\sin\frac{3\pi h}{2} - \sin\frac{\pi h}{2}\right)\left(\sin\frac{3\pi k}{2} - \sin\frac{\pi k}{2}\right)\sin\frac{\pi l}{2}\right) \end{cases} \quad (s2.1)$$

The number 4 as the first term in both sums derives from four atoms located on the first fcc lattice centered at 000. The second term is caused by the second fcc lattice shifted by ¼ ¼ ¼. A detailed analysis reveals that the first case (a) leads to an aggravation of the extinction conditions for certain hkl and, therefore, more reflections suffer from extinction. It is finally found for the structure factor

$$F_{diamond}(hkl) = \begin{cases} 8f & \text{all even and } h+k+l = 4n \\ 0 & \text{all even and } h+k+l = 4n+2 \\ (4+4i)f & \text{all odd} \\ 0 & \text{mixed} \end{cases} \quad (s2.2)$$

There are accordingly less reflections to observe in the diffraction pattern of diamond-structured materials than that of fcc-structured ones.

conditions, etc. Details on line width, reflection profiles, etc., are not specified because these depend on the instrumental setting and on the microstructure of the specimen.

Since reflex positions in the model pattern depend on the probing x-ray wavelength, the peak positions are often stored as d_i values rather than as scattering angles $2\theta_i$. The transformation of one quantity to the other proceeds by applying the Bragg equation $d = \lambda/(2\sin\theta)$. In this form the interplanar spacing d is just like the scattering angle 2θ considered as a continuous quantity. And in fact, diffraction patterns are frequently found with $1/d$ as abscissa against which the measured intensity is plotted. Storing the peak positions of a model substance in the ($1/d_i$, I_i) form has the advantage that the histogram may be converted into ($2\theta_i$, I_i) for any arbitrary wavelength. The user should be cautious, however, with respect to the specified intensities in this case. In principle, it would be possible to also transform the intensity by applying Eq. (1.43) to the appropriate wavelength. This is, however, rarely done and one should be aware of this fact. Quite often, the patterns are stored as measured for Cu Kα radiation and the relative intensities then only hold for the wavelength of $\lambda = 0.154$ nm.

Structure Box 3:

Sphalerite and Wurtzite Structure

The chemical compound zinc sulfide (ZnS) crystallizes in two distinct modifications: in the cubic sphalerite structure and in the hexagonal wurtzite structure (Table s3.1). The former is closely related to the diamond structure and may easily be derived from it. It has been demonstrated that the diamond structure comprises two fcc sublattices shifted by ¼ ¼ ¼. Whereas both sublattices are occupied by the same atoms in the diamond structure they are occupied by different types of atoms in sphalerite. Again, the zinc and sulfur atoms are tetrahedrally coordinated, but Zn solely has nearest S neighbors and vice versa. The metal atom is usually placed at the origin 000 of the unit cell, while the sulfur atom resides on the ¼ ¼ ¼ and symmetry-related positions. Each sort of atoms then represents an individual sublattice with fcc structure. Zinc sulfide is the prototype compound of this structure, which is sometimes also named the zinc blende structure (Fig. s3.1). The square of the structure factor reads

$$F^2_{sphalerite}(hkl) = \begin{cases} 4(f_A + f_B)^2 & \text{all even and } h+k+l = 4n \\ 4(f_A - f_B)^2 & \text{all even and } h+k+l = 4n+2 \\ 4(f_A^2 + f_B^2) & \text{all odd} \\ 0 & \text{mixed} \end{cases} \qquad (s3.1)$$

It can be seen that some reflections that are fully extinct in the diamond structure are now recovered. This effect is caused by the occupation of the two sublattices with different types of atoms and distinct atomic form factors $f_A \neq f_B$. Destructive interference as observed for the diamond structure will thus be lifted for some reflections. However, the reflection intensities remain rather small for these reflections and will increase the more the atoms differ by atomic number Z, i.e. the effect is more pronounced in CdS than in CdTe. It is concluded that the diffraction patterns of sphalerite compounds exhibit the same Bragg reflections hkl as elemental lattices with diamond structure plus some additional peaks. Many other binary compounds apart from ZnS crystallize in the sphalerite structure, with a large subset of them from the group of II–VI and III–V compounds. Roman numbers indicate the group in the periodic system from which the elements stem.

Also the sphalerite structure can be visualized as stacking sequence along the [111] direction. Because both the Zn and the S sublattice represent separate fcc lattices they both follow the ...ABCABC... sequence. However, the interpenetration of both lattices has to be indicated by the new sequence ...(AA'BB'CC')$_n$... where the primed and unprimed characters stand for those atom planes that are exclusively built up from S or Zn atoms, respectively. The subscript n indicates that the lattice is composed from a large sequence of this stacking.

At elevated temperatures ZnS transforms from the cubic structure to the hexagonal wurtzite structure. In the hexagonal form, Zn and S atoms are stacked according to the sequence ...(AA'BB')$_n$... (see Fig. s3.2). The square of structure factor and the extinction condition of the wurtzite structure read

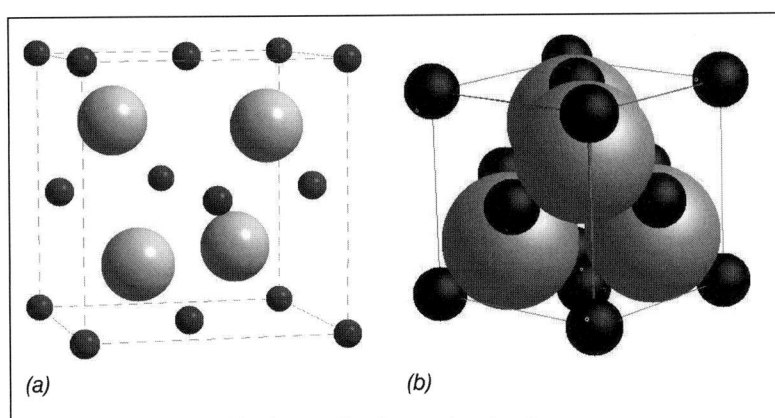

Figure s3.1 Cubic zinc blend unit cell with (a) reduced and (b) appropriate atomic dimensions.

Table s3.1 Structure and symmetries in the sphalerite and wurtzite structure.

Strukturbericht designation	Space group (number)	Atomic positions		Examples (lattice parameters a, c (nm))
Sphalerite structure (B3)	$F\bar{4}3m$ (216)	000	0½½	ZnS (0.5409)
		½0½	½½0	ZnSe (0.5669)
		¼¼¼	¾¾¼	ZnTe (0.6104)
		¾¼¾	¼¾¾	CdTe (0.6481)
Wurtzite structure (B4)	$P6_3mc$ (186)	000	⅓⅔½	ZnO (0.3250, 0.5207)
				ZnS (0.3819, 0.6256)
		00u	⅓⅔½+u	CdS (0.4137, 0.6716)
				CdSe (0.4298, 0.701)

$$F^2_{\text{wurtzite}}(hkl) = \left(f_A^2 + f_B^2 + 2f_A f_B \cos(2\pi ul)\right) \begin{cases} 1 & h+2k=3n\pm1 \; l \text{ even} \\ 3 & h+2k=3n\pm1 \; l \text{ odd} \\ 4 & h+2k=3n \; l \text{ even} \\ 0 & h+2k=3n \; l \text{ odd} \end{cases} \quad (s3.2)$$

Wurtzite may be considered as a structure with two free parameters, which are c/a and the positional parameter u. For a perfect close packing u would amount to 3/8, but slight deviations are typically observed.

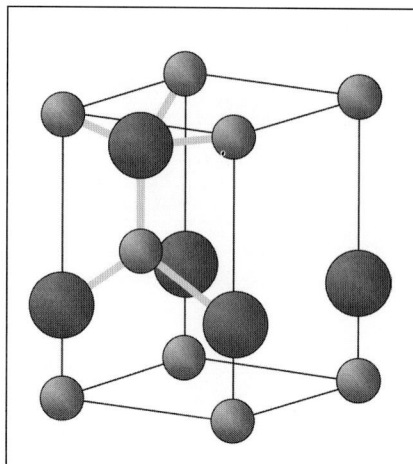

Figure s3.2 Hexagonal wurtzite unit cell with tetrahedral bonds for inner-cell atoms.

Bibliography

H. Hartmann, R. Mach, B. Selle, *Wide Gap II–VI Compounds as Electronic Materials*, North-Holland, Amsterdam, 1982.

W. Hirschwald (ed.), *Current Topics in Materials Science: Vol. 7. Zinc Oxide*, North-Holland, Amsterdam, 1981, pp. 143–482.

Like a fingerprint the powder diffraction pattern of a crystalline substance exhibits a characteristic appearance due to the distinct scattering angles where reflections occur and due to the pronounced variation of intensities. The identification of chemical phases thus proceeds by comparing the measured diffractogram with the set of possible patterns. This may be easily performed with modern computer technology by superimposing the model pattern over the measured one on a monitor and judging the agreement. An unconditional prerequisite to be fulfilled is that Bragg reflections occur. This might not always be obeyed, for instance in thin films from amorphous substances that only exhibit one or a few extremely broad peaks. If Bragg reflections occur in an experimental diffractogram, the set of possible patterns must be reduced from a huge number to a sizeable one. In most cases, this reduction follows from the chemical elements that might occur within the film and that are known from the deposition process.

The database with the largest number of datasets and the largest dissemination is edited by the International Center for Diffraction Data (ICDD), Philadelphia, USA. This data collection – the so-called powder diffraction file (PDF) – was published first by the American Society for Testing and Materials (ASTM) and later by the Joint Committee on Powder Diffraction Standards (JCPDS) [1] the acronyms of which still circulate in the literature. Additional data collections of comparative significance have also been included in the PDF. These are the Inorganic Structure Data Base (ISDB) collected by the Fachinformationszentrum (FIZ), Karlsruhe, Germany, the Cambridge Structure Database (CSD) from the Cambridge Crystallographic Data Centre (CCDC) and from the US National Institute of Standards and Technology (NIST). In 1999 the ISDB contained 40 000 powder patterns calculated from inorganic structure data, while the CSD focuses on large molecular structures. The source of a pattern in the powder diffraction file can be deduced from its nine-digit code number xx-xxx-xxxx. The first two digits are 00, 01, 02 or 03 if the pattern originates from ICDD, FIZ, CCDC or NIST, respectively.

The ICDD continuously exploits the published crystallographic literature, where so far unknown crystal structures are released. Also grant-in-aid studies are issued to research institutions for evaluating high-quality reference patterns. The data obtained from these sources are regularly included into the database, which is published every year. The ICDD is financed by selling the newest versions to research institutes in science and industry. In 2003 the PDF contained close to 300 000 diffraction patterns, divided into subsets of organic and inorganic substances with the majority in the latter category.

In addition to the ICDD there exist further providers of smaller databases for x-ray diffraction patterns like the International Union of Crystallography (IUCr), which distributes the Inorganic Structure Data Base at reduced costs to individuals and academic institutions. In principle, x-ray analysts have also the possibility of calculating model patterns of known structures for themselves. For this purpose the structure has to be inserted into the structure factor, Eq. (1.23), the Bragg equation, Eq. (1.15), and the formula for the integrated intensity, Eq. (1.45), to deduce the $(2\theta_i, I_i)$ histogram. Also computer programs like Powder Cell [2] and others are available that can perform the calculation of powder diffraction patterns when fed with the space group and unit cell coordinates of the atoms. These procedures might be appropriate for phase identification whenever only a small number of chemical phases regularly reoccur in investigations.

The phase identification by pattern assignment is detailed for the example of thin ZnSe films. Such films shall act as emitter layers in large-area solar cells in order to establish a built-in electrical field and to separate spatially the photogenerated charge carriers. In Fig. 2.1 the $\theta/2\theta$ scans of two ZnSe films deposited on glass at different substrate temperatures are displayed [3]. The samples were prepared by chemical vapor deposition (CVD) facilitating the transport of Zn and Se by hydrogen precursors of both elements [3]. The composition of the films was thus restricted to the elements Zn, Se and H. Moreover, in the preliminary process development H_2 was chosen as transporting agent because thermodynamic simulations predicted it not to be contained in the ZnSe films. The semiconducting layers could thus be expected with high probability to contain only Zn and Se, which restricted the number of possible phases to a limited number.

Figure 2.1 also shows the model patterns of cubic ZnSe (ICDD 00-037-1463) and hexagonal ZnSe (ICDD 01-080-0008) and it can be seen that both films comprise a mixture of the two allotropic phases of ZnSe. Under thermodynamic equilibrium conditions ZnSe would have only occurred in the cubic phase while the phase transition to the hexagonal phase should have only occurred at much higher temperatures than the deposition temperatures. However, it is a characteristic of thin-film growth that nonequilibrium phases and metastable phases are formed more often

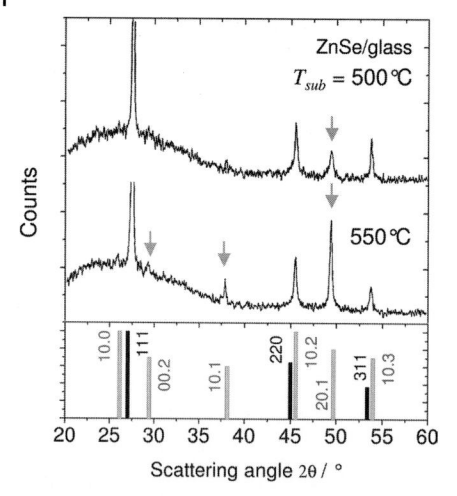

Figure 2.1 $\theta/2\theta$ pattern of thin ZnSe films on glass deposited by CVD at substrate temperatures of 500 and 550 °C. The histograms at the bottom give the ICDD stick patterns of cubic (black) and hexagonal (gray) ZnSe with their respective Miller indices (according to Ref. [3]).

than might be expected. The user should accordingly not restrict the number of possible patterns too strong by equilibrium considerations. In the case considered here it was intended to grow only the cubic sphalerite phase of ZnSe, but the wurtzite phase also occurred. The concentration of the first is seen to increase with decreasing deposition temperature. This example nicely demonstrates the valuable information that can be obtained from an x-ray diffraction pattern and how the results obtained may be used in the optimization of the deposition process.

It is not an uncommon situation, however, that the interpretation of the $\theta/2\theta$ pattern is not this simple. Some of the following observations are often made when comparing measured patterns with model ones:

1. Some reflections of the model pattern are missing.
2. Additional reflections occur that are not present in the model pattern.
3. A reliable agreement between measured data and one model file is achieved; however, the Bragg peaks appear to be shifted.
4. Relative intensities as predicted for the Bragg reflections by the model pattern are not obeyed.
5. In addition to the Bragg reflections the experimental pattern exhibits broad bumps that might be interpreted as reflections with an extremely large line width (compare Fig. 2.1 around $2\theta = 25°$).

In fact, it is the rule rather than the exception in the x-ray scattering analysis of polycrystalline thin films that one of these features occurs than that all of them are absent. These effects might appear disturbing at first sight, because they hinder the unique assignment of one crystallographic phase to the layer. On the other hand these features may reveal important information on the microstructure and morphology of the samples and on the mechanisms of thin-film growth. The points listed above and possible explanations for them are considered in the following.

It should finally be pointed out that various techniques have been developed to index diffraction patterns of unknown crystal structures. As has already been mentioned, it is rather unlikely that an unknown phase would be prepared in modern thin-film work on inorganic substances. Indexing procedures will thus not be presented here and the interested reader is referred to the literature [4].

2.2
Linear Attenuation Coefficient μ

The linear attenuation coefficient μ is of central importance for the identification of chemical phases and their quantitative determination by x-ray scattering techniques. Compared to powder work the attenuation coefficient is even more significant in thin-film analysis, because of the damping of scattered intensity caused by the absorption factor $A_{\theta 2\theta}(t)$ (see Eq. (1.32)). In order to improve our understanding of the attenuation coefficient μ, the various causes that determine its magnitude and dependency on wavelength λ and atomic number Z are discussed here in more detail.

Tabulated values of the linear attenuation coefficients μ are rarely given, because μ is proportional to the density ρ. Therefore, most tables specify the mass absorption coefficient μ_m, which is obtained by dividing by the mass density

$$\mu_m = \frac{\mu}{\rho} \tag{2.1}$$

The dimensions of μ_m are m² kg⁻¹, which can be understood as a cross-section – for the interaction with x-rays – per unit mass. For a discussion of the physical causes of attenuation or absorption the μ_m coefficient is advantageously further reduced to a cross-section σ that is specified for a single atom. The normalization is achieved by multiplying by the atomic mass M and dividing by Loschmidt's number N_L

$$\sigma = \frac{\mu}{\rho} \frac{M}{N_L} \tag{2.2}$$

In the SI system the dimensions of σ are m² per atom.

It has already been mentioned in chapter 1 that the interaction of x-rays with matter can be subdivided into photo-ionization, inelastic (Compton) scattering and elastic scattering. The elastic component is named after Thomson or Rayleigh according to whether the scattering from a single electron or an atom is considered. Its cross section will thus be abbreviated by the subscript R here. The total interaction cross-section σ_{tot} will decompose into the three terms according to

$$\sigma_{tot} = \sigma_{PE} + \sigma_C + \sigma_R \tag{2.3}$$

The three single components and the total cross-section σ_{tot} are shown as a function of energy for a silicon atom in Fig. 2.2. It can be seen from this plot that the energy loss of the interacting x-ray due to the generation of photoelectrons is the dominating effect up to about 10 keV. The σ_R and σ_C components only amount to a fraction of the total cross-section σ_{tot} at about 10 keV and beyond the three components are seen to attain comparable magnitudes. As a general tendency the total cross-section is seen to decline with increasing energy. This tendency can be understood in terms of the atoms becoming more transparent because the photon energy increasingly deviates from the resonance energies. Above 1022 keV a further loss channel due to pair production opens, but this will not be considered here.

As a further interesting feature the plot depicts a discontinuity in both the σ_{PE} component and for the total cross-section. Such absorption edges are also observed in the spectra of mass absorption coefficient μ_m that are shown for C, Si and Ge in Fig. 2.3 in the energy range from 1 to 100 keV. The plots were calculated for free, unbound atoms according to Hubbel and Seltzer [5]. For certain energies or wavelengths, characteristic edges occur at which the μ_m coefficients increase by about an order of magnitude. These edges are termed absorption edges. They occur whenever the energy exceeds the value above which electrons from inner shells may be excited to the vacuum level, i.e. when new processes of photoionization are enabled. The excitation from the innermost level is denoted as K edge and the numbering proceeds in alphabetical order to the more outlying levels, L, M, etc. Evidently, the K edge energy will increase with increasing Z. In addition, every chemical element can only exhibit as many edges in $\mu_m(E)$ as there are inner core levels. In Si, for instance, the K edge is found at 1.839 keV, while Ge exhibits a K edge at 11.1 and a first L edge at 1.217 keV. To give a further illustration of these general

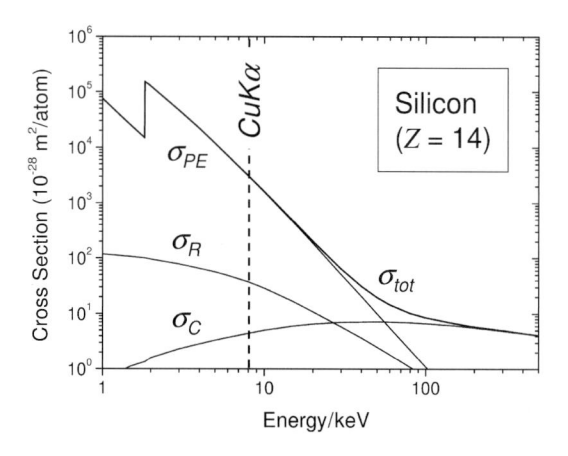

Figure 2.2 Interaction cross section for x-rays with silicon as a function of energy.

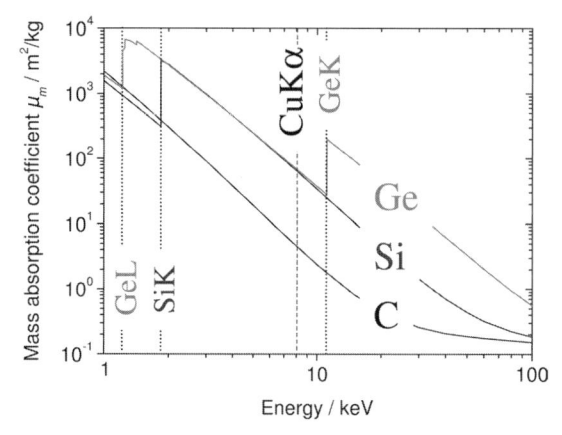

Figure 2.3 Mass absorption coefficient μ_m of C, Si and Ge in the energy range 1 to 100 keV.

rules the mass absorption coefficients of other elements relevant to this chapter are presented in Fig. 2.4.

In can be seen for the higher-Z elements that the L edge exhibits a fine structure in the absorption coefficient. The single L edges are thus denoted by roman numerals L_I, L_{II}, etc. It is found that the precise appearance of the fine structure is influenced by the chemical bonding of the atoms. Moreover, the K edges also exhibit this fine structure if the atoms are bonded to a solid or molecule. The variations of μ_m in the vicinity of the absorption edge are used as a valuable tool for the investigation of chemical bonding by the method of x-ray absorption spectroscopy (XAS), which is also termed EXAFS (extended x-ray absorption fine structure spectroscopy) or XANES (x-ray absorption near-edge spectroscopy) according to the energy range investigated. Since this book is restricted to the elastic scattering of x-rays these techniques will not be considered, but the reader is suggested to check whether these techniques might be applied for the solution of his analytical problem. The fine structure variations may in some cases be of relevance also for elastic scattering, because they extend some 100 eV in excess of the edge energy. If the probing x-ray wavelength is close to one of the absorption edges of the material under investigation, the analyst should be aware that the precise value of μ_m can be severely affected by edge variations.

The mass absorption coefficients μ_m of the chemical elements up to $Z = 92$ for the x-ray energy of Cu Kα radiation (8.04 keV, 0.154 nm) are listed in Table 2.1 and plotted in Fig. 2.5. It can be seen that the coefficients vary over three orders of magnitude. It is evident that the chemical composition of a sample may thus significantly influence the absorption factor $A(t)$, because μ_m appears in the exponent in the expression for $A(t)$ (see Eq. (1.32)). Numerical values of the mass absorption coefficients for other wavelengths can be found in Ref. [6].

Table 2.1 gives the mass absorption coefficients for one-elemental samples. If the linear absorption coefficient of a chemical compound is required, it must be

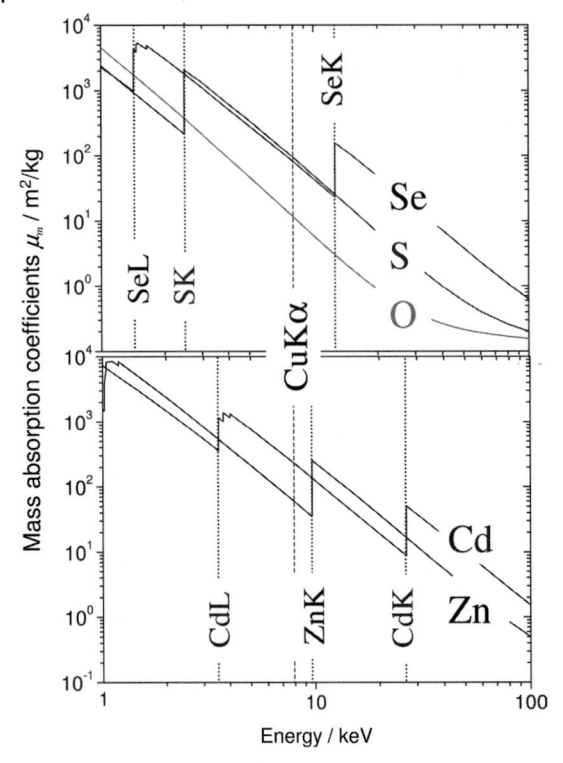

Figure 2.4 Mass absorption coefficients μ_m of (top) O, S, Se and (bottom) Zn and Cd in the energy range 1 to 100 keV.

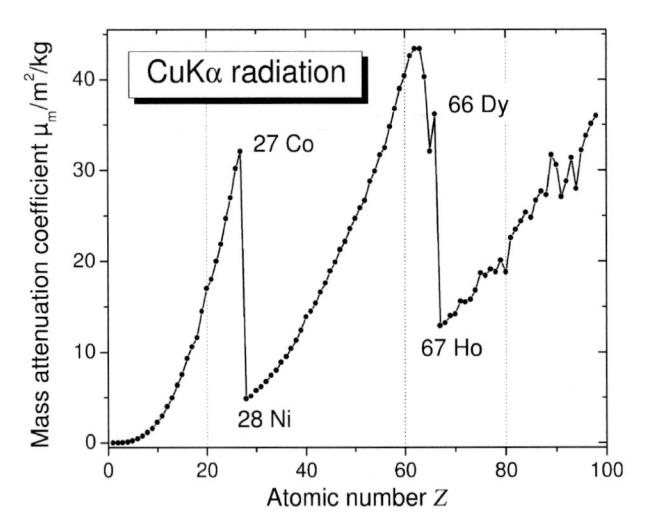

Figure 2.5 Linear attenuation coefficients the elements for Cu Kα radiation.

Table 2.1 Mass absorption coefficients in units of m² kg⁻¹ of all elements up to Z = 92 for Cu Kα radiation [6].

Z		μ_m	Z		μ_m	Z		μ_m	Z		μ_m
1	H	0.0391	24	Cr	24.7	47	Ag	21.3	70	Yb	14.2
2	He	0.0292	25	Mn	27.0	48	Cd	22.2	71	Lu	15.6
3	Li	0.05	26	Fe	30.2	49	In	23.6	72	Hf	15.5
4	Be	0.111	27	Co	32.1	50	Sn	24.7	73	Ta	15.8
5	B	0.231	28	Ni	4.88	51	Sb	25.9	74	W	16.8
6	C	0.451	29	Cu	5.18	52	Te	26.7	75	Re	18.7
7	N	0.744	30	Zn	5.79	53	I	28.8	76	Os	18.4
8	O	1.15	31	Ga	6.21	54	Xe	29.9	77	Ir	19.1
9	F	1.58	32	Ge	6.79	55	Cs	31.7	78	Pt	18.8
10	Ne	2.29	33	As	7.47	56	Ba	32.5	79	Au	20.1
11	Na	2.97	34	Se	8.00	57	La	34.8	80	Hg	18.8
12	Mg	4.00	35	Br	8.90	58	Ce	36.8	81	Tl	22.6
13	Al	4.96	36	Kr	9.52	59	Pr	39.0	82	Pb	23.5
14	Si	6.37	37	Rb	10.4	60	Nd	40.4	83	Bi	24.4
15	P	7.55	38	Sr	11.3	61	Pm	42.6	84	Po	25.4
16	S	9.33	39	Y	12.4	62	Sm	43.4	85	At	24.8
17	Cl	10.6	40	Zr	13.9	63	Eu	43.4	86	Rn	26.7
18	Ar	11.6	41	Nb	14.5	64	Gd	40.3	87	Fr	27.7
19	K	14.5	42	Mo	15.4	65	Tb	32.1	88	Ra	27.3
20	Ca	17	43	Tc	16.6	66	Dy	36.2	89	Ac	31.7
21	Sc	18	44	Ru	17.6	67	Ho	12.9	90	Th	30.6
22	Ti	20	45	Rh	18.9	68	Er	13.2	91	Pa	27.1
23	V	21.9	46	Pd	19.9	69	Tm	14.0	92	U	28.8

derived from its density ρ and the sum over μ_m of the component elements weighted by their mass fraction g_i/G. Here, g_i denotes the atomic mass of an element and G is the mass of the molecular unit, $G = \Sigma\, g_i$. The single-element μ_m then has to be substituted by a sum over all elements in accordance with

$$\mu_m \rightarrow \sum_i \frac{g_i}{G} (\mu_m)_i \tag{2.4}$$

Due to the large increase of μ_m with atomic number Z, the absorption coefficient of a compound or an alloy will in most cases be dominated by the heaviest elements in the sample.

Table 2.2 lists the linear absorption coefficients μ of some solids discussed in this chapter for Cu Kα radiation. Also the reciprocal values μ^{-1} are given that have the meaning of a penetration depth. For normal incidence of an incoming x-ray beam the μ^{-1} values specify the depth at which the x-ray beam intensity declines to $1/e$ of its initial value. It is evident that the penetration depths may significantly exceed the thickness of typical thin-film systems, which would remain valid even under inclined x-ray exposure. An x-ray beam may thus partially reach the deeper lying layers of the sample such that structural information from the substrate is also contained in the diffraction pattern.

Table 2.2 Atomic masses of selected elements; and density ρ, linear absorption coefficient μ and penetration depth μ^{-1} of selected solids for Cu Kα radiation.

Element	Atomic mass (u)	Modification	Structure type	Density, ρ (g cm^{-3})	μ (mm^{-1})	μ^{-1} (μm)
C	12.01	Diamond	Diamond	3.51	1.58	632
		Graphite	Graphite	2.25	1.01	985
Si	28.086	c-Si	Diamond	2.32	14.8	67.7
Zn	65.38	Cubic ZnS	Sphalerite	4.10	31.7	31.6
S	32.064	Hexagonal ZnS	Wurtzite	3.98	30.8	32.5
Se	78.96	Cubic ZnSe	Sphalerite	5.42	43.9	22.8
		Hexagonal ZnSe	Wurtzite	5.39	43.7	22.9

It is a further feature of thin-film analysis that the theoretically calculated μ_{cal} values are often found to be smaller than those actually observed in the layer, $\mu_{eff} < \mu_{cal}$. This fact may partially be explained by the porosity that may affect μ_{eff} through the inclusion of air or vacuum. In addition, the preparation of thin films in vacuum-based or wet-chemical processes may be accompanied by the inclusions of contaminants on inner surfaces, such as water, OH groups, and carbohydrate compounds, etc., having a lower density than the main part of the film. However, also the reverse situation, $\mu_{eff} > \mu_{cal}$, can occur. In plasma-assisted processes like physical vapor deposition (PVD) or plasma-enhanced CVD (PECVD), for instance, atoms from the sputter gas may be included in the growing film for which Ar is often used. Depending on process parameters like the substrate bias, the amount of Ar can rise to several at% that may increase μ above the value that is calculated from the ideal film composition. The general statement can be made that for any calculation of μ according to Eq. (2.4) the real chemical composition of the layer has be checked carefully – which may differ from the intended one. Among other techniques this may be done by the determination and interpretation of the μt product.

2.3
Determination and Interpretation of the μt Product

Probably the most important quantities in the research and development of thin films are the film thickness and its chemical composition. The thickness is required for the interpretation of a large number of characterization techniques and most of the layer properties sensitively depend on it. In Chapter 1 it is shown how the intensity of Bragg reflections is affected by the absorption factor A, which is a function of the diffraction angle θ, attenuation coefficient μ and film thickness t. The fact that these quantities influence the $\theta/2\theta$ pattern means that we have a

2.3 Determination and Interpretation of the µt Product

method at hand to determine the film thickness by evaluating the absorption factor. To be precise, the actual measurement quantity is not the film thickness t, but the product of attenuation coefficient μ and thickness, the μt product. This allows for various analytical approaches: (a) by making reliable assumptions about μ the thickness may be deduced or (b) further information on film morphology and composition may be derived when t is known.

One technique to determine the μt product of a layer is to measure the intensity loss of a substrate reflection caused by the deposition of the layer. This method presupposes that one or more Bragg reflections from a crystalline substrate or a crystalline bottom layer are available [7]. If this condition is fulfilled two $\theta/2\theta$ scans have to be recorded, one from the uncoated substrate and a second from the coated one. The integral intensity of the substrate peak at position $2\theta_{B,S}$ then has to be determined in both patterns yielding $I_{S,unc}$ and $I_{S,coa}$. According to the formulas presented in Chapter 1 the two integral intensities are related via the $(\mu t)_{TF}$ product by

$$I_{S,coa} = I_{S,unc} \exp\left(-\frac{2(\mu t)_{TF}}{\sin\theta_{B,S}}\right) \qquad (2.5)$$

Figure 2.6 displays a diffraction pattern of a molybdenum-coated glass substrate with a thin silicon film on top of it (Si/Mo/glass). The coating of glass by polycrystalline semiconductors is an active field of research and development for large-area electronic applications. In particular, the preparation of thin-film solar cells on low-cost, industrially relevant substrates may significantly contribute to cost reduction in the photovoltaic utilization of solar energy. The diffraction pattern in Fig. 2.6 shows a well-developed 110 reflection of the bcc-structured Mo layer. Also the position of the Mo 200 is indicated, but at the same scattering angle a peak due to the glass substrate occurs. This is one of the few cases where a glass substrate yields a pronounced reflection, which is seldom observed due to the amorphous glass structure (see below). In addition, all diamond-structure reflections 111, 220 and

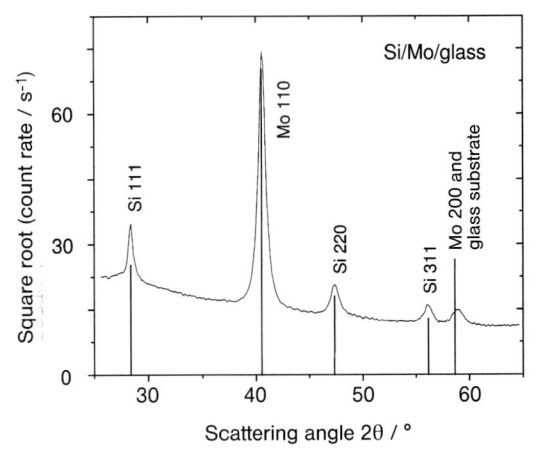

Figure 2.6 $\theta/2\theta$ pattern of a Mo-coated glass substrate with a thin silicon film on top.

311 of the Si layer arise in the scattering angle range investigated. The ordinate is given in units of square root of intensity to emphasize smaller reflections. Although the thickness of the Mo layer of 600 nm was less than that of the 1.8 µm Si layer, the reflections of the latter are significantly smaller than that of the Mo 110. This can be understood from the higher electron density ρ_e of elemental Mo of 2695 nm^{-3} compared to 699 nm^{-3} for Si. Because the x-ray scattering power scales with the square of ρ_e, the Mo reflection is higher in spite of the damping by the Si layer on top. The measurement of the Mo 110 integral intensity prior to and after Si deposition enabled the determination the $(\mu t)_{TF}$ product of the Si layer by applying

$$(\mu t)_{TF} = \frac{\sin\theta_{B,S}}{2} \ln\left(\frac{I_{S,unc}}{I_{S,coa}}\right) \qquad (2.6)$$

The layer thickness t is then determined by dividing the $(\mu t)_{TF}$ product by the value for the absorption coefficient of Si for Cu Kα. One decisive advantage of this technique is that the thickness is obtained without destruction of the sample or any small portion of it as would be necessary in other methods like profilometry or electron microscopy. Equation (2.6) may also be applied for the thickness determination of amorphous layers because it makes use of the substrate reflection and does not require a crystallized layer.

The technique can also be used to remove the substrate peak from the diffraction pattern when a thorough analysis of the thin-film Bragg reflections is required [8]. There is a point to be borne in mind, however, when deducing thickness values from this method when highly perfect single crystals like Czochralski-grown silicon wafers (CZ-Si) are used as substrates. In such cases the integral intensity formula of Eq. (1.44) from the kinematical theory does not apply, but instead intensity formulas from the dynamical theory have to be used. This is outlined in detail in Chapter 7. Regarding the sensitivity of this technique to numerical errors it has to be realized that the logarithm of an intensity ratio is calculated (Eq. (2.6)). Ideally, the two intensities should differ by a factor of two at least, but a much higher value would be preferred. The substrate peak intensity of the coated sample, however, must not become too small in order to operate with an integral peak intensity of not too large an estimated standard deviation.

The concept is easily generalized to multilayer systems as is shown in Fig. 2.7. The layer stack is assumed to comprise $J - 1$ separate layers labeled by j with the topmost layer being number 1 and the substrate labeled J. The integral intensity of each Bragg peak from one layer can be described by Eq. (1.45). However, the intensity of a Bragg reflection of the jth layer is affected by the absorption of the layers above. This is accounted for by a different form of the absorption factor, which reads for the jth layer of a multilayer system [9]

$$A_{\theta 2\theta}^{ML,j}(2\theta) = \left(1 - \exp\left[-\frac{2\mu_j t_j}{\sin\theta}\right]\right) \exp\left[-\sum_{p=1}^{j-1} \frac{2\mu_p t_p}{\sin\theta}\right] \qquad (2.7)$$

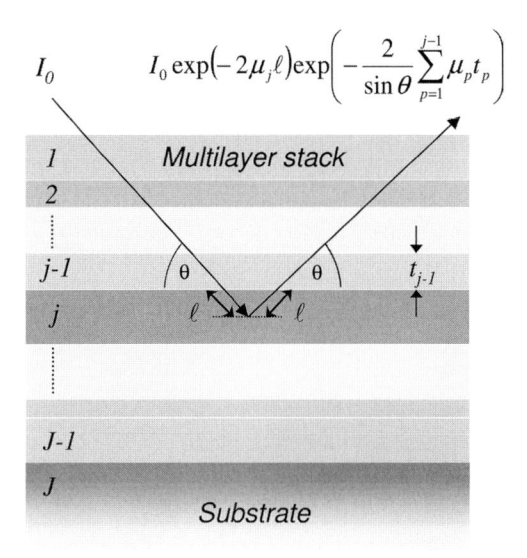

2.3 Determination and Interpretation of the μt Product

Figure 2.7 Schematic representation of absorption in a multilayer system.

This expression compares to the absorption coefficient of a single layer, see Eq. (1.32), except for the second factor that describes the damping of the incoming and outgoing beams by the layers on top of layer j.

A further technique to obtain the μt product of a thin film makes use of so-called harmonic reflections. Such reflections are characterized by an integer n through which both Miller index triples are related, i.e. $h'k'l' = n(hkl)$. In many cases only the fundamental and the first harmonic reflection, $n = 2$, are available from the measurement, but, evidently, the presented procedure will improve with the number of included reflections. Assume that the two peaks are found at θ_1 and θ_2 and have integral intensities I_1 and I_2 as quantified by Eq. (1.45). From the ratio of the two expressions one obtains

$$\left(1 - \exp\left[-\frac{2(\mu t)_{TF}}{\sin\theta_1}\right]\right) = \frac{I_1}{I_2}\frac{(Lp)_2}{(Lp)_1}\frac{F_2 F_2^*}{F_1 F_1^*}\left(1 - \exp\left[-\frac{2(\mu t)_{TF}}{\sin\theta_2}\right]\right) \quad (2.8)$$

which can be solved for $(\mu t)_{TF}$ numerically by inserting the measured intensities and calculating the other factors. It should be emphasized that the two intensities in the equation now account for reflections from the thin film, in contrast to Eq. (2.6) where the intensities stemmed from the substrate. In the ratio of Eq. (2.8) all other factors like multiplicity, etc., have canceled because they are equal for harmonic reflections. It should be noted that the structure has to be known in order to calculate the structure factors. The decisive point, however, is the that the texture factors T_h vanish, because both Bragg peaks monitor the same family of lattice planes or the same group of crystallites, which is not valid for nonharmonic reflections. In an ideal powder sample $T_h = 1$ would hold for all reflections, but in polycrystalline thin films one almost always encounters preferentially oriented grain ensembles. Equation (2.8) may accordingly not be applied for pairs of reflections other than harmonic ones.

Harmonic reflections are rather far separated in the diamond structure due to the severe extinction conditions and the concentration of the scattering intensity to only a few Bragg reflections. The first harmonic pair is identified as 220 and 440, which would arise at 47.3° and 106.7° in a $\theta/2\theta$ scan of a thin Si film recorded with Cu Kα radiation. In sphalerite- and wurtzite-type compounds the extinction conditions of the diamond structure are partially lifted allowing for more harmonic combinations. In the cubic sphalerite structure, for instance, the first harmonic pairs are 111/222 and 200/400, while the hexagonal wurtzite structure shows the harmonic 10.0/20.0 and 00.2/00.4 couples. These reflections might well be elucidated for the determination of the μt product according to Eq. (2.8). The main obstacle will probably be the low intensity of the second reflection caused by the increased damping at higher scattering angles 2θ. Moreover, in some harmonic pairs the second peak is additionally modulated by variation of the structure factor which holds, for instance, for the sphalerite 111 and 222 peaks. After deriving μt from Eq. (2.8) the final calculation of t would then proceed in the same way as given above by division of μt by a reliable value for the absorption coefficient μ of the layer.

In effect, everything has been said as to how the thickness might be derived from integral intensities of a diffraction pattern. One should be aware, however, of the fact that a highly accurate thickness value is not easily obtained for polycrystalline layers and rarely is an estimated standard deviation of less than 1% achieved. Obtaining high accuracy is impeded, for instance, by surface or interface roughness and thickness inhomogeneities, the latter constraining the analyst to define carefully what is meant by thickness and at what position on the specimen it has been measured. The systematic errors connected with the thickness value from the procedures outlined above may best be discussed by rewriting μt as the product of quantities to which the experimental value is theoretically related

$$\left(\mu t\right)_{\exp} = \rho \mu_m t \tag{2.9}$$

Here, ρ and μ_m are again the density and the mass absorption coefficient of the layer. In the case where a chemical compound or mixture is considered, μ_m has to be substituted by the weighted sum of the elemental coefficients $\mu_{m,i}$. It will become transparent from an analysis of this equation which structural information may be derived from the determination of μt.

In the main application of Eq. (2.9) the thickness is determined by dividing $(\mu t)_{\exp}$ by $\rho \mu_m$. The systematic errors introduced are related to the inaccuracy in both quantities ρ and μ_m. Typically, the density for macroscopic bulk material ρ_{bulk} is inserted which may be the first source of error. Very often in thin-film preparation the obtained density is smaller by at least a few percent, because voids and microcavities are formed during growth. Also the inclusion of contaminants from the growth medium – mainly light elements – can make ρ_{bulk} a larger quantity than the true density of the layer. The geometrical thickness t_{geo} determined with the inclu-

2.3 Determination and Interpretation of the µt Product

sion of these systematic errors will then be underestimated. This thickness value $t_{pro} = (\mu t)_{exp}/(\rho\mu_m)$ would correspond to a kind of "projected" density the film would have when its morphology is compared to bulk material (Fig. 2.8). If the geometrical thickness t_{geo} value is available from another technique like profilometry that specifies the spatial extension of the layer from the interface to its surface, the ratio t_{geo}/t_{pro} can be calculated. Such a ratio might be useful to derive the porosity of a layer like porous silicon, although the free volume in the layer should probably exceed at least 10% to yield reliable results.

Another, less often applied use of Eq. (2.9) consists in deriving an areal density by dividing the experimental quantity by μ_m only

$$\rho t = \frac{(\mu t)_{exp}}{\mu_m} \tag{2.10}$$

The dimension of ρt is kg m^{-2} and it can thus be considered as a mass thickness, where the mass is that of the solid of which the layer is comprised. The reliability of any presupposed composition inserted into Eq. (2.10) again determines the accuracy of the obtained results. It is intriguing to compare the usefulness of the two quantities thickness t and mass thickness ρt in specifying the amount of deposited material on a substrate. In general, one has become accustomed to applying the geometrical thickness t_{geo}, but in some investigations, for instance on growth rate, the use of mass thickness ρt might be more appropriate [10, 11]. The advantage of ρt over t is that it does not rely on morphological properties of the film like porosity, roughness, phase gradients, etc. Mass thicknesses are often determined as areal densities by Rutherford backscattering [10, 11], but Eq. (2.10) demonstrates that they may also be determined by x-ray diffraction procedures.

As a final application of the μt product the monitoring of thickness inhomogeneities is mentioned. An inhomogeneous covering of a substrate by a thin film is frequently observed and may be caused by the geometric constraints of the deposition process. Thickness variations may be monitored by measuring a substrate peak as outlined above in the sample area where Δt has to be specified. This must be performed by a spatial scan of the substrate peak and, evidently, the sample holder used for this application would require an xy lateral translation stage. The smaller the sample area of interest is the more measures have to be taken to reduce the x-ray beam focus on the sample surface to an appropriate size. This might be

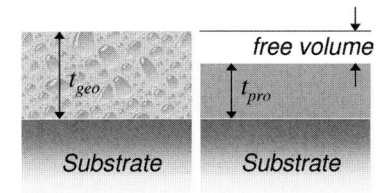

Figure 2.8 Visualization of geometrical thickness (left) and projected thickness (right). The difference between them accounts for the free volume of the layer.

achieved by choosing a high scattering angle reflection of the substrate, by setting appropriate slits or even by changing the x-ray tube from line focus to point focus mode. Minimum beam spot sizes in the range of square millimeters can be achieved with a conventional diffractometer, while even smaller areas can be resolved with so-called microdiffraction equipment.

It is concluded from this section that the determination of the μt product directly leads to the two most important structural parameters of a thin film, i.e. its thickness and composition. It turns out that the proper interpretation of the μt product is intimately related to perturbations of the ideal thin-film structure, i.e. one has to consider thickness inhomogeneity, inclusion of contaminants, porosity and other real world effects. In many R&D projects these "side effects" are identified as the essential features with which the thin-film grower has to cope in order to prepare materials with the intended properties.

2.4
Analysis of Phase Mixtures

Consideration is now given to a single-layer thin film that contains P different crystalline phases, each of which having a weight fraction w_j/W. The sum over all phase weights w_j gives the total weight W of the film,

$$\sum_{j=1}^{P} w_j = W.$$

The $\theta/2\theta$ diffraction pattern of the sample may now be analyzed with respect to a quantitative determination of the different phases. The analysis makes use of the quantitative expression of the integral intensity I_h of Bragg reflections, given in Eq. (1.44). Some modifications have to be carried out, however, because the equation was derived for a single-phase sample. We are thus aiming at a formula for the integral intensity I_{hj} for reflection h of the jth phase with weight fraction w_j/W. In principle, one may proceed by replacing the subscript h by subscript hj in all appropriate quantities of Eq. (1.44). This would hold for the multiplicity of the reflection m_h, the texture factor T_h and the structure factor F_h, where the arguments for their appearance in Eq. (1.44) are independent of the number of phases. Also the Lorentz polarization factor Lp and the scaling factor SCF would not be affected, because they are related to the geometrical and instrumental constraints of the measurement. However, the $V/2\mu$ term and the absorption coefficient $A_{\theta 2\theta}(t)$ need to be reconsidered when a multiple-phase sample is under investigation.

Let us start with the attenuation coefficient μ or the two quantities of which it is composed, i.e. the mass attenuation coefficient μ_m and the density ρ. An averaging of both quantities has to be performed, because all the phases present in the sample contribute to the damping of beam intensity. The average mass attenuation coefficient $\overline{\mu_m}$ derives from the sum over μ_m coefficients of all phases weighted by their weight fraction

$$\overline{\mu_m} = \sum_j \frac{w_j}{W}(\mu_m)_j \qquad (2.11)$$

This formula compares to Eq. (2.4), but now the sum does not extend over individual atoms, but over chemical phases. Next, the average density is simply the ratio of the sum over all phase weights and the total volume

$$\bar{\rho} = \frac{1}{V}\sum_j w_j \qquad (2.12)$$

The product of both quantities gives the average attenuation coefficient $\bar{\mu} = \bar{\rho}\overline{\mu_m}$ of a phase-mixed sample which has to be inserted in Eq. (1.44) instead of μ. Of course, this also holds for the absorption coefficient $A_{\theta 2\theta}(t)$, see Eq. (1.32), in which μ must be replaced by $\overline{\mu_m}$.

The illuminated sample volume V is seen from Eq. (1.44) to appear in the expression for the integral intensity of a Bragg reflection. Because the sample considered now comprises different phases this volume has to be replaced by the volume occupied by the jth phase only. Each phase volume V_j is related by $V_j = w_j/\rho_j$ to the weight and density of the appropriate phase (this should not be confused with the volume of the unit cell of phase j, denoted by $V_{uc,j}$). After insertion of these results in order to modify Eq. (1.44) we arrive at the expression

$$I_{hj} = SCF \cdot Lp \cdot m_{hj} \cdot T_{hj} \cdot \frac{|F_{hj}|^2}{V_{uc,j}^2} \cdot \frac{w_j}{\rho_j \bar{\mu}} \cdot A_{\theta 2\theta}(t) \qquad (2.13)$$

that describes the integral intensity of the hth reflection of phase j in a $\theta/2\theta$ diffractogram of a phase-mixed thin film.

The first interesting point of this equation is that the reflection intensity scales with the weight fraction of the phase. In consequence, a $\theta/2\theta$ diffraction pattern of a polycrystalline phase mixture may strongly be dominated by that compound that is comprised of elements of the highest Z. The effect is caused by the concomitant increase of atomic weight, number of electrons and x-ray scattering power in the periodic table. Moreover, it is enhanced by the F^2/V^2 term as was pointed out before. The general tendency is consequently observed that heavy-element compounds will dominate in phase-mixed patterns. Hence, a fully numerical analysis of the diffraction pattern is recommended whenever a quantitative determination of phase content is to be performed. In a study on mixed-oxide films the accuracy that might be achieved with this procedure has been investigated [12]. The authors decomposed the $\bar{\mu}t$ product of the layers into a product of mass attenuation coefficient $\overline{\mu_m}$ and average mass thickness $\bar{\rho}t$, in accordance with Eq. (2.10). For various degrees of phase mixing the integral intensities from the crystalline substrate and from each of the two phases were measured and subsequently $\overline{\mu_m}$, $\bar{\rho}t$ and w_j/W could be deduced. It was concluded in this work that mean accuracies of about 3% for $\bar{\rho}t$, 1% for $\overline{\mu_m}$ and 4% for w_j/W can be achieved [12]. One may add that $\bar{\mu}t$ products in this study were in the range 0.1–0.6 – corresponding to film thicknesses of the order of 10 µm – and that achievable accuracies may drop for nonoptimal $\bar{\mu}t$ products.

The next important factor that may influence the application of Eq. (2.13) is the occurrence of preferred orientation in a sample. This is often the case and may severely affect the integral intensities collected in a $\theta/2\theta$ scan as evidenced by the T_h term in Eq. (2.13). If a phase-mixed sample exhibits a strong texture, the quantitative determination of chemical composition can only be obtained in combination with a texture investigation. A whole-pattern fitting procedure should then preferably be applied [9]. Caution should be taken when using Eq. (2.13) without disposing of information on the state of preferred orientation.

An interesting example for the procedures given above is shown in Fig. 2.9 [13]. The $\theta/2\theta$ diffraction pattern was measured from the multilayer system ZnS:Mn/Bi/W/glass as shown in the inset. Mn-doped ZnS films are investigated with respect to their suitability in electroluminescent displays and the structural questions of interest are related to phase purity, microstructure and texture that all affect the functionality of devices. As for other large-area optoelectronic systems, low-cost materials like glass are the substrates of choice. The Bi and W metal layers were deposited to contact electrically the active ZnS:Mn layer and to improve the adhesion to the substrate. All three layers are seen to cause Bragg peaks, which for one reason is due to the low absorption of the ZnS:Mn layer and for another reason due to the high electron density of the metal layers. For some peaks a severe overlap occurs that complicates the quantitative phase analysis. The Mn content in ZnS is only of the order of 0.1 at% and the top-layer pattern is thus indistinguishable from a pure ZnS pattern. In addition to the cubic ZnS peaks a small freestanding h-ZnS 103 peak is also observed, while the h-ZnS 100 peak is completely overlapping with Bi 210. Such severe overlapping in a $\theta/2\theta$ pattern can only be modeled by a whole-pattern fitting algorithm that makes use of multilayer absorption and phase-mixing effects (see Eqs. (2.7) and (2.13)). The application of these tech-

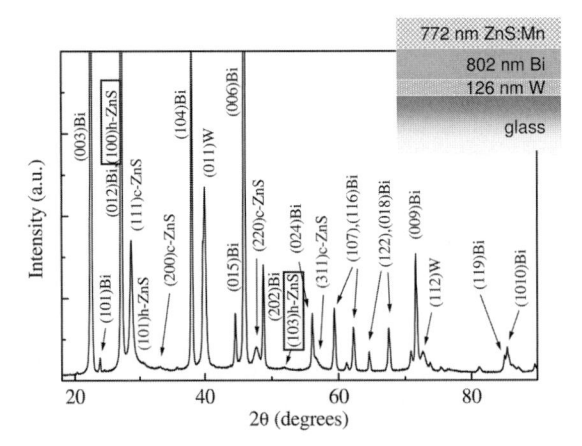

Figure 2.9 $\theta/2\theta$ diffraction pattern of ZnS:Mn on Bi/W/glass (from Ref. [13]).

niques revealed that the hexagonal phase was contained in the layer at a weight fraction of $w_j/W = 3.3$ wt%, which is intolerably high for proper device functionality. This concentration might not have been expected from the small peak intensity in the $\theta/2\theta$ pattern and emphasizes the necessity to analyze carefully the measured pattern by quantitative fitting procedures. The study also investigated the alternative growth on Sn-coated glass, where no h-ZnS could be identified and concluded that the growth of ZnS:Mn sensitively depends on the substrate used [13].

An instrumental feature to be considered in any quantitative analysis of a $\theta/2\theta$ pattern is the improvement by sample rotation or spinning. The rotation of the thin film during data collection may significantly improve the signal-to-noise ratio of the measured intensity. The geometry of the movement is shown in Fig. 2.10 in the hemisphere of the sample reference frame. In this system the x-ray tube and the detector rotate around the sample normal s_3; however, in the laboratory setup the sample is in fact rotating on the sample holder. The angle of rotation around the sample normal is termed azimuth and denoted by ϕ – which will apply throughout all following chapters.

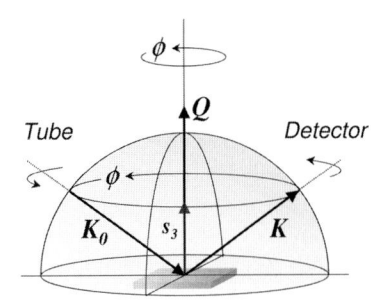

Figure 2.10 Sample rotation seen from the sample coordinate system.

Sample rotation causes an averaging of the Bragg peak intensity. The effect cannot be understood by only considering the central beam, but relies on the beam divergence δ and the registration of crystallites with reflecting lattice planes that are not quite parallel to the surface plane. These crystallites are rotated into and out of the diffraction condition. As has already been pointed out, it is only a subset of crystallites that causes the Bragg reflection, but this subset might be significantly enlarged by rotating the sample. It has to be emphasized that sample rotation will not enhance the measured peak intensity, but is associated with reducing the noise. The improvement in signal-to-noise ratio will more be pronounced the larger the crystallites. When the size of coherently scattering domains is of the order of nanometers the effect might not be too severe, but for coarse grains in the micrometer range sample spinning may lead to a significant averaging of otherwise noisy reflections.

2.5
Amorphous Thin Films

A typical phenomenon in thin-film growth at lower temperatures is the formation of amorphous phases. These are, in principle, difficult to determine by x-ray diffraction techniques because Bragg reflections rely on a spatially periodic structure. An important example is represented by amorphous silicon, a-Si. The structural arrangement of Si atoms in a-Si is characterized by the absence of a long-range crystalline order, a high density of vacancy sites and, consequently, the formation of uncoordinated chemical bonds, so-called dangling bonds. Layers of amorphous silicon may be prepared by CVD processes through plasma-assisted decomposition of gaseous silane (SiH_4) [14], which causes a high hydrogen concentration of up to several at% to be contained in the film, then described by the acronym a-Si:H. The hydrogen acts by saturating the dangling bonds and reduces the density of electronically active defects enabling the use of thin a-Si:H films as active layers in solar cells, photocopier drums and thin-film transistors of liquid crystal displays (LCDs).

In an amorphous solid the relative arrangement of atoms is not fully random, but due to their self-volume the interatomic distances between nearest neighbors compare to those in crystalline structures. The distinction between an amorphous and a crystalline configuration can be visualized by considering the atoms to be arranged on bonding spheres around a central atom. In a crystal the radii of bonding spheres r proceed in a definite fashion and the arrangement of atoms on the spheres is in accordance with the point symmetry group of the central atom. In an amorphous solid only the radii of the first very few bonding spheres are well defined whereas with increasing distance from the central atom both the distance r and the symmetric arrangement of atoms upon the spheres become washed out. It is said that an amorphous solid is only characterized by short-range order; the long-range order as in a crystallite is missing. Interestingly, the local ordering of atoms in an amorphous solid may cause a characteristic diffraction pattern and Fig. 2.11 displays the $\theta/2\theta$ diffractogram of a 2 µm thin a-Si:H film [15]. It can be seen that instead of sharp Bragg peaks large bumps occur at around 27.6° and 51° when Cu Kα radiation is applied. On the scale of interplanar spacing this corresponds to d values of 0.325 and 0.179 nm.

In 1915 Debye devised a scattering equation for short-range ordered atomic ensembles like gases and liquids that are comprised of molecules of random orientation. In the derivation he assumed that the geometrical arrangement of atoms can be described by definite interatomic distances r_{mn}, but that other structural features like the angular arrangement between various r_{mn} are lost in the diffraction experiment. The geometry is visualized in Fig. 2.12 with γ denoting the angle between the scattering vector Q and the bond vector r_{mn}. The bond vectors are equally distributed on a sphere of radius r_{mn}. If f_m and f_n are the atomic scattering factors of atoms m and n the scattering intensity of the atomic ensemble scales with

$$\sum_m \sum_n f_m f_n \exp(-iQr_{mn}) \tag{2.14}$$

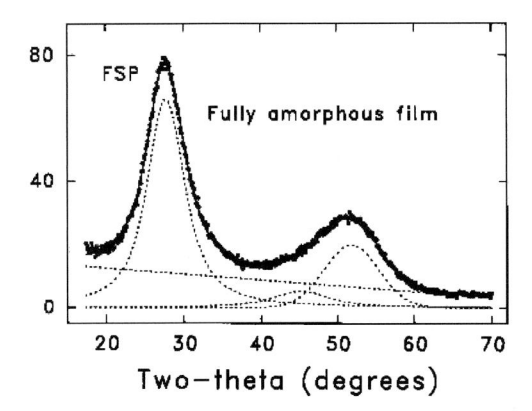

Figure 2.11 θ/2θ pattern of a 2 μm silicon film on stainless steel deposited by CVD [15].

The measured intensity is due to an averaging over all configurations of r_{mn} and $\mathbf{Q}$

$$\sum_m \sum_n f_m f_n \langle \exp(-i\mathbf{Q}\mathbf{r}_{mn}) \rangle \quad (2.15)$$

which may be evaluated by virtue of the angle γ between both vectors

$$\langle \exp(-i\mathbf{Q}\mathbf{r}_{mn}) \rangle = \frac{1}{4\pi r_{mn}^2} \int_0^\pi \exp(-iQr_{mn}\cos\gamma) 2\pi r_{mn}^2 \sin\gamma \, d\gamma \quad (2.16)$$

Performing the integration for each mn term yields the Debye scattering equation for the interference function $I(Q)$.

$$\Im(Q) = \sum_m \sum_n f_m f_n \frac{\sin(Qr_{mn})}{Qr_{mn}} \quad (2.17)$$

This formula has been successfully applied to model the diffraction pattern of many short-range ordered objects like molecules in liquids and gases [4]. It may also be applied to the modeling of amorphous substances like glass [16] and other amorphous solids. For these simulations the number of atoms in the scattering entities is usually restricted to a finite number N and fixed values r_{mn} are assumed for the different bonds. The number of bonds in the cluster is $(N^2 - N)/2 = N(N-1)/2$ and all bond lengths have to be included if the scattering is to be calculated.

The case of amorphous silicon is considered as an example of a monatomic amorphous solid. For this purpose, the solid is assumed to consists of a large number of 17-atom Si clusters that are constructed from the diamond structure by only taking the first two bonding spheres (see Fig. 2.13). The central Si atom at (000) is surrounded by a tetrahedron of four Si atoms that constitute the first bonding sphere with a radius of $\sqrt{8}a/4$. The second bonding sphere of radius $\sqrt{3}a/4$ is occupied by the 12 next-nearest neighbors that reside on the faces of the cubic unit cells. There are, however, many more interatomic distances r_{mn} than only between the central atom and those on the two bonding spheres. Because each cluster con-

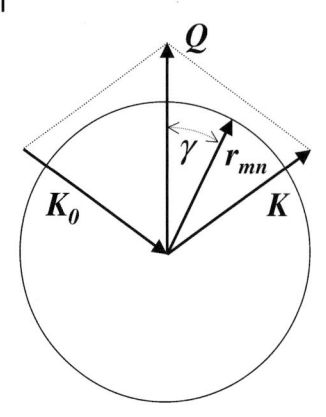

Figure 2.12 Relative arrangement of scattering vector Q and interatomic bond vector r_{mn} in an atomic cluster for the derivation of the Debye scattering equation.

tains 17 atoms we have to extend the sum of Eq. (2.17) over $17 \times 16/2 = 136$ different bond lengths r_{mn}. A numerical analysis reveals that these bonds can be grouped into nine different bond lengths r_m with frequency c_m. Values for both are given in Table 2.3. In terms of this very simplified a-Si model the Debye scattering equation takes the form

$$\mathfrak{J}(Q) = 17 f_{Si}^2 + \sum_m c_m f_{Si}^2 \frac{\sin(Qr_m)}{Qr_m} \qquad (2.18)$$

The first term is seen to scale with the number of atoms in the cluster and the square of their atomic form factor. This is a general property of Eq. (2.17). The second term has the appearance of a harmonic sum over rational multiples of the product of scattering vector Q and the smallest bond length $\sqrt{3}a/4$. In order to apply the equation for the calculation of the scattering power of an amorphous silicon film, the product $Qa = 4\pi a \sin\theta/\lambda$ has to be evaluated by inserting the Cu Kα wavelength and the Si cell edge. The course of the scattering intensity for the 17-Si-atom cluster is shown in Fig. 2.13.

Two very broad peaks are seen to occur in the same scattering angle range as shown in Fig. 2.11. The intensity is seen to reach local maxima at around $2\theta = 27.4°$ and 49° which compares with the corresponding a-Si:H pattern. Various differences between both patterns in Fig. 2.11 and 2.13 have to be stated, however, which are related to the precise maxima positions, the relative intensity of the "peaks" and their width. In particular, the width of the so-called first scattering peak (FSP) in the experimental pattern is considerably smaller than that in the simulation. The effect is assigned to the so-called medium-range order (MRO) in the amorphous network. The precise nature of MRO in amorphous silicon is still under discussion and various structural models like the ordering of vacancies, nonrandom orientation relations between individual Si clusters, etc., are considered. MRO turns out as a highly relevant issue for the functionality of the material since its improvement

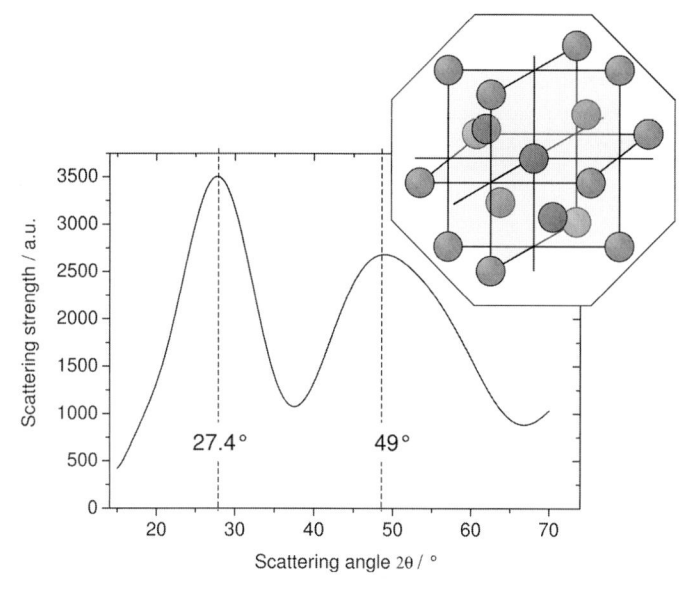

Figure 2.13 Simulation of the scattering of 17-atom Si clusters comprising the central atom and the first two bonding spheres as in the diamond structure unit cell as shown in the inset.

was shown to correlate with better light stability [17]. As a further point of concern, the inclusion of hydrogen has to be introduced in any proper modeling, which was neglected here. Finally, the pattern shown in Fig. 2.13 would have to be corrected by including the Lorentz polarization factor Lp and the absorption factor $A_{\theta 2\theta}$ to make a proper comparison. The interesting aspect in applying Debye's scattering equation is that it represents a good starting point to the structural modeling of amorphous solids. And, in fact, it is often found to yield the most characteristic features of a $\theta/2\theta$ diffraction pattern of an amorphous solid, but a subsequent improvement is required to arrive at more adequate structural models.

A central issue in applying Eq. (2.17) to amorphous substances is the assumption of well-defined bond lengths r_{mn}. The approach is fully justified for molecules, but in an amorphous network the interatomic distances are subject to fluctuations that increase with increasing magnitude of r_{mn}. The effect is considered in a more appropriate modeling by introducing the so-called pair distribution function $P(r)$. This function describes for a central atom the probability P with which other atoms will be found at a distance r. For a perfect crystalline substance $P(r)$ will exhibit sharp maxima whenever a bonding sphere at r_m is crossed. These peaks will become washed out in the $P(r)$ of an amorphous solid and will reach a saturated value $\bar{P}$ for large r. In fact, amorphous networks may also be characterized by variations of bond distances r_{mn} whereas in the original derivation of the Debye scattering equation only angular bond variations were considered. A further property of the Debye scattering equation is that it does not distinguish between (a) the random orientation of a large number of scattering clusters with respect to Q and (b)

Table 2.3 Interatomic distances r_m and their number of appearances c_m in a 17-Si-atom cluster in the cubic diamond structure. Bond lengths r_m are given in units of the Si cell edge a.

m	0	1	2	3	4	5	6	8
r_m/a	0	$\sqrt{3}/4$	$\sqrt{8}/4$	$\sqrt{11}/4$	$\sqrt{16}/4$	$\sqrt{19}/4$	$\sqrt{24}/4$	$\sqrt{32}/4$
c_m	17	32	84	48	24	12	48	12

a possible random arrangement of atoms on their bonding spheres. Recently, the approach has been extended by Wieder and Fuess by introducing a bond orientation distribution function that allows for a generalization of Debye's scattering equation [18]. The authors arrived at a general expression from which both the Laue equation or the Debye scattering equation result as limiting cases.

A typical observation made in thin-film technology is that when the preparation of polycrystalline material at low temperatures is intended, quite often only a partially crystallized layer is obtained that comprises an amorphous fraction. If a diffractogram of such a film is to be analyzed sufficiently long measurement times are recommended to achieve a good signal-to-noise ratio, because the amorphous scattering is not concentrated into a few Bragg peaks but is equally distributed over a large range of scattering angles 2θ. Various techniques might be applied to determine quantitatively the amorphous fraction – which is not a simple task. If both the amorphous and the crystalline phase are thoroughly dispersed into each other the intensity $I(Q)$ can be conceived as a superposition of distinct Bragg peaks and an amorphous share described by Debye scattering. Preferably, one uses for this analysis two standard patterns of both a fully crystallized and a fully amorphous film. These patterns can be combined numerically to model the experimental pattern of a layer comprising both phases [19]. Another approach was introduced by LeBail, who simulated an amorphous solid by small nanometer-sized crystallites the Bragg peaks of which were subjected to severe line broadening [20] (see Chapter 3).

2.6
Accurate Determination of Lattice Parameter

A further possibility of chemical phase identification or stoichiometry determination is related to the precise measurement of lattice parameters like the unit cell edge a of a cubic structure. The method may advantageously be applied to ternary compound systems such as $Si_{1-x}Ge_x$, $Si_{1-x}C_x$ or $(Cd_{1-x}Zn_x)Te$. Thin films of these materials are prepared to adjust the electronic band gap energy E_g to the intended functionality. In some cases, like the introduction of C into a Ge lattice, it is not only the gap energy but even the nature of the fundamental electronic transition that may be varied from an indirect to a direct one. The alloying of different semiconductors can thus be considered as band gap engineering. The stoichiometry variation will not only affect the electronic properties of the material but also the aver-

2.6 Accurate Determination of Lattice Parameter

age dimension of the crystallographic unit cell. Restricting the discussion to cubic structures the unit cell edge $a(x)$ of an alloy $A_{1-x}B_x$ is found between the unit cell edges of the individual structures of elements A and B. For most cubic semiconductors $a(x)$ is described with sufficient precision as depending linearly on x. A linear relation between unit cell length a and stoichiometry has been described by Vegard to hold for $KBr_{1-x}Cl_x$ bulk material [21] and any linear relationship

$$a(x) = a_0 + a_1 x \qquad (2.19)$$

between both is generally termed Vegard's rule.

If the stoichiometry of a ternary $A_{1-x}B_xC$ is to be determined one may proceed by examining the unit cell edge a from the peak positions in a $\theta/2\theta$ diffraction pattern via the Bragg equation. This will yield the stoichiometry coordinate x by use of Eq. (2.19). However, caution should be taken in applying Vegard's rule with respect to various sources of possible misinterpretations. One source of error is related to the use of the constants a_0 and a_1. These values have in most cases been determined for bulk material and their application to the thin-film system under investigation at least has to be critically considered. Also the standard deviations Δa_0 and Δa_1 should be considered with respect to their effect on the stoichiometry error through the laws of error propagation. Additionally, Bragg peak positions from thin-film specimens might be shifted due to intrinsic stresses and, thereby, intrinsic strains. The peak shift is then due to mechanical reasons and erroneous results are obtained by interpreting the shift in terms of chemical composition.

For stoichiometry determinations via Eq. (2.19) the analyst is recommended to perform a careful calibration of the instrument as a first step. In general, all procedures as outlined in the manual for the instrument should be thoroughly carried out. As a second step, a standard sample should be measured for which the unit cell is known with high accuracy. A widely used standard is a silicon powder sample distributed by the National Institute of Standards and Technology (NIST), denoted as standard reference material SRM 640. Its cell edge is given with an accuracy of better than 2×10^{-6}. The powder pattern of this sample should be measured over the whole accessible range of scattering angles. Cell edges a_h should then be calculated from each reflection h and plotted as a function of scattering angle 2θ. In most cases it will be seen from this plot that the first three digits of the cell edge for each reflection a_h compare to the NIST-specified value of $a_{640c} = 0.54311946(92)$ nm. However, a closer inspection of the course of $a_h(2\theta)$ will probably reveal systematic variations of a_h over 2θ. Finally, the procedures given below should be applied to derive a cell edge value of the NIST sample. The difference with respect to the NIST value gives an estimate of how accurate unit cell edges – and via Vegard's rule, stoichiometries – might be determined with the instrument used. The significance of measuring standards in x-ray scattering cannot be overemphasized. The reader is strongly recommended to do so on a regular basis and whenever the configuration of the instrument is changed. In particular, after a new instrument has been delivered to a laboratory or when the analyst is a newcomer to the field the measurement and thorough analysis of standards will be helpful to realize the instrumental limitations to which the results are subjected.

It is considered in the following which instrumental errors might be introduced when attempting to determine the cell edge a with high precision. The discussion is restricted to cubic structures as group IV and II–VI semiconductors typically adopt them. Unit cell length a and Bragg angle θ_B are related by

$$a = \frac{\lambda\sqrt{h^2 + k^2 + l^2}}{2\sin\theta_B} \tag{2.20}$$

This formula suggests that the cell edge a might be derived with high accuracy from any reflection $\mathbf{h}$. In fact, it is relatively easy to measure the centroid θ_0 of a reflection, i.e. its center of mass, with a precision of the order $\Delta(2\theta)/2\theta = 10^{-5}$. Also, the Cu K$\alpha$ wavelengths are known with an accuracy of better than 10^{-6}. However, this does not mean that the cell edge a can be derived with the same accuracy by simply evaluating Eq. (2.20). The loss in accuracy is due to instrumental shifts $(\Delta 2\theta)_j$ to which the peak is subjected in the parafocusing configuration. Although these shifts are small, the analyst should be aware of them.

The effect of the instrumental peak shifts on the determination of the unit cell edge can be considered by replacing θ_B in Eq. (2.20) by the sum of the measured centroid position θ_0 minus the shifts

$$\sin\left(\theta_0 - \frac{1}{2}\sum\Delta(2\theta)_i\right) = \frac{\lambda\sqrt{h^2 + k^2 + l^2}}{2a} \tag{2.21}$$

The sum has to be extended over the various shifts associated with the actual configuration of the instrument. A question arises: what is the nature of the shifts and can they be analytically be described? Wilson [22] and Parrish et al. [23] reviewed the aberrations of peak positions in a powder diffractometer and categorized them into those of geometrical and those of physical nature. The most relevant aberrations are now presented. For their derivation the reader is referred to the work of Wilson [22].

1. *Zero shift.* The inaccuracies of the gear system and detector arm movement may be associated with a constant shift of Bragg peak positions over the whole range of scattering angles

$$\Delta(2\theta)_1 = c \tag{2.22}$$

Relying on the occurrence of harmonic reflections, techniques have been proposed by which the zero shift in the pattern may be eliminated if this is the dominant instrumental shift [24, 25]. These procedures might be applied to obtain a starting value for $(\Delta 2\theta)_1$ for the full regression of all shifts (see below).

2. *Height misalignment.* The deviation of the sample surface from the θ and 2θ axes is considered as a height misalignment b and is associated with a shift of the reflection by

$$\Delta(2\theta)_2 = -2b\cos\theta / R \tag{2.23}$$

where R is the radius of the diffractometer. The shift is to higher (lower) scattering angles for a sample position above (below) the tangent on the focusing circle and the shift approaches zero for $2\theta \to 180°$. In order to tackle this source of error a dial gauge is available for many diffractometers or the sample is mechanically pressed by a spring against a defined position to adjust the sample height to a precision in b of several micrometers. For an assumed misalignment of $b = +8$ μm, with $R = 280$ mm and $2\theta = 28.4°$ a shift of $(\Delta 2\theta)_2 = 0.003°$ is calculated for the Si (111) reflection, yielding $(\Delta 2\theta)_2/\theta_B = 10^{-4}$ and thus an absolute error in cell length of $\Delta a \approx 5 \times 10^{-5}$ nm. The height misalignment may be dominant for low scattering angles and, in general, a highly precise cell edge value a cannot be expected from evaluating a single low-angle reflection. With the recent introduction of graded multilayer mirrors a conventional powder diffractometer in parafocusing configuration can be converted into a parallel beam configuration (see Instrumental Box 5 in Chapter 4). One of the important advantages of parallel beam geometry is that a peak shift by height misalignment is systematically avoided.

3. *Sample transparency.* X-rays are not directly reflected at the sample surface, but the beam enters into the sample volume up to a depth that is governed by the linear attenuation coefficient μ. The peak shift due to beam penetration may be evaluated by considering each scattering lattice plane to be subjected to a height misalignment z that is the distance to the sample surface. The total shift then derives from the averaging over all depths

$$\Delta(2\theta)_3 = -\frac{2}{R}\langle z \rangle \cos\theta \tag{2.24}$$

The averaging is achieved as usual by integrating over all depths from 0 to t

$$\langle z \rangle = \frac{\int_0^t z \exp(-2\mu t/\sin\theta)dz}{\int_0^t \exp(-2\mu t/\sin\theta)dz} = \frac{\sin\theta}{2\mu}\left\{1 + \frac{2\mu t/\sin\theta}{1 - \exp(2\mu t/\sin\theta)}\right\} \tag{2.25}$$

This yields in the limit of a thin sample, having $t \ll 1/\mu$

$$\Delta(2\theta)_3 = -t\cos\theta/R \tag{2.26}$$

Equation (2.25) may also be approximated for thick samples ($t \gg 1/\mu$) but for thin-film work only Eq. (2.26) will be of relevance. For films less than 1 μm thick the shift becomes negligibly small for conventional diffractometers.

4. *Flat sample.* In the parafocusing Bragg–Brentano geometry the focusing circle must intersect the goniometer circle at any measuring point $(\theta, 2\theta)$. This condition is only strictly obeyed by a sample that is bent with a radius of the focusing circle. However, a flat specimen is only aligned tangentially and the focus of the reflection is thus shifted by an amount that scales with the square of the beam divergence in the scattering plane. If a divergence slit of opening angle δ_{ds} is used this shift becomes

$$\Delta(2\theta)_4 = -\frac{\delta_{ds}^2}{6}\cot\theta \tag{2.27}$$

Also this shift takes its largest values for small 2θ and approaches zero with increasing scattering angle.

5. *Axial divergence.* In addition to the scattering plane divergence the x-ray beam exhibits an axial divergence δ_a accounting for the angular spread of the beam in direction of the θ and 2θ axes. The axial divergence δ_a is preferably restricted by inserting Soller slits in the beam path. The two cases of using either one or two Soller slits have to be distinguished. In a setup with Soller slit aperture angle δ_{Sol} the peak shift becomes

$$\Delta(2\theta)_{5,\text{oneSol}} = -\frac{1}{12}\left(\delta_{Sol}^2 + \frac{B_a^2}{R^2}\right)\cot 2\theta \tag{2.28}$$

when one slit system is either inserted in the incident or the diffracted beam path. Here B_a accounts for the width of the focal line on the sample and the receiving slit. A typical value for B_a would be the length of the x-ray focus, i.e. 12 mm in the long fine-focus mode. In the case of two Soller slits the peak shift due to the axial divergence reads

$$\Delta(2\theta)_{5,\text{twoSol}} = -\frac{\delta_{Sol}^2}{6}\cot 2\theta \tag{2.29}$$

In both cases the condition for narrow slits, $R\delta_{Sol} < B_a$, must hold in order for the formulas to apply, but the condition will probably be obeyed in any high-precision work. Other cases were considered by Wilson [22]. It is realized in the case of two Soller slits that the axial divergence shift can be made arbitrarily small by reducing δ_{Sol}, which would remain unattainable with only one set of slits due to the finite axial extension of the focus (B_a/R term). Therefore, the use of two Soller slits is recommended if highly accurate cell parameters are desired. The choice of divergence δ_{Sol} should be made in accordance with intensity optimization, which may significantly decline when two Soller slits are inserted.

The above listed aberrations are the most relevant ones that might occur in a symmetrical $\theta/2\theta$ scan in parafocusing geometry. Further shift components were investigated by Wilson and others, but these appear to be of minor importance in modern diffractometer systems. The various aberrations $\Delta(2\theta)_j$ are depicted in Fig. 2.14, where the instrumental parameters were set to some typical values. It should be noted that the shifts appear as their negative values in Eq. (2.21). The plot shows that the shift due to height misalignment, sample transparency and setting of the divergence slit all tend to zero for $2\theta \rightarrow 180°$. The shift due to axial divergence, however, tends to increase for both $2\theta \rightarrow 0°$ and $180°$ and has a minimum around $2\theta = 90°$. If a highly accurate lattice parameter is to be derived the a values of each reflection are advantageously plotted as a function of scattering angle 2θ. The course of the function should then be subjected to a numerical regression to fit Eq. (2.21) where the unit cell edge a and the unknown instrumental parameters serve as variables that have to be determined. The technical point should be mentioned that all shift formulas are given in radians and thus have to be converted if θ or 2θ are specified in degrees.

In the so-called internal standard method in powder diffractometry a standard powder with known cell edge is dispersed into the powder with cell parameters to be determined. Since both diffraction patterns superimpose in a $\theta/2\theta$ scan of the mixture, the unknown cell edge can be gauged with the known one. This is performed by determining the total shift $\Delta(2\theta)$ from each reflection of the standard and inserting interpolated values into Eq. (2.21) to correct for the unknown unit cell edge. This technique may be applied to thin films when their preparation proceeds by wet-chemical procedures into which a standard powder may be inserted. Possible effects on the standard's cell edge by the preparation must be excluded, however.

Finally, the important distinction between precision and accuracy has to be emphasized. The point is of general importance in x-ray work and can be illustrated by an example from the evaluation of unit cell length a. Assume that a single reflection hkl is measured with a small step width and a long counting time to derive the centroid of the reflection $2\theta_0$ very carefully. The application of Bragg's equation in the form of Eq. (2.20) will then yield a precise value of the unit cell edge a, but not an accurate one. A highly accurate value of a is only obtained by making use of Eq. (2.21) and the related procedures. Therefore, if real physical quantities are required that have to be compared with textbook values, for instance, high accuracy is needed, for which high precision is only a presupposition to be obtained. In general, one may associate precision with reproducible measurement results and accuracy with the correction for possible systematic errors [23]. In practice the dis-

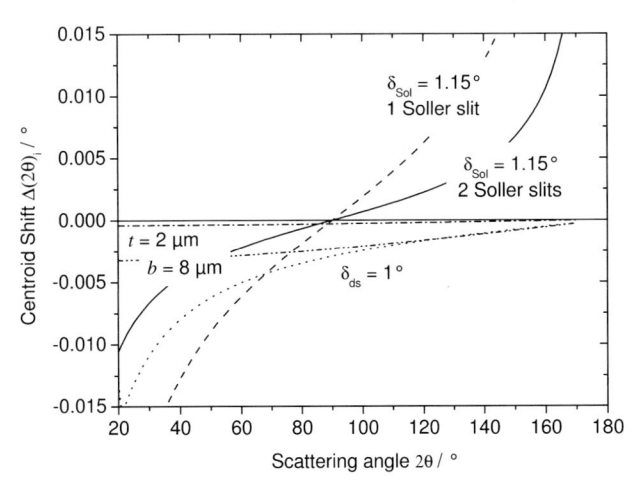

Figure 2.14 Instrumental peak shifts due to height misalignment ($b = 8$ μm), sample transparency ($t = 2$ μm) and beam divergence $\delta_{ds} = 1°$ and $\delta_{Sol} = 1.15°$. The parameters $R = 280$ mm and $B_a = 12$ mm were used.

tinction becomes apparent when not only a value for the cell edge a or the stoichiometry coordinate x is to be derived from the $\theta/2\theta$ scan but also their estimated standard deviations are required according to the laws of error propagation. This is the main task in the accurate determination of unit cell edges and it is strongly recommended that it be performed.

2.7
Applications

The separation of a thin-film $\theta/2\theta$ scattering pattern into fractions from different chemical phases might be complicated due to one of the following reasons. Firstly, the dispersion of phases might not be a homogenous one, but a kind of multilayer stacking may appear with one phase layer adjacent to the substrate. Secondly, the phases might exhibit a texture with most of the crystallites orienting along a selected direction h parallel to the substrate normal s_3. Thirdly, the dispersion of phases or the degree of texture may exhibit a pronounced depth dependency within the thin film. In order to decide on how strongly these phenomena affect the thin film under investigation a thickness series should be examined, which is a set of samples that were all prepared under the same deposition conditions but with varying deposition times. The obtained set of $\theta/2\theta$ diffraction pattern measured for this series gives an estimation of how strong a possible thickness dependence may be. For instance, an amorphous interlayer can be uncovered in such a series, because its broad scattering peaks would continuously diminish with increasing thickness. If the amorphous phase is only located at the interface the modeling of the pattern by a multilayer approach is recommended comprising a bottom amorphous layer and a top crystalline layer. Such spatial phase separations have indeed been observed to occur during the growth of a large variety of thin films like those from the Si–H [10, 26] system and many other. From a more general point of view the initial occurrence of metastable phases is often observed, which can be amorphous, but can also be of polycrystalline nature in a metastable modification that changes to a thermodynamically stable phase with increasing thickness, for instance in chemical vapor-deposited aluminum oxide [27]. These growth phenomena have been assigned to equilibrium processes between bulk and surface energies. The explanation is based on the rule of stages of Ostwald [28] as was probably first realized by Buckel for low-temperature-deposited metallic films [29, 30] and recently applied to μc-Si thin films [10]. The modeling should be performed in all these cases by use of a multilayer model instead of a phase-mixture one. Also a possible depth dependence of preferred orientation will be revealed by a thickness series.

The accurate determination of cell edges has, for instance, been applied to powders of iron pyrite to elucidate the constants a_0 and a_1 in Vegard's rule for sulfur-deficient FeS_{2-x} [31]. Pyrite has been investigated with respect to various applications in thin-film technology like energy storage and solar cells. The stoichiometry deviation is of particular importance, since a too large a density of defect states in the band gap would prohibit the material's integration into electronic devices. The determination of the stoichiometry deviation is only possible in high-precision work by taking into account all shift components mentioned in Section 2.6, because the a_1 coefficient in Vegard's rule is of the order of 1.5×10^{-4} nm for sulfur-deficient iron pyrite. A stoichiometry deviation of $x = 0.02$, which would be highly relevant for device functionality of pyrite layers, would relate to a cell edge variation $\Delta a/a_0 = xa_1/a_0$ of only 5.5×10^{-6}. In the cited study the investigations were performed on powder material, which appears to be a justified strategy whenever basic material questions have to be considered. It has to be emphasized that the application of these procedures to polycrystalline films may be a challenging task. For one reason this is due to the reduced material volume available in thin films and translates into higher counting times to obtain a comparable signal-to-noise ratio. For another reason, there are more systematic peak shifts possible in thin films, as mentioned above. For polycrystalline films the $\Delta a/a_0$ variations intended to be observed for an application of Vegard's rule should probably exceed the range 10^{-3}–10^{-4} in order to derive reliable x values for the determination of either stoichiometry deviations or chemical composition.

Exercises

2.1 The mass absorption coefficient μ_m of N and O are 0.74 and 1.104 m²/kg for Cu $K\alpha$ and 0.079 and 0.116 m²/kg for Mo $K\alpha$ radiation. Calculate the linear attenuation coefficient of a gaseous mixture of N_2 and O_2 (78.4 vol% over 21.6 vol%) with a density of 1.205 kg/m³ and depict I/I_0 as a function of an x-ray path length from 0 to 1 m for both wavelengths.

2.2 The outermost tissue of the human skin comprises of O, C, H, N, Ca, Cl, P, K and S with relative weights of 56.1, 28, 9.3, 2, 1.5, 1, 1, 0.25, 0.2%. Calculate the absorption coefficient and penetration depth of Cu $K\alpha$ by assuming a specific gravity of $\rho_{skin} = 1.1 \times 10^3$ kg/m³. It is in this upper layer where the total beam energy is deposited. X-rays may cause severe skin burnings when exposing your body to them. Manual adjustments of the sample with an opened tube shutter can cause ulceration that might only be treated by amputation. Make 100% use of radiation protection measures and be aware of possible harms.

2.3 Assume an elemental sample is illuminated by Cu $K\alpha$ or Mo $K\alpha$ radiation. Elucidate by searching the appropriate tables as to which elements would yield maximum fluorescence for these wavelengths. Which radiation tube might alternatively be applied to avoid the strong fluorescence background from these elements?

2.4 Calculate the distances of the first four bonding spheres for a C atom in the diamond lattice and give their respective occupation numbers. How do the latter relate to the point symmetry group T_d of the C atom position?

2.5 Next to the diamond structure carbon adopts other structural modifications, from which the graphite structure is the most important one. In the hexagonal graphite unit cell (space group $P6_3/mmc$, $a = 0.2464$ nm, $c/a = 2.723$) the carbon atoms occupy (000), (00½), (⅓ ⅔ 0) and (⅔ ⅓ ½) positions. What is the graphite structure factor?

2.6 The interatomic distances in the O_2 and N_2 molecules are 0.1207 and 0.1098 nm, respectively. Calculate the $I(Q)$ diffraction pattern of both molecules according to the Debye scattering equation and the appropriate superposition of both for air. How would the pattern of He gas under the same ambient conditions compare?

2.7 Glass is a widely used substrate material for thin-film deposition. It is used in laboratory experiments and in industrial practice. Typical usages are related to optoelectronic thin films for solar cells, thin-film transistors, IR-protective coatings, etc. Glass comprises mainly amorphous silicon dioxide into which other metal oxides are included. Pure amorphous SiO_2 is denoted as quartz glass, but its use is restricted due to its high price to very few applications. The atomic short-range order in quartz glass is characterized by a tetrahedral coordination of silicon atoms to four oxygen neighbors, exhibiting a bond length of 0.162 nm. Calculate the $I(Q)$ pattern according to the Debye scattering equation for a single Si_2O_4 cluster with a central SiO_4 tetrahedron and an additional Si linked to one of the oxygen atoms. At which scattering angle would the first two peaks occur when Cu Kα radiation is applied? Where are the peak maxima and what is their full width at half maximum if one performs a symmetric $\theta/2\theta$ scan of a glass substrate?

2.8 The thermal expansion coefficient of silicon is 4×10^{-6} K^{-1} at room temperature. How much could the cell edge a of the NIST SRM 640c powder vary during the measurement, if the temperature of the sample was stable to ±1 K.

2.9 Friedel's law states that the intensity of a single crystal reflection is the same for lattice planes (hkl) and $(\bar{h}\bar{k}\bar{l})$. Give the general prove for the principle and neglect anomalous scattering.

2.10 Friedel's law may be violated when anomalous dispersion is included. In order to illustrate the phenomenon calculate the structure factor of ZnO in wurtzite structure for (00.2) and (00.$\bar{2}$) by using the form factors $f'_{Zn} + if''_{Zn}$ and $f'_O + if''_O$. What will the ratio $F^2(00.2)/F^2(00.\bar{2})$ look like?

2.11 How has the Lorentz polarization factor Lp to be modified when a secondary monochromator is introduced into the beam path that operates under Bragg angle θ_{SM}? Hint: only the polarization factor is affected.

2.12 Some ternary metal-chalcogenide compounds ABC_2 crystallize in the tetragonal chalcopyrite structure, space group $I\bar{4}2d$, with the prototype compound being chalcopyrite $CuFeS_2$. The structure may be imagined by stacking two sphalerite unit cells onto each other and elongating or shrinking the new cell along the stacking direction. A tetragonal cell then results having $c/a \neq 2$. The chalcogen positions are slightly shifted in the xy plane as accounted for by fractional coordinates (¼, u, ⅛), ($\bar{u}$, ¾, ⅛), (¼, ½ + u, ⅜), (¾, ½ − u, ⅜), (½ + u, ¼, ⅝), (½ − u, ¼, ⅝), (¼, $\bar{u}$, ⅞), (¾, u, ⅞), with positional parameter u. The metal atoms are distributed in an alternating fashion over metal lattice sites with A on (000), (0½¼), (½½½) and (½0¾) and B on the remaining ones (for instance, in CuGaSe$_2$ $c/a = 1.966$ and $u = 0.243$). Can you derive the structure factor? For which values c/a and u does the chalcopyrite structure merge into the sphalerite structure for equal A and B atoms?

References

Monographs relevant to this chapter

L. E. Alexander, H. P. Klug, *X-ray Diffraction Procedures for Polycrystalline and Amorphous Materials*, 2nd ed., Wiley-Interscience, New York, 1974.

A. J. C. Wilson, *Mathematical Theory of X-Ray Powder Diffractometry*, Philips Technical Library, Eindhoven, 1963.

Special papers

[1] D. K. Smith, R. Jenkins, ICDD and the Powder Diffraction File-Past, Present and Future, published by the International Center for Diffraction data; www.icdd.com/products/ICDD.PDF

[2] W. Kraus, G. Nolze, Powder Cell: a program for the representation and manipulation of crystal structures and calculation of the resulting X-ray powder patterns, *J. Appl. Cryst.* **29** (1996) 301.

[3] M. Toplak, Chemische Gasphasendeposition von Zinkselenidschichten für die Anwendung als Puferschicht in Chalkopyrit-Dünnschichtsolarzellen, Diplomarbeit am Fachbereich Physik, Freie Universität, Berlin, 1998.

[4] L. E. Alexander, H. P. Klug, *X-ray Diffraction Procedures for Polycrystalline and Amorphous Materials*, 2nd edn, Wiley-Interscience, New York, 1974.

[5] S. M. Seltzer, Calculation of photon mass energy-transfer and mass energy-absorption coefficients, *Rad. Res.* **136** (1993) 147; physics.nist.gov/PhysRefData/XrayMassCoef/cover.html.

[6] D. C. Creagh, J. H. Hubbel, X-ray absorption (or attenuation) coefficients, in *International Tables for Crystallography*, Vol. C, ed. A. J. C. Wilson, E. Prince, Kluwer, Dordrecht, 1999, p. 220.

[7] W. G. Sloof, B. J. Kooi, R. Delhez, T. H. de Keijser, E. J. Mittemeijer, Diffraction analysis of nonuniform stresses in surface layers: Application to cracked TiN coatings chemically vapor deposited on Mo, *J. Mater. Res.* **11** (1996) 1440.

[8] J.-D. Kamminga, R. Delhez, T. H. de Keijser, E. J. Mittemeijer, A tool for X-ray diffraction analysis of thin layers on substrates: substrate peak removal method, *J. Appl. Cryst.* **33** (2000) 108.

[9] P. Scardi, L. Lutterotti, XRD characterization of multilayered systems, *Thin Solid Films* **236** (1993) 130.

[10] M. Birkholz, B. Selle, W. Fuhs, S. Christiansen, H. P. Strunk, R. Reich, Amorphous-crystalline phase transition during the growth of thin films: the case of microcrystalline silicon, *Phys. Rev. B* **64** (2001) 085402.

[11] F. Fenske, B. Selle, M. Birkholz, Preferred orientation and anisotropic growth in polycrystalline ZnO:Al films prepared by magnetron sputtering, *Jap. J. Appl. Phys. Lett.* **44** (2005) L662-L664.

[12] S. Battaglia, M. Franzini, L. Leoni, Simultaneous determination of layer thickness, composition, and mass absorption by X-ray diffraction, *Powder Diffract.* **7** (1992) 194.

[13] M. A. Tagliente, M. Penza, M. Gusso, A. Quirini, Characterization of ZnS:M thin films by Rietveld refinement of Bragg–Brentano X-ray diffraction patterns, *Thin Solid Films* **353** (1999) 129.

[14] G. Bruno, P. Capezzuto, A. Madan, *Plasma Deposition of Amorphous Silicon-Based Materials*, in *Plasma-Materials Interactions* (Academic Press, Boston, p. 324, 1995).

[15] D. L. Williamson, Medium-range order in a-Si:H below and above the onset of microcrystallinity, *Mater. Res. Soc. Symp. Proc.* **557** (1999) 251.

[16] B. E. Warren, *X-Ray Diffraction*, Addison-Wesley, Reading, MA, 1969, chapter 11.

[17] S. Guha, J. Yang, D. L. Williamson, Y. Lubianiker, J. D. Cohen, A. H. Mahan, Structural, defect, and device behavior of hydrogenated amorphous Si near and above the onset of microcrystallinity, *Appl. Phys. Lett.* **74** (1999) 1860.

[18] T. Wieder, H. Fuess, A generalized Debye scattering equation, *Z. Naturf.* **A52** (1997) 386.

[19] L. Houben, M. Luysberg, P. Hapke, R. Carius, F. Finger, H. Wagner, Structural properties of microcrystalline silicon in the transition from highly crystalline to amorphous growth, *Phil. Mag. A* **77** (1998) 1447.

[20] A. Le Bail, Modelling the silica glass structure by the Rietveld method, *J. Non-Cryst. Solids* **183** (1995) 39.

[21] L. Vegard, Die Konstitution der Mischkristalle und die Raumfüllung der Atome, *Z. Phys.* **5** (1921) 17.

[22] A. J. C. Wilson, *Mathematical Theory of X-Ray Powder Diffractometry*, Philips Technical Library, Eindhoven, 1963.

[23] W. Parrish, A. J. C. Wilson, J. I. Langford, 5 Determination of lattice parameters; 5.2 X-ray diffraction methods: polycrystalline, in *International Tables for Crystallography*, Vol. C, ed. A. J. C. Wilson, E. Prince, Kluwer, Dordrecht, 1999, p. 485.

[24] S. Popovic, An X-ray diffraction method for lattice parameter measurements from corresponding $K\alpha$ and $K\beta$ reflexions, *J. Appl. Cryst.* **4** (1971) 240.

[25] C. Dong, F. Wu, H. Chen, Correction of zero shift in powder diffraction patterns using the reflection-pair method, *J. Appl. Cryst.* **32** (1999) 850.

[26] M. Birkholz, B. Selle, E. Conrad, K. Lips, W. Fuhs, Evolution of structure in thin microcrystalline silicon films grown by electron-cyclotron resonance CVD, *J. Appl. Phys.* **88** (2000) 4376.

[27] O. Zywitzki, G. Hoetzsch, Correlation between structure and properties of reactively deposited Al_2O_3 coatings by pulsed magnetron sputtering, *Surf. Coat. Technol.* **94/95** (1997) 303.

[28] W. Ostwald, Studien über die Umwandlung fester Körper, *Z. Phys. Chem.* **22** (1897) 289.

[29] W. Buckel, The influence of crystal bonds on film growth, in *Elektrische en Magnetische Eigenschappen van dunne Metallaagies*, ed. IUPAP, Palais der Academien, Leuven, 1961, p. 264.

[30] W. Buckel, Liquid-like Amorphous Thin Films, in: *Diffraction Studies on Non-Crystalline Substances*, I. Hargittai and W. J. Orville-Thomas (eds.), Akademiai Kiado, 1981 p. 711.

[31] M. Birkholz, S. Fiechter, A. Hartmann, H. Tributsch, Sulfur deficiency in iron-pyrite (FeS_{2-x}) and its consequences for band-structure models, *Phys. Rev. B* **43** (1991) 11926.

3
Line Profile Analysis

Soon after the invention of x-ray diffraction it became clear that the method was not only suited to resolve crystalline structures, but could also reveal information on a sample's microstructure. In 1918 Scherrer devised a formula describing how the width of a Bragg reflection increases with decreasing crystallite size. The formula became very popular in the investigation of polycrystalline samples and made the crystallite size–peak width relation an ongoing issue in x-ray diffraction. It has to be emphasized, however, that the crystallite size derived from x-ray peak profiles accounts for domain sizes that scatter the incoming x-rays coherently. The size of coherently diffracting domains is generally smaller than the crystallite size as obtained by other techniques like transmission electron microscopy. It thus has to be carefully specified as to which experimental quantity is exactly meant whenever crystallite sizes are compared that have been obtained by different techniques.

The width of Bragg reflections was also realized to be affected by microstructural features other than size broadening, like the crystallite shape, the shape distribution and any distortion of the crystal structure like microstrain, dislocations, twin planes and stacking faults. The general question in the microstructural analysis of polycrystals relates to how the physical properties of the specimen are affected by its microstructure. It is evident that the microstructural richness of a polycrystalline thin film can become a hard task to be fully elucidated. The investigations are thus often restricted to certain aspects like the density of certain lattice faults.

Microstrain and dislocations play a prominent role in the various distortions of the crystal lattice. In the vicinity of dislocations the atoms reside on equilibrium positions distinct from those in the unperturbed lattice and cause the surrounding bond lengths to contract and expand. The strain fields introduced by dislocations may extend over many hundreds of unit cells in the crystal. The interplanar spacing d as it appears in the Bragg equation is thus subjected to a variation and may not be accounted for by a fixed value d_0, but by a distribution of d values. The strain fields thus cause a smearing of scattered x-ray intensity around d_0 and a broadening of reflections. These strain fields are denoted as microstrain, because they appear on a length scale that is small when compared with the inverse linear attenuation coefficient, $1/\mu$, of the probing x-ray beam. Although this definition is rather unsatisfactory, because a sample property is related to the technique by which it is

Thin Film Analysis by X-Ray Scattering. M. Birkholz
Copyright © 2006 WILEY-VCH Verlag GmbH & Co. KGaA, Weinheim
ISBN: 3-527-31052-5

probed, it is a practical approach and has become widely used. In consequence, the microstrain fields rise and fall within the x-ray illuminated sample volume. This contrasts with so-called macrostrains which exceed the $1/\mu$ scale and can cause a complete shift of a Bragg reflection from d_0 to a new lattice spacing $d_0 + \Delta d$. In this chapter only the reflection broadening due to microstrains is considered, whereas peak shifts due to macrostrains are postponed to Chapter 6.

Line profile analysis (LPA) endeavors to derive microstructural features in the sample from the shape and broadening of Bragg reflections. Because broadening due to crystallite size and microstrain typically occur together, techniques were to develop that allow for the separation of both effects. These developments were mainly performed by investigations of bulk material from metals or simple inorganic compounds. LPA ideally requires diffraction patterns with a high signal-to-noise (S/N) ratio and freestanding reflections with negligible overlap. In many cases the usage of Kα_1 monochromatized radiation is recommended to obtain reliable results. It is evident from this listing that the applications of LPA to thin films may not be straightforward. Some of the following examples will therefore relate to powder samples, for which the techniques were originally developed. It has to be emphasized that the crystallite orientation distribution is assumed in many LPA approaches to be a random one. This presupposition is not generally fulfilled in polycrystalline thin films: rather, preferred orientation or texture is typically observed. This does not invalidate the application of LPA techniques in principle, but it should carefully be checked as to how far the interpretation of the data would be affected by texture effects.

This chapter follows the historical development of LPA in so far as size effects are presented first, while the analysis of concomitant size and strain broadening is outlined subsequently. It should be kept in mind, however, that thin-film samples typically exhibit both microstructural properties. Instrumental Box 4 is devoted to numerical methods and software techniques that are used in the analysis of x-ray scattering data. Nanocrystalline and nanocomposite materials are chosen as the material class for which the concepts of LPA are illustrated. In these materials new functionalities may be elicited, since the ratio of surface over bulk atoms may achieve unusually large values. Structure Box 4 covers crystalline lattice faults.

3.1
Model Functions and Peak Parameters

The first step in LPA is most often the numerical regression of the measured Bragg reflections by standardized model functions. This procedure can be considered as a data reduction technique: instead of handling N data points $(2\theta_j, I_j)$ measured over a certain range of 2θ this step reduces the full data set to a few parameters. In the case of a Bragg reflection these parameters account for the peak position, $2\theta_0$, usefully specified as the peak centroid, the peak intensity at the centroid, I_0, and one or more parameters that relate to the peak width and peak shape. In the following, $2w$ is always associated with the full width at half maximum (FWHM).

Instrumental Box 4:

Numerical Data Analysis

A measured reflection profile or diffraction pattern essentially consists of a set of N data couples of intensities I_i collected at scattering angles $2\theta_i$. All physical parameters derived from it, like unit cell edges, crystallite size or average microstrain, are based on the adaptation of the data to a physical or a microstructural model. The aspects of structural modeling are shown schematically in Fig. i4.1. Often the experimentalist is interested in the preparation of thin films with optimized physical properties (lower right box), which are governed by the structural and microstructural properties of the layer–substrate system (upper box). To the same extent the x-ray scattering pattern (lower left box) is determined by the specimen's structure. In order to elicit certain material properties the structure has to be deduced from the pattern, which can be considered as the inverse problem of thin-film analysis by x-ray scattering.

The model should allow for the formulation of a theoretical intensity function $y(2\theta, \boldsymbol{x})$ depending on a set of P parameters x_j. The Cauchy function with parameters $\boldsymbol{x} = (I_0, 2\theta_0, 2w)$ or any other of the profile functions of Section 3.1 may serve as examples. It is realized from their analytical expressions that the parameters may be linearly (I_0) or nonlinearly ($2\theta_0$, $2w$) related to the model function. A model function for microstructural analysis by whole pattern fitting may comprise the parameter vector $\boldsymbol{x} = (a_0, \mu t, D_0, \ln\gamma, \rho_d)$. If only one nonlinear parameter occurs in $\boldsymbol{x}$, nonlinear techniques are required for the solution. The task to be solved is the identification of parameter values x_j that optimally describe the set of measured intensities I_i. If the errors of the measurement are equally distributed "optimum" parameters follow from the minimization of the chi-square function

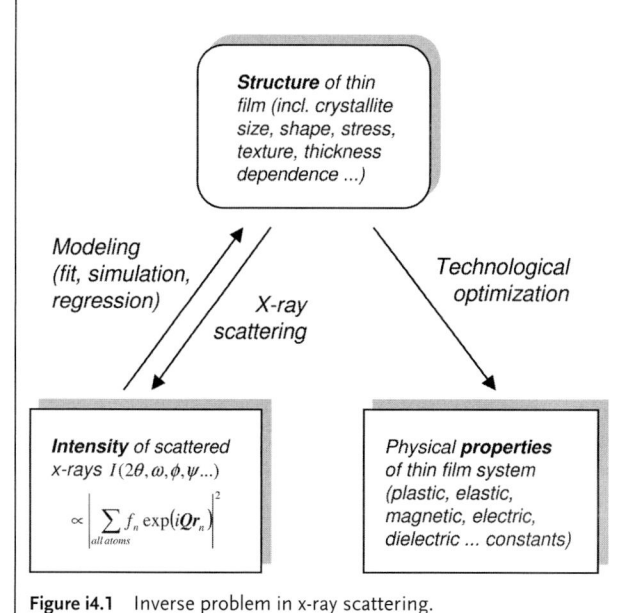

```
                    ┌─────────────────────┐
                    │ Structure of thin   │
                    │ film (incl. crystallite
                    │ size, shape, stress,│
                    │ texture, thickness  │
                    │ dependence ...)     │
                    └─────────────────────┘
                      ↗               ↘
       Modeling                          Technological
       (fit, simulation,                 optimization
       regression)       X-ray
                         scattering
                      ↙               ↘
   ┌──────────────────────┐      ┌──────────────────────┐
   │ Intensity of scattered│      │ Physical properties  │
   │ x-rays I(2θ,ω,φ,ψ...)│      │ of thin film system  │
   │                      │      │ (plastic, elastic,   │
   │  ∝ |Σ fₙ exp(iQrₙ)|² │      │ magnetic, electric,  │
   │     all atoms        │      │ dielectric ... constants)│
   └──────────────────────┘      └──────────────────────┘
```

Figure i4.1 Inverse problem in x-ray scattering.

that is typically given by the sum of squared differences that are normalized to the standard deviation σ_i of the measurement point

$$\chi^2(x) = \sum_{i=1}^{N} \left| \frac{I(2\theta_i) - y(2\theta_i, x)}{\sigma_i} \right|^2 \to \text{Min} \tag{i4.1}$$

Various techniques have been developed for the solution of this problem. A widely used technique is the nonlinear regression using the Levenberg–Marquardt algorithm. Its application requires the derivatives of the model function with respect to parameters x_j. Other numerical techniques are the methods of downhill simplex, Monte Carlo and simulated annealing. Recently, so-called evolutionary or genetic algorithms have been increasingly applied. Nonlinear regression may yield misleading results when the shape of the chi-square function $\chi^2(x)$ in parameter space x allows for various local minima to occur. The problem can be reduced or circumvented by restricting the number of model parameters P or their allowed ranges $[x_{j,b}, x_{j,t}]$ as strongly as possible. It may also be useful to start the fitting with estimates that are mostly close to the final refined values.

There are different measures in use to quantify the adequacy of the fit to the measured data. The goodness-of-fit indicator is given by

$$GoF = \left(\frac{1}{N - P + C} \sum_{i=1}^{N} \left(I(2\theta_i) - y(2\theta_i, x) \right)^2 \right) \tag{i4.2}$$

Here, C is the number of constraints by which the model parameters might be related. $N - P + C$ can be considered as the degree of freedom of the regression. Other widely used quality parameters are the so-called R values, of which a large variety exist. The R_p value

$$R_p = \frac{\sum_i |I_i - y_i(x)|}{\sum_i I_i} \tag{i4.3}$$

specifies the average deviation between the measured intensity and the intensity values predicted by the model function ("p" for pattern). A second variant is the weighted R_w value

$$R_w = \left(\frac{\sum_i w_i |I_i - y_i(x)|^2}{\sum_i w_i I_i^2} \right)^{1/2} \tag{i4.4}$$

in which statistical weights w_i are included. A reliable measure for the adequacy of a fitted pattern is also provided by the plot of the difference data and their optical inspection (see Fig. 3.3).

Performing one of the different nonlinear regression techniques yields the parameters x_j, which are in most cases the quantities of interest to the x-ray scattering analyst. However, the specification of a parameter x_j without any error margin is useless, because it does not allow one to estimate the reliability of either the underlying model function or the parameter value obtained. The techniques of numerical regression therefore determine also a set of estimated standard deviations σ_j for each x_j. If too large σ_j values are obtained from a fit either the measurement should be improved (e.g. better signal-to-noise ratio) or the underlying model function has to be doubted or replaced by a more appropriate one.

In addition, Fourier transformation techniques are also often applied in numerical data analysis as is demonstrated for the deconvolution of instrumental and sample-broadened line profiles. Also Laplace transforming of measured data is of relevance in thin-film analysis as is discussed in Chapters 4 and 6 for the simulation of depth-dependent properties. It can be stated that the development of adequate software techniques for the analysis of x-ray diffraction data is of the same significance as the development of improved diffractometer hardware.

Ready-to-use computer subprograms to be integrated into various codes are found, for instance, in Ref. [1]. It has to be emphasized, however, that a large number of extended software packages for the analysis of x-ray diffraction patterns are available. An important internet site for powder diffraction analysis software is provided by the Collaborative Computational Project no. 14, the so-called CCP14 suite [2]. The project was founded by the UK Engineering and Physical Science Research Council and accommodates a large set of x-ray diffraction computer programs and manuals. For thin-film investigations the chosen program should always be checked as to whether it enables the inclusion of thickness effects (absorption correction, μt product, etc.), since many program tools have been developed for powder applications.

References

[1] W. H. Press, B. P. Flannery, S. A. Teukolsky, W. T. Vetterling, *Numerical Recipes*, Cambridge University Press, Cambridge, 1989.
[2] CCP14 Homepage: Single Crystal and Powder Diffraction (freely available crystallographic software for students and academia), available at: www.ccp14.ac.uk/index.html.

These definitions allow one to decompose a reflection profile $I(2\theta)$ into the product of I_0 and a profile function f related to the centroid position at $2\theta_0$

$$I(2\theta) = I_0 f(2\theta - 2\theta_0, 2w, ...) \tag{3.1}$$

Some model profiles make use of more shape parameters than only the FWHM $2w$ and thereby allow for the fitting of more complicated line profiles. (Since atomic form factors are only rarely considered in the following the symbol f stands for the profile function throughout this chapter.)

A rather general and important result achieved during the development of profile analysis was the observation of the different effects that are caused either by small crystallite size or by microstrain. It turned out that small sizes are associated with a Cauchy-type profile, while pure microstrains basically lead to a broadening described by a Gauss function. (The Cauchy profile is equally termed a Lorentz profile in LPA, but in order to avoid confusion with the correlation length L use will be made of the Cauchy notation and its abbreviation C.) The phenomenon is related to the fact that the Kα emission line of an x-ray tube is Lorentz/Cauchy shaped due to lifetime effects of the atomic transitions. In contrast, microstrain fields often ex-

Structure Box 4:

Crystalline Lattice Faults

Crystallographic lattices are never perfect, but contain lattice faults that may be characterized by their spatial extension as zero-dimensional point defects, one-dimensional line defects or two-dimensional area defects. A defect is thereby considered as a deviation from the ideal lattice periodicity. In the framework of this categorization the grain size of a polycrystalline sample may also be understood as a defect, namely a three-dimensional one, with the grain boundaries restricting the lattice periodicity to the grain volume. Defects can have a significant impact on the intended thin-film properties. Here, some background information on one- and two-dimensional defects is given. The density of defects may in some cases be deduced from the analysis of x-ray diffraction line profiles.

The most relevant one-dimensional defects are dislocations that occur in different forms like screw dislocations and edge dislocations. In Fig. s4.1 the effect of an edge dislocation on the stacking of unit cells in a cubic lattice is visualized. Let us imagine moving in a circle around the line defect in a plane perpendicular to it and adding all lattice vectors r_{mnp} along this way. Any such circle in a perfect lattice would cause the sum of vectors to vanish. It can be seen from the figure, however, that the sum over all r_{mnp} around the line defect will yield a vector different from zero: $\Sigma r_{mnp} = b$. The vector b is called the Burgers vector and its direction and magnitude specify in a very condensed manner the type of dislocation.

A dislocation causes the neighboring atoms to shift from their ideal lattice sites and thereby introduces a strain field $u(r)$ in the surrounding volume. The strain field may extend over a distance of some micrometers from the core of the dislocation and the defect line thus affects a large volume. It is this defective volume that causes the broadening of the Bragg reflection. The most severe reordering occurs in the dislocation core, which has a radius R_c of typically 1 nm and is frequently termed the outer cut-off radius. Dislocations are introduced into a crystal to accommodate the atomic positions when the strain exceeds a critical limit. The largest number of dislocation studies has been per-

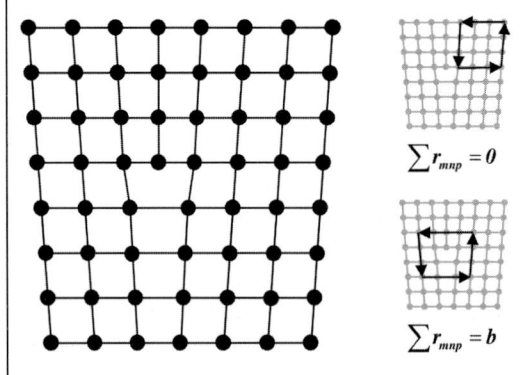

Figure s4.1 Schematic representation of an edge dislocation in a simple cubic lattice.

formed for one-elemental metals and semiconductors; there is much less knowledge concerning alloys and compounds.

The occurrence of dislocations in solids is difficult to avoid. Single-crystalline silicon is one of the few commercially available materials where a zero dislocation density can be certified. The density of dislocations ρ_d is measured by counting the points where they pierce through the sample surface and ρ_d thus has the dimensions of m^{-2}. Metal single crystals typically exhibit dislocation densities of the order of 10^6 to 10^{12} m^{-2}, but this value can increase up to 10^{18} m^{-2} after severe plastic deformation. The so-called dislocation configuration parameter M is the product of the core radius and the square root of the density, $M = R_c \rho_d^{1/2}$.

In real solids dislocations do not occur in the ordered state as depicted in Fig. s4.1, but are found to be bent, twisted and kinked. The defect lines may also multiply by branching out, albeit they never end in the volume of the crystallite, but always extend to a surface or form a closed loop. Dislocations can also move through the crystallite, which may be excited by thermal gradients or mechanical loading. Interestingly, there exists no thermal equilibrium concentration of dislocations as for the density of vacancies in a solid. Rather dislocations enlarge the entropy term in the thermodynamic state function.

Dislocations play a decisive role in plasticity, because the gliding of dislocations is one of the basic mechanisms of plastic deformation. The process is visualized in Fig. s4.2, where a dislocation travels from the right to the left side of a crystal. After the process is finished the upper part of the crystal has been shifted with respect to the lower part. In fact, dislocation glide or slip is one of the mechanisms by which the deformation of a solid proceeds under mechanical loading. Slip processes are categorized according to the crystallographic plane (hkl) and direction [uvw] along which they proceed. It turns out that some slip systems (hkl)[uvw] are energetically favored over others and the choice of the activated slip system depends on the structure of the solid. In fcc-structured metals, for instance, (111)[10$\bar{1}$] and their symmetry-related analogues are the preferred slip systems. The combination of four slip planes having each three possible slip directions results in 12 possible slip systems in the fcc structure that are given in Table s4.1. In bcc metals the two slip systems (112)[1$\bar{1}$1] and (110)[1$\bar{1}$1] are typically activated by mechanical loading.

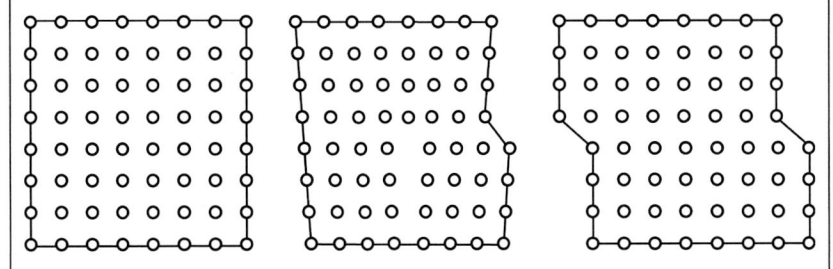

Figure s4.2 Dislocation glide via translation of an edge dislocation.

Table s4.1 The 12 slip systems (hkl)[uvw] of fcc lattices.

Slip plane (hkl)	(111)	(11$\bar{1}$)	($\bar{1}$1$\bar{1}$)	($\bar{1}$11)
Slip direction [uvw]	[$\bar{1}$10]	[$\bar{1}$10]	[110]	[110]
	[0$\bar{1}$1]	[011]	[011]	[0$\bar{1}$1]
	[$\bar{1}$01]	[101]	[$\bar{1}$01]	[101]

Two-dimensional crystal lattice faults are always characterized by a plane, along which an abrupt change in the structural order occurs. An important example is the twin plane. This two-dimensional defect acts like a mirror plane at which the continuous stacking of unit cells is disrupted. It should be emphasized that the twin plane is a mirror plane on the length scale of the grain size rather than on the length scale of the unit cell. Some other prominent two-dimensional crystal defects are stacking faults and the formation of anti-site domains. Stacking faults may occur for instance in a hexagonal lattice, when the usual ...ABAB... sequence is interrupted by the inclusion of a C plane ...ABAB-CABAB... . Anti-site domains, however, cannot arise in monatomic lattices, but only in binary or higher multinary compounds. In their case the site occupation of atoms is changed when the defect plane is passed, for instance, Zn atoms would occupy S positions and vice versa.

Bibliography

J. Bohm, *Realstruktur von Kristallen*, Schweizerbart'sche Verlagsbuchhandlung, Stuttgart, 1995.

hibit a normal distribution of d values around an average value d_0. The Cauchy and Gauss functions $f_C(2\theta)$ and $f_G(2\theta)$ thus became the prototype model functions in LPA. They are used here in the form

$$f_C(2\theta) = \left[1 + (2\theta - 2\theta_0)^2 / w^2\right]^{-1}$$
$$f_G(2\theta) = \exp\left[-\ln 2 (2\theta - 2\theta_0)^2 / w^2\right] \tag{3.2}$$

where the parameters have the meaning introduced above. The factor ln 2 occurs in the exponent of the Gauss function because the relation FWHM = $2w$ should be obeyed. A Bragg reflection profile is then obtained by multiplying $f_C(2\theta)$ or $f_G(2\theta)$ by the intensity at the centroid I_0. Figure 3.1 depicts both model function for a peak with centroid at 32.5°, a FWHM of 0.8° and $I_0 = 1$. It is seen from the plot that for a Cauchy-type function the peak tails are more pronounced than for a Gauss function and the flanks within the FWHM are slightly steeper.

The so-called integral breadth β of a diffraction line describes the width a peak would have if it were modeled by a rectangular box of the same integral intensity, but height I_0 and width β. This translates into the formula

$$\beta = \frac{1}{I_0} \int_{-\infty}^{\infty} I(2\theta) d2\theta \tag{3.3}$$

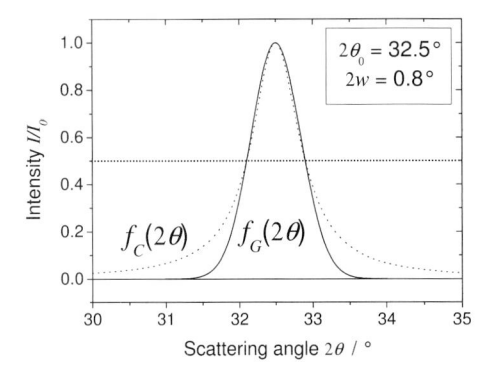

Figure 3.1 Gauss and Cauchy model functions $f_G(2\theta)$ and $f_C(2\theta)$ of the same FWHM and I_0 for a Bragg reflection with centroid at $2\theta_0 = 32.5°$ and $2w = 0.8°$.

The concept of the integral breadth was introduced by Laue [1] and is visualized in Fig. 3.2. It is evident from the plots of the Cauchy and Gauss functions in Fig. 3.1 that the β parameter of the former is greater than that of the latter because of the larger intensity in the tails of the Cauchy profile. The Cauchy and Gauss functions may be reformulated using the integral breadth and then become

$$f_C(2\theta) = \left[1 + (2\theta - 2\theta_0)^2 \pi^2 / \beta_C^2\right]^{-1}$$
$$f_G(2\theta) = \exp\left[-(2\theta - 2\theta_0)^2 \pi / \beta_G^2\right] \quad (3.4)$$

It should be noted that the integral breadth does not extend the number of parameters of a chosen model profile, but that β derives from other line parameters and may be used to substitute for one of them.

The regression of a measured reflection by either a pure Cauchy function or a pure Gauss function only rarely yields satisfying results. Rather often the flanks or the tails of the Bragg peak deviate significantly from the $f_C(2\theta)$ or the $f_G(2\theta)$ prototype profile. Figure 3.3 displays the 10.0 reflection of a hollow-cathode sputtered

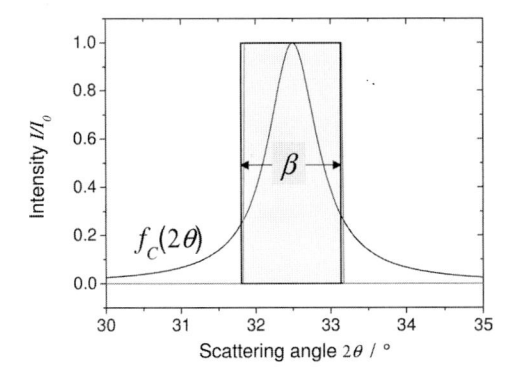

Figure 3.2 Visualization of the concept of integral breadth.

thin titanium film of 300 nm thickness. Hollow cathodes operate as gas aggregation sources, where nanoparticles are directly formed in the gas phase and which are currently under development for applications in nanocomposite thin films [2, 3]. The two upper plots in Fig. 3.3 depict the fit of the Ti 10.0 reflection by a Gauss and Cauchy function. The difference between the measured and the modeled profile, $\Delta I = I_{meas} - I_{cal}$, is plotted in the box below each profile. It can be seen that the measured reflection is in better agreement with a Gauss than with a Cauchy profile and that systematic deviations between the measured and the model profile occur. In general, there is typically only little agreement between a measured reflection profile and either a pure Gauss or a pure Cauchy profile.

Various mathematical functions have thus been investigated with respect to their ability to model adequately Bragg reflections. Three of them turned out to yield the most convincing results and they can all be considered as mixtures of Gauss and Cauchy functions. The first of these functions derives from a generalization of the Cauchy profile in which the weight of intensity between center and tails is adjusted by the parameter m. The function is generally termed Pearson VII [4, 5], but may

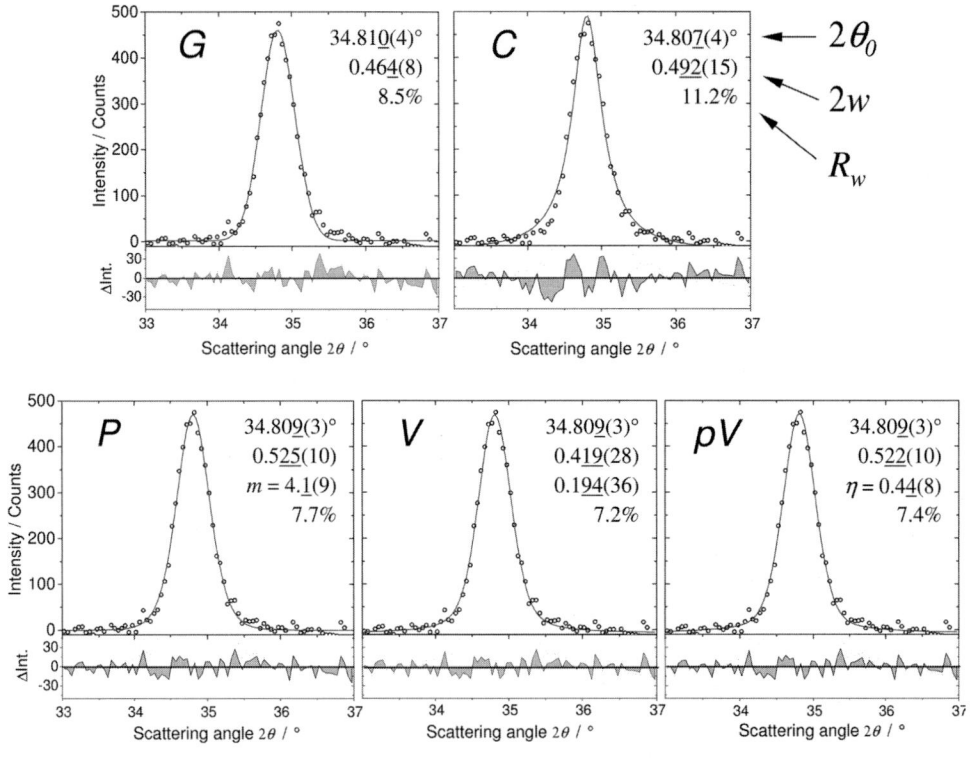

Figure 3.3 10.0 Bragg reflection of a hollow-cathode sputtered thin titanium film and its simulation by Gauss (G), Cauchy (C), Pearson (P), Voigt (V) and pseudo-Voigt (pV) functions. Top and bottom numbers in the upper right of each plot give the peak centroid $2\theta_0$ and R_w value of the fit, while the numbers between indicate $2w$ values or other profile parameters that split into $2w_G$ and $2w_C$ for the V profile.

also be named the Pearson function, because Pearson functions other than the seventh are only rarely considered. The Pearson function reads

$$f_P(2\theta) = \left[1 + (\sqrt[m]{2} - 1)(2\theta - 2\theta_0)^2 / w^2\right]^{-m} \quad (3.5)$$

The relationship with the Cauchy function is clearly recognized and $f_C(2\theta)$ is identically obtained for $m = 1$. It may not as easy be recognized that the Gauss function $f_G(2\theta)$ is achieved in the limit of $m \to \infty$ (Exercise 2).

The Voigt function is the second model function to reliably fit the shape of Bragg reflections in a $\theta/2\theta$ pattern [6, 7]. The function follows from the mathematical operation of a convolution of a Cauchy and a Gauss function, i.e. by integrating over the Cauchy profile and weighting with a Gauss function

$$f_V(2\theta) = \frac{\beta}{\beta_C \beta_G} \int_{-\infty}^{+\infty} f_C(\xi) f_G(2\theta - \xi) d\xi \quad (3.6)$$

Here, β, β_C and β_G denote the integral breadths of the convoluted and underlying basis functions f_C and f_G. The various integral breadths in the Voigt function are related through

$$\beta = \frac{\beta_G \exp(-k^2)}{1 - \mathrm{erfc}(k)} \quad (3.7)$$

where erfc accounts for the complementary error function, see Ref. [8], and k scales with the ratio of both integral breadths

$$k = \frac{\beta_C}{\sqrt{\pi}\beta_G} \quad (3.8)$$

In the form of Eq. (3.6) the normalization $f_V(0) = 1$ holds. One should be aware of the fact that the shape of $f_V(2\theta)$ depends only on β_C and β_G, since β is a function of these two quantities. The fitting of diffraction peaks by the Voigt function was circumvented for many years because of the comparatively large computing efforts required. These restrictions have now significantly relaxed with the progress in computing power, but due to the historical constraints a simplified version of the Voigt function has found much wider use, which is known as the pseudo-Voigt function [9]. The pseudo-Voigt function is simply obtained by a weighted superposition of Cauchy and Gauss functions

$$f_{pV}(2\theta) = \eta f_C(2\theta) + (1 - \eta) f_G(2\theta) \quad (3.9)$$

where the weight parameter η describes the amount of Cauchy share in the mixed profile [9].

The use of any of these three functions instead of a pure Cauchy or Gauss function extends the number of adjustable fitting parameters. The number of extra parameters is one for the Pearson function, which is the additional shape parameter m. In case of the Voigt and the pseudo-Voigt functions two different values for the FWHM $2w$ may be introduced, one for the Cauchy fraction of the profile and the second for the Gauss part. Rather often both 2w are constrained to the same value, $2w_C = 2w_G$, in order to minimize the profile complexity [10]. In most practical cases the use of the Pearson, the Voigt or the pseudo-Voigt model function causes an improvement of the regression. Figure 3.3 demonstrates the effect for the 10.0 reflection from a 300 nm thin Ti film. The use of the three model functions as shown in the lower row of the figure is associated with a significant reduction of the difference intensity $\Delta I = I_{meas} - I_{cal}$ and the R_w value of the fit (see Instrumental Box 4).

The integral breadths of all five model functions are listed in Table 3.1 as a function of shape parameters 2w, m, k or η. The integral breadth β of the Voigt function can be approximated to better than 0.16% by the rational function given in the table [11].

Table 3.1 Integral breadths of the five most relevant model functions in LPA, where $\Gamma(x)$ is the gamma function, $A = 0.9039645$, $B = 0.7699548$, $C = 1.364216$, $D = 1.136195$ and $E = 2(\ln 2/\pi)^{1/2} = 0.9394372$.

Model function	Equation	Integral width β
Cauchy	(3.2)	πw
Gauss	(3.2)	$\sqrt{\pi/\ln 2}\, w$
Pearson	(3.5)	$\dfrac{\pi 2^{2(1-m)} \Gamma(2m-1)}{(2^{1/m}-1)\left[\Gamma(m)\right]^2} w$
Voigt	(3.6)	$\dfrac{2(1+Ck+Dk^2)}{E(1+Ak+Bk^2)} w$
Pseudo-Voigt	(3.9)	$\left(\eta\pi + (1-\eta)\sqrt{\pi \ln 2}\right) w$

The adjustment of a measured Bragg peak to one the above given functions can be achieved by numerical procedures that may be named nonlinear regression, mathematical fit or model simulation. In fact, these three expressions are frequently used interchangeably in this book. The distinction between simulation on the one hand and fit or regression on the other, however, is that a simulation might be performed without any experimental data by just calculating a Bragg peak from a theoretical model. In contrast, fitting procedures follow the rules of least-square minimization by variation of the adjustable parameters as outlined in Instrumental Box 4. In general, the determined line parameters $2\theta_0$, $2w$, I_0, etc., should always be given with their estimated standard deviations (esd) from the covariant matrix obtained from the numerical regression. An important point in peak fitting is the reliable choice of a background function that accounts for the scattering of the substrate, fluorescence radiation of the sample or other factors (Exercise 3). In most cases the assumption of a linear background in the vicinity of one peak allows for a reliable modeling, but in some cases – for instance for a low-angle peak from a thin film on glass – it might become necessary to assume a more complex description of the background function. Technically speaking, the background causes the introduction of additional parameters into the regression procedure and might lead to larger estimated standard deviations for the fitting parameters of interest, i.e. the line parameters $2\theta_0$, $2w$, I_0, etc.

Also the effect of asymmetry should be considered in the profile fit. Peak asymmetries are introduced by instrumental effects, but are also related to the course of the scattered intensity as a function of scattering angle. An inspection of Eq. (1.45) shows that the Lorentz polarization factor Lp and the absorption factor A depend on 2θ. Also the atomic form factors and therefore the structure factor F vary with $2\sin\theta/\lambda$. Broad peaks that extend over large 2θ ranges may thus already be distorted by the influence of $Lp(2\theta)$ and $A(2\theta)$. The course of the $Lp \cdot A$ function continuously declines over the whole range of 2θ from 0 to 90°. It can thus be concluded that for broader peaks the correction of the measured intensity by division by $Lp(2\theta) \cdot A(2\theta)$ should be performed prior to the fitting of reflections. If the variation of the structure factor $F(2\sin\theta/\lambda)$ also has to be considered to deal with peak asymmetry, the use of whole-pattern fitting as outlined in Section 3.5.4 might be recommended.

3.2
Instrumental Line Profile

Peak profiles from Bragg reflections are always affected by the instrumental settings like emission profile of the tube, widths of the divergence and receiving slit, aperture of Soller slits, filters and monochromators in the beam path and other factors that are independent of the sample microstructure. In fact, it is a practical guiding rule to choose the diffractometer setting such that the instrumental width compares with the sample broadening to arrive at a reliable compromise between resolution and intensity. The instrumental line width then is of the same magnitude as the broadening from the sample and inevitably has to be considered in LPA. If the instrumental profile is symbolized by $g(2\theta)$ and that of the sample by $f(2\theta)$, Bragg reflections in a $\theta/2\theta$ scan are the convolution $h(2\theta)$ of both, which is indicated by the $\otimes$ symbol

$$h(2\theta) = f(2\theta) \otimes g(2\theta) = \int_{-\infty}^{\infty} f(2\theta) g(2\theta - \xi) d\xi \qquad (3.10)$$

Now discussed are the relevant instrumental features that determine the instrumental line profile in order to subsequently consider the deconvolution of the sample profile from the measured profile, i.e. the inverse operation of (3.10).

The first instrumental factor to be mentioned is the profile of the Cu Kα emission line itself. The line is composed of the α_1 and α_2 components that both can be modeled by Cauchy functions, since the natural shape of x-ray emission lines is of the Cauchy type. Figure 3.4 depicts the emitted intensity of a copper tube as a function of energy in the vicinity of Cu Kα around 8.04 keV. According to this splitting every $\theta/2\theta$ scan will thus consists of two separate diffraction patterns comprising two sets of peaks with the second set shifted by

$$\theta_2 = \arcsin\left(\frac{\lambda_{\alpha 2}}{\lambda_{\alpha 1}}\sin\theta_1\right) \tag{3.11}$$

relative to the first peak set. In the diffraction patterns shown in Chapters 1 and 2 the instrumental resolution was too low and the α_1 and α_2 lines merged for every reflection, but in a high enough resolution scan Bragg peaks are observed to split into two separate peaks. Equation (3.11) predicts the α_1–α_2 peak splitting to increase for increasing scattering angle 2θ and this effect is typically observed in $\theta/2\theta$ powder patterns. The intensity ratio R_{21} between the two components is close to 1:2, i.e. the α_2 component basically accounts for 50% peak intensity I_0 compared to the α_1 component. Caution should be taken, however, when a secondary monochromator or a multilayer mirror is introduced in the beam path, which might cause the α_2/α_1 intensity ratio to adopt smaller values than 1:2 [13, 14]. The precise ratio should then be determined by measuring a standard sample.

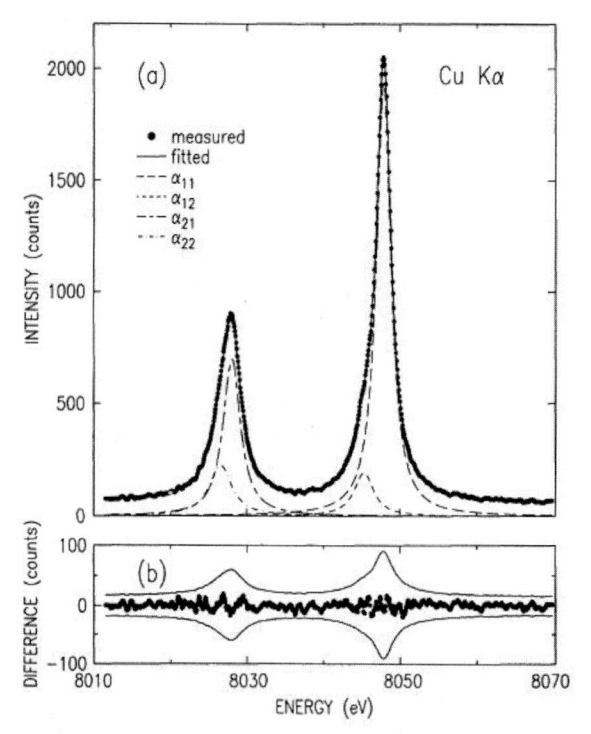

Figure 3.4 (a) CuKα emission spectrum from a laboratory x-ray tube. The dashed lines indicate the decomposition into four Lorentz/Cauchy functions. (b) Difference intensity between measured and modeled spectrum with the thin solid line accounting for the 2σ limits as determined by the count statistics of the measured spectrum. (According to Ref. [15].)

Recent studies of emission lines from laboratory x-ray tubes revealed the Cu Kα line to be composed of four Cauchy functions rather than only two. The effect can be seen from the underlying Cauchy-shaped lines in Fig. 3.4 and the values given in Table 3.2. The splitting into four instead of two emission lines is due to the multi-electron nature of the Cu Kα transition in a solid, i.e. the participation of atomic 3d levels in the basically 2p → 1s transition [12]. 3d hole states or so-called spectator states have been assigned to contribute to the double splitting of the Cu Kα line. Also for other transition metals x-ray emission lines asymmetries caused by simultaneous multielectronic transitions are observed [15]. These investigations also revealed the α_2/α_1 intensity ratio to deviate slightly from the ideal 1:2 value, having $R_{21} = 0.51$. Although these results are of fundamental importance for an understanding of the basic nature of the transition, for most x-ray diffraction investigations of thin films the analytical description of the Cu Kα line by two Cauchy functions α_1 and α_2 with parameters given in Table i1.1 will yield sufficiently precise results.

Table 3.2 Line parameters of the phenomenological decomposition of Cu Kα into four Lorentz/Cauchy functions [15].

Component	$\alpha_{1,1}$	$\alpha_{1,2}$	$\alpha_{2,1}$	$\alpha_{2,2}$
Energy (eV)	8047.837(2)	8045.367(22)	8027.993(5)	8026.504(14)
Intensity I_0	0.957(2)	0.090(1)	0.334(1)	0.111(1)
Width (eV)	2.285(3)	3.358(27)	2.666(7)	3.571(23)

It is generally recommended to perform line profile analyses using Kα$_1$ monochromatized radiation, which reduces the overlap between neighboring peaks and the number of reflections by one half. This recommendation may not always be followed in thin-film work, because α_1–α_2 monochromators cause a reduction in incident beam intensity and thin-film experiments already lack from reflected intensity compared to powder diffraction. The use of α_1–α_2 monochromators is also partly inhibited by the laborious calibration procedures that are necessary for the installation and re-installation of some of them. An alternative approach is based on the numerical correction of the measured pattern by subtracting that part of intensity that is due to Kα$_2$ radiation. In accordance with Eq. (3.11) this may be performed by a recursion formula introduced by Rachinger [16]

$$I_1(2\theta) = I_{1+2}(2\theta) - R_{21}I_{1+2}\left(2\theta - 2\arcsin\left(\frac{\lambda_{\alpha 2}}{\lambda_{\alpha 1}}\sin\theta\right)\right) \tag{3.12}$$

Here, I_1 is the pure α_1 intensity and I_{1+2} is the measured intensity composed of α_1 and α_2 radiation. This formula operates through all scan steps by subtracting intensity at the 2θ positions as calculated from the intensity at lower 2θ values. The recursion has to be begin for 2θ values that exceed the starting value by $2\arcsin(\lambda_{\alpha 1}/\lambda_{\alpha 2}\sin\theta)$ and the first few data points in the full 2θ range will thus remain uncorrected. The Rachinger correction has in fact been applied to the Ti 10.0

reflection shown in Fig. 3.3 that was originally measured with α_1–α_2 unmonochromatized radiation (Ref. [7] of Chapter 1).

To summarize, there are two strategies to deal with the α_2 component: (a) in the first approach the α_2 component is eliminated by use of a primary monochromator and (b) in the second approach it is numerically corrected for by application of the Rachinger correction of Eq. (3.12). Evidently, the first procedure will yield more accurate results, but only at the expense of increasing the integration time and it may be difficult to apply to thin films.

In addition to the spectral composition the instrumental profile $g(2\theta)$ is affected by the various beam shaping effects through the optical components of the diffractometer. The discussion proceeds along the same line as in Chapter 2, where the shift of the peak centroid due to instrumental settings was discussed, but this time the shape of the instrumental profile has to be considered. The procedure would lead to a – rather complex – formula of $g(2\theta)$ including all beam shape parameter like the width of the x-ray focus B_a, the various opening angles like δ_{ds} and δ_{Sol} and other instrumental parameters [17]. Such a strategy has been followed in the so-called fundamental parameter approach [13] (see Section 3.5.4). Another strategy relies on the measurement of a standard sample exhibiting neither size broadening nor strain broadening.

A standard sample for LPA should preferably fulfill the conditions: (a) large enough crystallite size, (b) polycrystalline untextured ensemble of a large number of crystallites and (c) absence of microstrains, which might be achieved by high-temperature annealing. It is evident from this list that the conditions are partially conflicting. The most common line profile standards are oxide compounds or other ceramics with crystallite sizes in the micrometer range that are prepared by standard ceramic processing procedures. The measurement of a standard yields a set of line parameters like the integral width β or FWHM $2w$ as a function of scattering angle 2θ that will enter into the process of deconvolution. In general the use of a standard of the same material as the specimen is preferred. In some cases, however, the procedure is unfeasible and the deconvolution has to be obtained with line parameters from a standard of another material. Then, the Bragg peaks of sample and standard occur at different 2θ and an interpolation of the standard's line parameter becomes necessary. Various interpolation formulas have been developed and tested. Well known, for instance, is the Caglioti formula that approximates the square of $2w$ by a sum over $\tan^2\theta$ and $\tan\theta$ and a constant [18]. The formula was developed, however, for neutron diffraction patterns, where reflection profiles can reliably be modeled by Gauss functions.

For laboratory diffractometers a sufficient approximation for the integral breadth may reliably be accounted for [19] by the function

$$\beta_{instr}^2 = A\tan^2\theta + B + C\cot^2\theta + D\sin^2 2\theta \tag{3.13}$$

where A, B, C and D are constants that depend on the actual diffractometer settings. The constants have to be determined in a first measurement run of the standard. In a subsequent scan, the shape parameters of the reflections from the sample under investigation are determined. Having arrived at this point the analyst has two sets of shape parameters $\beta(2\theta_i)$, one for the reflection of the specimen under investigation and another one for the standard. This result does not enable a full deconvolution of the sample profile $f(2\theta)$ according to Eq. (3.10), but it allows for some first interpretations – if the line profiles from standard and sample both are either purely Gauss or Cauchy type.

The simple rule holds that the convolution of two Cauchy functions yields a Cauchy function and the convolution of two Gauss functions yields a Gauss function. It can also be shown that the integral breadth of the convoluted peak increases by

$$\beta_h = \beta_f + \beta_g \text{ (pure C)} \quad (3.14)$$

in the case of convoluting two Cauchy functions and

$$\beta_h = \sqrt{\beta_f^2 + \beta_g^2} \text{ (pure G)} \quad (3.15)$$

in the case of two Gauss functions. Therefore, the β_f of the specimen functions may easily be derived in these cases and inserted into the Scherrer equation or the pure microstrain equation to reveal either pure crystallite size or pure microstrain.

From the three combined model functions presented above the Voigt function offers the great advantage that it enables a simple deconvolution when both the structure-broadened profile of the sample $f(2\theta)$ and the instrumental profile $g(2\theta)$ may be reliably modeled by a Voigt function. In that case the six integral breadths of the three functions g, h and f are related via

$$\beta_C^f = \beta_C^h - \beta_C^g$$
$$\left(\beta_G^f\right)^2 = \left(\beta_G^h\right)^2 - \left(\beta_G^g\right)^2 \quad (3.16)$$

The mathematical process of deconvolution may be performed, however, for any arbitrary line profile. This point might be of relevance for Bragg peak profiles that withstand any reliable refinement by analytical model functions and is considered in the next section.

3.3
Deconvolution by Fourier Techniques

A deconvolution technique that makes use of the decomposition of line profiles into a Fourier sum was developed rather early to derive the broadened sample profile $f(2\theta)$ from the measured profile $h(2\theta)$ [20]. The significance of this approach, however, exceeds by far that of a numerical procedure, because it paved the way for the separation of size and strain broadening [21]. Hence, first are introduced the method and the physical meaning of the Fourier coefficients are outlined in later sections.

The first basic step in the procedure is the transition to a new abscissa, for which the intensity is given. Instead of the scattering angle the magnitude of the scattering vector is used for this purpose. The abscissa is thus transformed from 2θ to $Q = 4\pi\sin\theta/\lambda$. Accordingly, the transformation of the intensity is performed by multiplying by $\lambda/2\pi\cos\theta$

$$I(Q) = \frac{\lambda}{2\pi\cos\theta} I(2\theta) \tag{3.17}$$

because $dQ/d(2\theta) = 2\pi\cos\theta/\lambda$ is valid and the infinitesimal relation $dI(Q) = [d(2\theta)/dQ]dI(2\theta)$ must be obeyed. Equivalent relations for the integral breadth and the FWHM of a reflection can be derived from

$$\beta_Q = \beta_{2\theta} \frac{dQ}{d(2\theta)} = \beta_{2\theta} \frac{2\pi\cos\theta}{\lambda} \tag{3.18}$$

The transformation to the scattering coordinate Q is reasoned from the following considerations. It was seen in Chapter 1 that a powder sample's signal reception by the detector in the symmetric $\theta/2\theta$ diffraction experiment is a highly selective one. For any Bragg reflection only those grains are observed for which the lattice planes are parallel or nearly parallel with the surface plane (see Fig. 1.9). One arrives from this at the general statement that the sample is always probed in the direction of the scattering vector $\mathbf{Q}$. This is intuitively understood, because this is the direction of momentum transfer, $\mathbf{K} - \mathbf{K_0} = \mathbf{Q}$, along which the beam interacts with the sample. The representation of the measured intensity as a function of Q can thus be considered as the "natural" coordinate of the scattering process. As for $\mathbf{K_0}$ and $\mathbf{K}$ the dimensionality of $\mathbf{Q}$ is that of a reciprocal length.

The operations of decomposing the measured line profile $h(2\theta)$ into its Fourier sum is visualized in Fig. 3.5. For reasons of simplicity the use of $K\alpha_1$ monochromatized radiation will be assumed. Firstly, the starting and end values on the abscissa $2\theta_S$ and $2\theta_E$ are chosen. The peak centroid should ideally fall close to the middle of the sampled range, near $(2\theta_E - 2\theta_S)/2$. Moreover, the sampling window should exceed the FWHM of the reflection by at least a factor of five, $(2\theta_E - 2\theta_S) > 10w$. The latter condition may sometimes not easily be obeyed due to peak overlap, which might particularly occur in nanocrystalline thin films. The $N + 1$ intensity data $I_j, j = 0, 1,...N$, are then collected in the step scan mode at $2\theta_S + j(2\theta_E - 2\theta_S)/N$. The intensities are normalized with respect to the maximum intensity of the reflection, i.e. $h(2\theta_j) = I(2\theta_j)/I_0$. Subsequently, the $h(2\theta_j)$ function is transformed to the new abscissa as shown in the upper part of Fig. 3.5. The profile is then limited by momentum transfer starting point Q_S and end point Q_E with individual data points separated by $\Delta_Q = (Q_E - Q_S)/N$.

The new coordinate to which the $h(Q)$ profile is Fourier transformed deserves special attention. Another situation might be more familiar to the reader, which is the Fourier transformation from a signal in the time domain into the frequency domain. Let us consider a signal transient $h(t'_i)$ of a total length of $(t'_E - t'_S) = 1$ s from which $N = 1000$ data points have been sampled at a time resolution of $\Delta_{t'} = 1$ ms. The Fourier transform projects the time window $(t'_E - t'_S)$ to a frequency range from

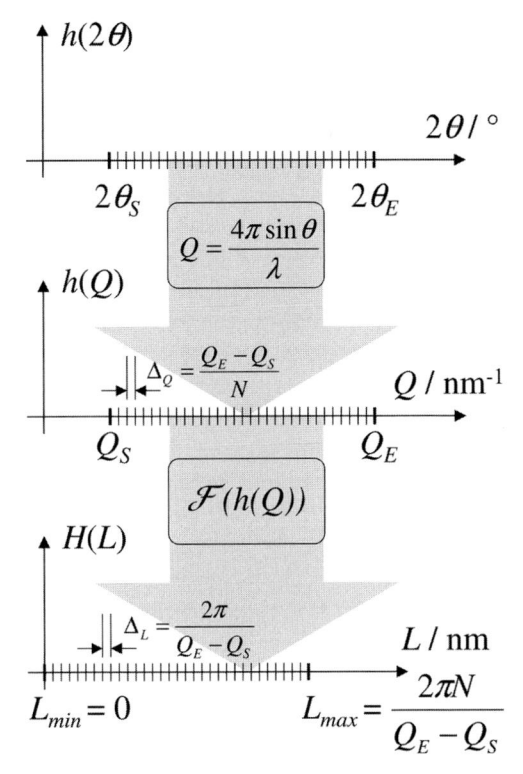

Figure 3.5 Schematic representation of abscissa transformation involved in the deconvolution of a sample profile. First, the abscissa is transformed from scattering angle 2θ to momentum transfer Q. Second, a discrete Fourier transform maps the set of $I(Q_i)$ data to $H(L)$.

$\nu_{min} = 0$ to $\nu_{max} = 1$ kHz, with adjacent points at a distance of $\Delta\nu = 1$ Hz. The resolution in the time domain thus defines the data window in the frequency domain, $(\nu_{max} - \nu_{min}) = 1/\Delta_t$, and the duration of the transient determines the resolution of the spectrum, $(t'_E - t'_S) = 1/\Delta_\nu$. Time and frequency are therefore denoted as Fourier-conjugated variables.

In the same sense the Fourier-conjugated variable of momentum transfer Q is defined and is denoted by L in the following. L might be considered to indicate a length, since the dimension of Q is that of a reciprocal length and, in fact, L is often denoted as Fourier length in the literature. As in the time–frequency example, the width of the L window follows from the resolution in the Q window, $(L_{max} - L_{min}) = 2\pi N/(Q_E - Q_S)$, and the resolution in L scales with the width of the Q window, $\Delta_L = 2\pi/(Q_E - Q_S)$ (see Fig. 3.5). What the discrete Fourier transform of $h(Q)$ yields is thus a number of N data points H_n given for an abscissa L defined by

$$L = \frac{2\pi n}{Q_E - Q_S} = \frac{n\lambda}{2(\sin\theta_E - \sin\theta_S)} \tag{3.19}$$

The L coordinate is a central quantity in LPA and the reader should be fully aware from where it stems. The number already includes various experimental constraints, i.e. the wavelength applied, the data window $2\theta_E - 2\theta_S$ and – indirectly by the resolution – the number of data points N used for the analysis. Hence, L is a quantity very closely related to the discrete numerical analysis.

The Fourier transformation of a reflection profile is illustrated by the study of Krill and Birringer, who investigated an fcc nanocrystalline Pd powder sample prepared by inert gas condensation [22]. The $\theta/2\theta$ powder pattern was measured using Mo Kα radiation. The pattern as processed by Rachinger's Kα_2 correction and background subtraction is shown in Fig. 3.6(a). Also a Pd reference sample was measured for comparison and the 111 peaks of sample and reference are shown in the inset.

After the abscissa change from 2θ to Q the Fourier transformation can finally be performed. Here, the reflections from the standard and the sample are assumed to be symmetric. If this assumption is invalid, the procedure has to be extended to include the appropriate sin terms as outlined in Ref. [23]. The discrete Fourier transformation proceeds by calculating coefficients H_n from $h(Q_j)$ data points according to

$$H_n = \sum_{j=0}^{N-1} h(Q_j)\cos(2\pi nj / N) \tag{3.20}$$

with the subscript j running over all measured data points. By virtue of the H_n coefficients the reflection profile $h(Q)$ might be expanded into a finite sum of harmonic functions

$$h(Q_j) = \frac{1}{N}\sum_{n=0}^{N-1} H_n \cos(2\pi jn / N) \tag{3.21}$$

where the $1/N$ factor is introduced for normalization. It should be noted that H_n could equally be symbolized by $H(L)$ due to the relation, Eq. (3.19), between n and L. While j is the number index of measured intensity data in 2θ space, n indexes the Fourier-conjugated data points, but both subscripts run from 0 to N. A Fourier transformation may thus be considered as linking the two-component vectors $(Q_j, h(Q_j))$ and $(L, H(L))$ via the back and forth transformations of Eqs. (3.20) and (3.21). The significance of the Fourier transform derives from the emphasis of periodic features in one of the data sets that cause enhanced amplitudes at the respective positions of the transformed variable.

Comparable formulas to Eqs. (3.20) and (3.21) also hold for the measured profile of the standard and for the – so far – unknown profile of the deconvoluted sample profile $f(Q)$.

$$G_n = \sum_{j=0}^{N-1} g(Q_j)\cos(2\pi nj / N)$$
$$A_n = \sum_{j=0}^{N-1} f(Q_j)\cos(2\pi nj / N) \tag{3.22}$$

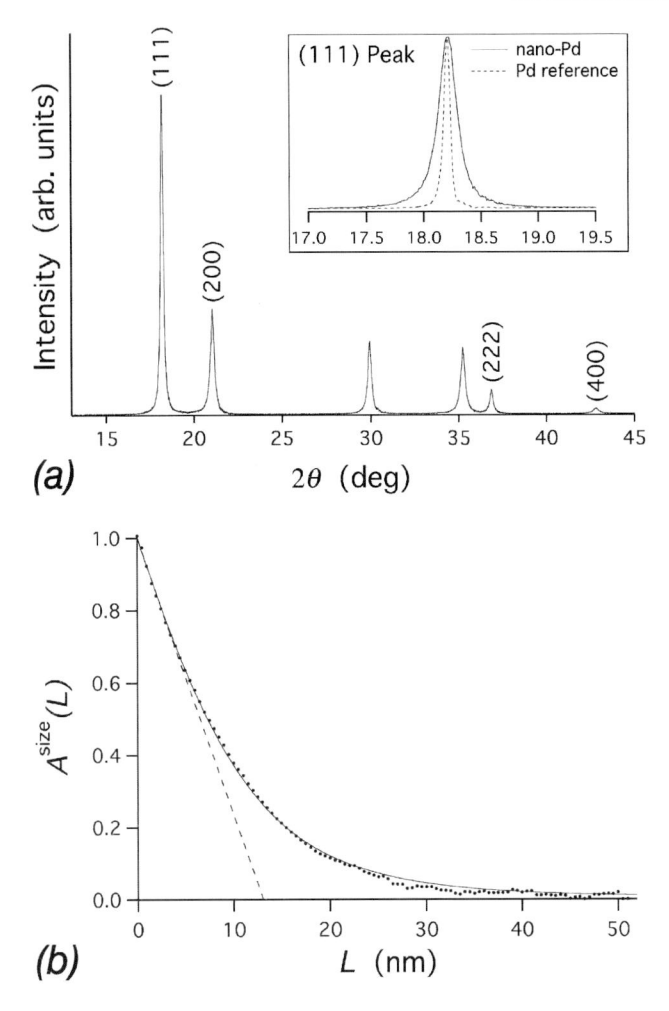

Figure 3.6 (a) $\theta/2\theta$ diffraction pattern of a nanocrystalline Pd powder sample measured with Mo Kα radiation. (b) Fourier coefficients $A(L)$ of the 111 reflection as a function of the Fourier-conjugated coordinate L (from Ref. [22]).

where the second set of coefficients is symbolized by A_n instead of F_n in order to avoid confusion with the structure factor.

The fundamental point now is that according to a theorem of Fourier analysis the coefficients A_n, H_n and G_n are related by

$$A_n = H_n / G_n \tag{3.23}$$

if the $h(Q)$ profile is the convolution of $g(Q)$ and $f(Q)$ according to Eq. (3.10).

Therefore, we have finally arrived at a method that allows to deconvolute the $f(Q)$ profile by expanding it in a harmonic sum

$$f(Q_j) = \frac{1}{N} \sum_{n=0}^{N-1} A_n \cos(2\pi jn / N) \qquad (3.24)$$

with the A_n coefficients known from the sample profile and from the standard profile. The determination of H_n and G_n coefficients from the measured h and g profiles thus allows for the calculation of A_n coefficients. The set of A_n coefficients can be used to generate the instrumental-free sample line profile $f(2\theta)$ by Fourier synthesis according to Eq. (3.24) and we have solved the deconvolution problem.

The information content obtained with this procedure will depend on the number of coefficients N that are determined with the analysis. This is recursively defined by the number of measured data points, since no information can be achieved for $L > N/(Q_E - Q_S)$. In other words, any uncertainties of the evaluation procedure are buried in the residual by which the finite Fourier sum of Eq. (3.24) differs from an infinite one.

As an example, the A_n coefficients of the Pd 111 peak are given in Fig. 3.6(b). A high resolution in L space of $\Delta_L = 0.5$ nm was essentially achieved by subtracting neighboring peaks and transforming the peaks from a large Q range [22]. Since the profile broadening is caused by the small size of the nanogranular sample the coefficients are denoted as A_n^S or A_n^{size} coefficients. The coefficients were normalized by setting $A(0) = 1$. It can be seen that the slope $dA(L)/dL$ is constantly negative. Moreover, an oscillation is seen for the higher L data points, which is a numerical artifact typically introduced by the discrete Fourier transformation of a finite data range.

An important constraint that decides on the content of analyzable physical information within the coefficients A_n is the degree of overlap of the reflection with neighboring ones. The application of the Fourier deconvolution ideally requires freestanding and only slightly overlapping reflections. Of course, this constraint is rarely fully obeyed and truncation errors are introduced into the discretization (Eq. (3.24)) of the peak. The errors turn out to take the largest values for coefficients A_n with small n and the phenomenon is termed the so-called hook effect in the literature.

It is concluded that (a) the model fitting of Bragg reflections (section 3.1) and (b) a discrete Fourier transformation (this section) both allow for the determination of the deconvoluted sample-broadened profile $f(2\theta)$. In case (a) the standard data enter via the instrumental profile function (Eq. (3.13)), while in case (b) the appropriate G_n coefficients have to be determined for every reflection. It appears useful to state the distinctions between both approaches. First, one ends up with few data points in (a), while as many coefficients A_n are obtained as there are measuring points in (b). Second, while the discrete Fourier transform may be subjected to truncation errors, these are avoided by modeling, but this is only achieved at the expense of regions that are not so well fitted, as seen in Fig. 3.3. Third, the deconvolution of the sample-broadened intensity $f(2\theta)$ is fully general in (b), while it relies on the validity of a model function to fit the observed reflections. In general, it might be stated that the modeling of individual peaks or whole diffraction patterns imposes the microstructure of the sample to be "fitable" by the model functions and the underlying assumptions. However, these presuppositions might not generally be obeyed. On the other hand, if the peaks behave well, i.e. if reliable fits are achieved by model functions, the results obtained for the sample microstructure might also be reliable.

3.4
Reflection Broadening by Small Crystallite Size Only

One source of line broadening is due to the finite dimensions of coherently scattering domains. The effect is illustrated by the example of the profile of an ε-phase Co film shown in Fig. 3.7(c) [24]. Ferromagnetic materials like Co, Fe, FePt, etc., with small crystallite sizes are currently investigated with respect to their application in ultrahigh-density recording. However, for crystallites that are too small the magnetic state could switch from ferromagnetic to superparamagnetic and the magnetic polarization within each magnetic domain would be governed by thermal fluctuations. Crystallite sizes thus have to be carefully controlled in the development of new thin-film recording media. The full $\theta/2\theta$ pattern shown in Fig. 3.7(c) has been measured for a sample with an average crystallite size of 11 nm, while the peaks in the inset were recorded for decreasing crystallite sizes of 9, 7, 5, and 3 nm. The Bragg peaks are seen to broaden significantly resulting in an overlap between neighboring peaks like the group of 221, 310 and 311. In this example the size distributions were determined by transmission electron microscopy (TEM; see Fig. 3.7(a) and (b)), and were found to be rather monodisperse, i.e. the standard deviation of the size distribution function amounted to only 7% of the average size.

Rather often, polycrystalline samples exhibit pronounced variations in crystallite size with grains including dislocation networks, stacking faults, twin planes and other lattice faults. These distortions can severely complicate the interpretation of the peak broadening. In such complex problems like the microstructure of solids the interpretation of the results strongly depends on underlying model assumptions. A set of formulas is presented in the following, which relate microstructure parameters and line profile parameters. When applying these formulas, it should always be kept in mind that they are only valid for some restricting assumptions on the microstructure, i.e. within the validity range of the underlying model. The models should be subjected to mistrust and results derived from them should only be given with simultaneously indicating the basic assumptions. This section is concerned with models that neglect any broadening due to microstrain and other lattice distortions, but solely account for small crystallite size. The presentation starts by assuming a polycrystalline sample to be comprised of crystallites having all the same shape and dimension. Although this is a purely hypothetical case, it allows for the introduction of the relevant quantities and the derivation of the Scherrer equation. The discussion is restricted to the case of cubic materials.

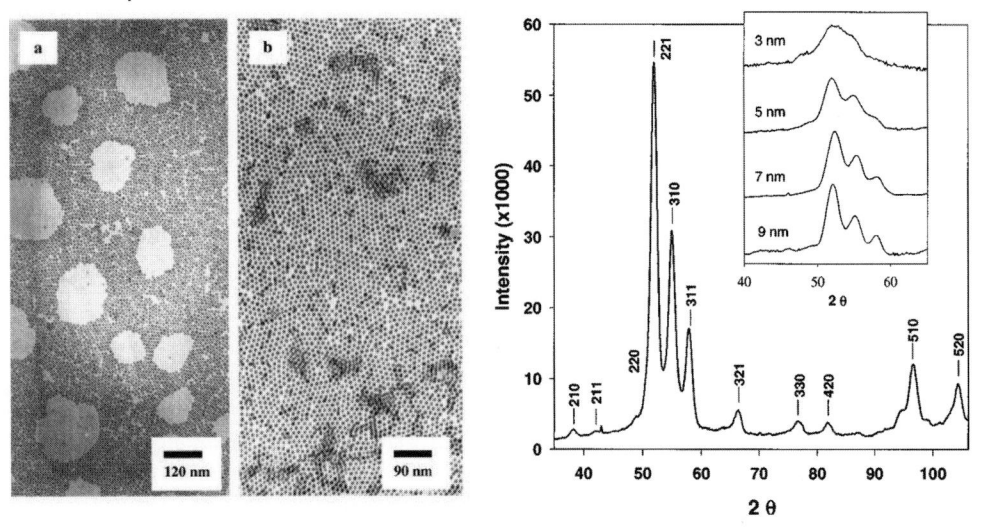

Figure 3.7 TEM images of Co nanoparticles with grain sizes of (a) 6 nm and (b) 9 nm. (c) Symmetric $\theta/2\theta$ diffraction pattern of 11 nm ε-phase Co nanocrystals measured with Co Kα radiation. Inset: Large-peak region for nanocrystals of decreasing size (from Ref. [24]).

3.4.1
Scherrer Equation

Elucidated first is the integral breadth of a Bragg reflection that is caused by a cube-shaped crystallite (see Fig. 3.8(a)). For this purpose, the intensity in the peak maximum $I_0(2\theta_0)$ and the integral intensity I_{int} have to be determined. The interference function $\Im(Q)$ introduced in Section 1.1 that contains the information on the line profile in angular space is considered. The derivation is simplified to the situation of a 00l reflection and the edges of the cubic crystallites are assumed to coincide with those of the cubic unit cell, $c_3 \parallel Q$. Within the appropriate range of scattering angles 2θ the Laue conditions (Eq. (1.10)) for constructive interference are fully valid or obeyed to a good approximation. The scattering vector Q then has a single component along c_3, while its projection on c_1 and c_2 vanishes. Insertion of $c_1 Q = c_2 Q = 0$ into Eq. (1.9) yields for the interference function dependent on momentum transfer $Q = c_3 Q$

$$\Im_{cub}(Q) = \frac{\sin^2(aN_3 Q/2)}{\sin^2(aQ/2)} \tag{3.25}$$

If the scattering coordinate $\xi = aQ/2\pi$ is introduced, we end up with the function $\sin^2 \pi N_3 \xi / \sin^2 \pi \xi$ that has to be investigated and which is frequently denoted as Laue interference function or Laue function. For a cube the relation $D_{cub} = N_3 a$ connects the crystallite height with the number and length of unit cells in one column along the direction of the scattering vector Q.

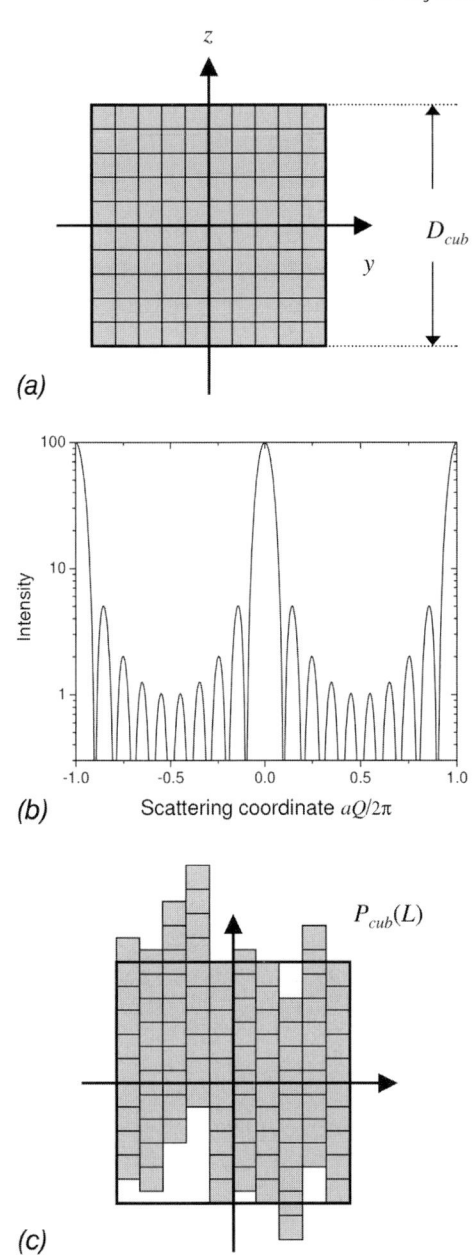

Figure 3.8 (a) Cube-shaped crystallite and (b) its associated Bragg peak line profile $I(Q)$. The crystallite is decomposed into columns of unit cells to visualize the intragrain column height distribution $P_{cub}(L)$. The column theorem predicts both crystallite shapes (a) and (c) to give the same reflection profile (b).

The course of $\Im_{cub}(Q)$ as a function of $\xi = aQ/2\pi$ is displayed in Fig. 3.8(b) on a logarithmic intensity scale for $N_3 = 10$. The fundamental properties of the Laue function are well known. First, main maxima arise whenever $aQ/2\pi$ becomes an integer, and, second, the maxima amount to N_3^2. A set of satellite peaks occurs next to the main peaks. It is assumed that the central peak extends from $\xi = -1/2$ to $+1/2$, i.e. half the satellite peak intensity is included in the central peak. Then, the integral of one of the main peaks becomes

$$\int_{-1/2}^{1/2} \frac{\sin^2 \pi N_3 \xi}{\sin^2 \pi \xi} d\xi = N_3 \tag{3.26}$$

In order to apply this result to the integral intensity of a Bragg reflection (Eq. (3.25)) a transformation from ξ to Q has to be performed, which is accomplished by $dQ = 2\pi d\xi/a$. The ratio of integral intensity over peak intensity can then be calculated and it is found for the integral breadth β_Q of a reflection in the Q representation

$$\beta_Q = \frac{I_{int}}{I_0} = \frac{2\pi N_3}{a} \frac{1}{N_3^2} = \frac{2\pi}{D_{cub}} \tag{3.27}$$

The transformation from β_Q to the more usual 2θ representation $\beta_{2\theta}$ is performed by virtue of Eq. (3.18) and we finally arrive at a relation between the integral breadth of the peak and the dimension of the cube-shaped crystallites

$$D_{cub,mono} = \frac{\lambda}{\beta_{2\theta} \cos\theta_0} \tag{3.28}$$

where θ_0 is the Bragg angle of the peak centroid. The size parameter D has been indexed by the subscript cub,mono to emphasize that the formula is valid only for cube-shaped monodisperse crystallites. A slightly modified version of this equation was first given by P. Scherrer in 1918 [25] and has been known as the Scherrer equation since then. Equation (3.28) was originally derived for the FWHM rather than for the integral breadth [25]. The transition to FWHM is performed by multiplying the right-hand side by a factor of $2\sqrt{(\ln 2)/\pi} = 0.9394$ as is shown in Ref. [16] of chapter 2. The technical point should be noted that integral breadth and FWHM have to be inserted in radians.

In general, the Scherrer equation yields too large values of the coherently diffracting domains as will be seen in the following. The equation established the first relation between microstructural properties of samples and their Bragg reflection profile. Today, the use of more advanced techniques like those presented in the following sections is recommended.

3.4.2
Column Height Distribution

In deriving the Scherrer equation, the crystal reference frame $\{c_i\}$ and the laboratory reference frame $\{l_i\}$ are assumed to be oriented parallel, $c_3 \parallel l_3 \parallel Q$. This assumption fully neglects shape anisotropy and allows Eq. (3.28) to be only applied to $00l$ reflections. In order to apply the concept of the Scherrer equation to crystallites other than cubes the formalism requires a more general approach.

So far in this book, diffraction peaks have been understood as being caused by parallel lattice planes occupied by atoms that form the solid. Another fruitful consideration of the process was introduced by Bertaut [26], who argued that for a constructive interference to occur the x-ray "sees" one crystallite being composed of columns of unit cells. The situation is visualized in Fig. 3.9 for a crystallite of arbitrary shape. According to Bertaut's column theorem, a diffraction peak at $2\theta_0$ is caused by the scattering of all unit cell columns that are oriented with respect to Q such that the Bragg equation is obeyed. If L denotes the height of one of the columns, it is the column height distribution function that is mapped into the line profile of the Bragg peak. One distinguishes here between two column height distributions: one for an individual crystallite $P(L)$ and another for the full polycrystalline sample $p(L)$. Bertaut's column theorem was originally formulated for the latter $p(L)$, but the description of individual crystallites by $P(L)$ also turns out to be a useful approach. For a distribution of crystallites of the same size D, a so-called monodisperse distribution, both column height distributions are equal.

The column theorem comes fully into effect when the scattered intensity of a crystallite of arbitrary shape is to be elucidated as shown in Fig. 3.9. The crystallite and its unit cells are oriented such that a Bragg reflection is excited. The scattering vector Q is normal to the (x, y) plane that passes through the crystallite's center of mass. In addition, the height of a single column, $L = L(x, y)$, depends on the position of the column within the (x, y) plane that cuts the grain perpendicular to Q, and thereby $P(L)$ also depends on x and y. One may thus write $P(L(x, y))$ for the column height distribution, but for reasons of brevity we will restrict ourselves to $P(L)$.

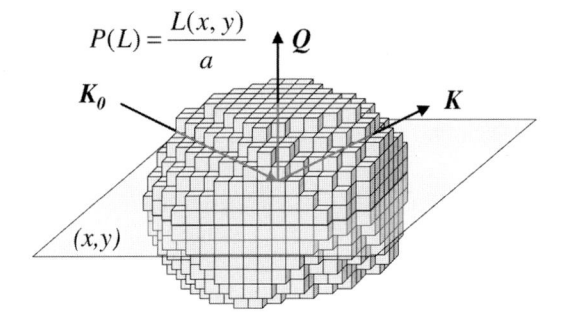

Figure 3.9 Visualization of Bertaut's column theorem. A crystallite is modeled by columns of unit cells with distribution $P(L)$.

In generalizing Eq. (3.25) the interference function is the sum over all Laue functions from individual columns normalized with respect to the projecting plane

$$\Im(Q) = \iint \frac{\sin^2(aP(L)Q/2)}{\sin^2(aQ/2)} dxdy \bigg/ \iint dxdy \tag{3.29}$$

where the integration has to be performed over all area elements $dxdy$, for which $P(L) \neq 0$. This is the general expression to be applied for deriving a reflection profile of a single crystallite with column height distribution function $P(L)$. It can be seen that inserting the column height distribution function of the cube-shaped grain, $P_{cub}(L) = N_3$, into Eq. (3.29) yields the special case of the interference function of Eq. (3.25) (Exercise 10).

In order to demonstrate fully the consequences of the column theorem the cube-shaped grain in Fig. 3.8(a) is rebuild to a crystallite of completely different shape, but of the same column height distribution $P_{cub}(L)$ (see Fig. 3.8(c)). Remarkably, this rather improbable grain shape would cause the same line profile shown in Fig. 3.8(b). It can thus be concluded that it is the column height distribution function $P(L)$ that is mapped into the x-ray diffraction line profile. Any crystallite size parameters D are only indirectly obtained from interpreting $P(L)$ or $p(L)$ as derived from the peak profile, i.e. by modeling.

3.4.3
Crystallite Shapes Other Than Cubes

Polycrystalline grain ensembles often exhibit shapes other than that of a cube. A more realistic model assumes the crystallites to have spherical shapes. The interference function and a Scherrer equation for spherical crystallites is thus derived in the following, by which the usefulness of Eq. (3.29) will become evident. In a mathematical continuous limit the column height distribution function of a sphere composed of cubic unit cells becomes

$$P_{sph}(L(x,y)) = \frac{2}{a}\sqrt{(D_{sph}/2)^2 - x^2 - y^2} \tag{3.30}$$

The prefactor of 2 ensures that the full height of columns is considered that extend in a sphere above and below the $z = 0$ plane. Insertion in Eq. (3.29) yields an integral that may be solved by transforming to polar coordinates

$$\Im_{sph}(Q) = \frac{2}{\pi R^2} \frac{2\pi}{\sin^2(aQ/2)} \int_0^R \sin^2\left(\sqrt{R^2 - r^2}Q\right) r dr \tag{3.31}$$

to give the intensity profile from spherical crystallites with a single diameter $D_{sph} = 2R$

$$\Im_{sph}(Q) = \frac{\sin^2 QR + QR(QR - \sin 2QR)}{Q^2 R^2 \sin^2(aQ/2)} \tag{3.32}$$

3.4 Reflection Broadening by Small Crystallite Size Only

An example of this profile function is shown in Fig. 3.10(b) for a diameter D_{sph} that compares through

$$D_{sph} = \sqrt[3]{\frac{6}{\pi}} D_{cub} = 1.2407 D_{cub} \tag{3.33}$$

with the cube edge of the previous example to attain the same illuminated sample volume. It can be seen that the $\mathfrak{I}_{sph}(Q)$ profile is broader than the $\mathfrak{I}_{cub}(Q)$ profile. This effect can be assigned to the outer columns of the spherical grain the height of which is significantly smaller than D_{sph}. A further distinction between both profiles is that the satellite peaks are significantly reduced in $\mathfrak{I}_{sph}(Q)$. In general, $\mathfrak{I}_{sph}(Q)$ compares much better to real-world Bragg reflections than $\mathfrak{I}_{cub}(Q)$. This underlines the enhanced appropriateness of modeling the shape in polycrystalline samples by spheres rather than by cubes. It should be noted that the strong satellites associated with the Laue function (Eq. (3.25)) are a typical feature of the scattering profile from parallelepipeds. These objects are characterized by the occurrence of only a single column height and it can thus be concluded that the variance of column height distribution leads to a reduction of the satellite peaks. The washing out of satellite peaks can also be understood by imagining a superposition of many Laue functions. Since each column length causes a strongly modulated intensity as in $\mathfrak{I}_{cub}$ the superposition of many of such signals with different minima positions averages out the strong extrema. The sphere profile $\mathfrak{I}_{sph}(Q)$ therefore becomes the smoother the more columns of different height are contained within the grain, i.e. the larger the diameter becomes.

In order to derive a Scherrer equation for spherical-shaped crystallites we can proceed as above by integrating over the intensity function

$$\int_{-a/2}^{a/2} \frac{\sin^2(aNQ/2) + (aNQ/2)\big((aNQ/2) - \sin aNQ\big)}{2\pi (aNQ/2)^2 \sin^2(aQ/2)} dQ = \frac{4}{3}\frac{N}{a} \tag{3.34}$$

Redistributing the formula and transforming again from β_Q to the 2θ representation yields for $\beta_{2\theta}$

$$D_{sph,mono} = \frac{4}{3}\sqrt[3]{\frac{\pi}{6}}\frac{\lambda}{\beta_{2\theta}\cos\theta_0} \tag{3.35}$$

A comparison with Eq. (3.28) reveals that the calculation of crystallite sizes by the original Scherrer equation yields values too small by a factor of 1/1.0747 when compared with the more realistic assumption of spherical shape. Since practically all samples investigated in x-ray diffraction are better approximated by spheres than by cubes, the use of Eq. (3.35) is a much better choice than Eq. (3.28). However, also in this case the validity of the formula is restricted to a monodisperse grain ensemble.

Other crystallite shapes like tetrahedrons, octahedrons, cylinders, etc., have already been evaluated [27–29]. If the geometrical shape of the grains is described by a single shape parameter D, the influence of shape may be included in the Scherrer formula scheme by introducing a Scherrer constant K_S

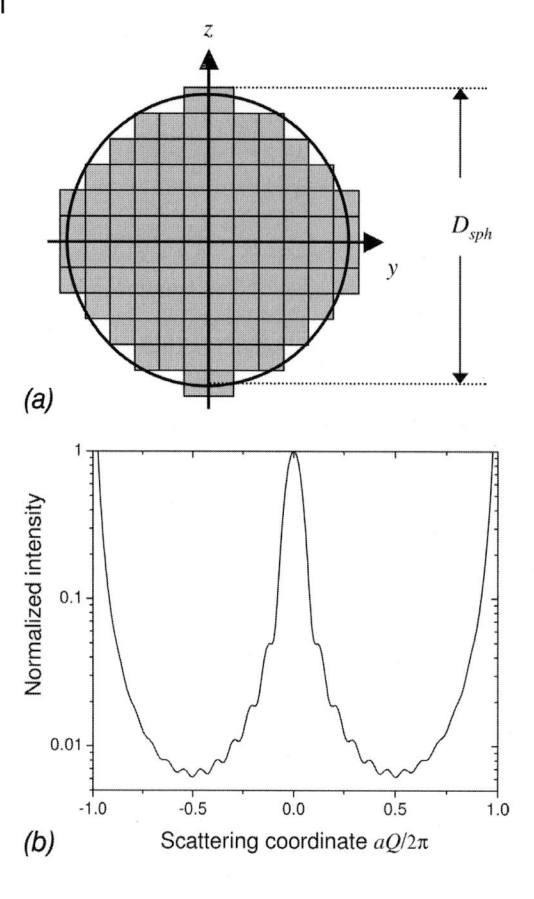

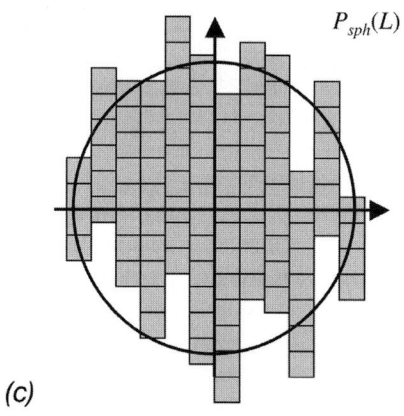

Figure 3.10 (a) Spherical crystallite and (b) its associated Bragg peak line profile $I(Q)$. The crystallite with the shape shown in (c) would yield the same line profile (b) if the column heights are distributed according to the same $P_{sph}(L)$.

$$D_{mono} = \frac{K_S \lambda}{\beta_{2\theta} \cos\theta_0} \tag{3.36}$$

For the case of a sphere $K_S = (4/3)(\pi/6)^{1/3}$ has been shown. Further values of Scherrer constants as calculated for some grain shapes are compiled in Table 3.3. The values specified are valid for general orientations between crystallographic and sample reference frame. In the special case of an $h00$ reflection for a cube exhibiting $c_3 \| Q$, Eq. (3.28) with $K_S = 1$ can be extracted from the table. In the general case of nonspherical grains, the anisotropy is included as an hkl dependence in K_S. The Scherrer constant takes numerical values of about unity. Compilations of K_S for various grain shapes and Miller indices can be found in Ref. [28].

Table 3.3 Scherrer constants K_S for some crystallite shapes. The conventions are obeyed that $h = |h|$, $k = |k|$, $l = |l|$, $h \geq k \geq l$, $S_1 = h + k + l$ and $S_2 = h^2 + k^2 + l^2$ (from Ref. [28]).

Shape	$h \geq k + l$	$h \leq k + l$
Sphere	$(4/3)(\pi/6)^{1/3}$	
Cube	$6h^3 / S_2^{1/2}(6h^2 - 2(k+l)h + kl)$	
Tetrahedron	$2h / 3^{1/3} S_2^{1/2}$	$S_1 / 3^{1/3} S_2^{1/2}$
Octahedron	$\dfrac{2S_1^3}{6^{1/3} S_2^{1/2}(S_1^2 + (k+l)S_1 + 2kl)}$	$\dfrac{2S_1^3}{6^{1/3} S_2^{1/2}(2S_1^2 - S_2)}$

Independent of the value of K_S it is a common property of the Scherrer equation that the integral breadth β (or FWHM) is inversely proportional to the size parameter D. The proportionality constant $K_S \lambda$ is of the order of magnitude of the wavelength and amounts to about 0.1 nm. Practically, it is found that reliable results are only obtained with the Scherrer formula for crystallite sizes of less than about 100 nm. Larger diameters are subject to strong instrumental errors due to the small breadth β associated with them. Modern procedures like the FPA, however, allow for the determination of size parameters in the micrometer range (see Section 3.5.4). It has to be emphasized that the integral breadth in the Scherrer equation has to be inserted in its instrumental corrected form (Eq. (3.16)). If the correction is neglected, too large a β value is inserted and D values are derived that are smaller than with the correction. The same shift to smaller D values is associated with neglecting any possibly existing microstrain.

3.4.4
Determination of the Column Height Distribution Function

So far, only monodisperse ensembles of crystallites have been considered, but in real samples the crystallite size will be subject to variations and has to be fully accounted for by a crystallite size distribution function. If the shape is described by only one geometrical parameter D, the size distribution function $g(D)$ will indicate the probability $g(D)dD$ that the shape parameter of an arbitrarily selected crystallite will fall in the range dD. The introduction of a size distribution $g(D)$ also alters the meaning of the column height distribution function. Whereas the function $P(L)$ was sufficient to describe the column height distribution for a monodisperse grain ensemble, a more complex function $P(L, D)$ has to be introduced when D becomes a distributed quantity. The distribution function $p(L)$ now accounts for the column height distribution in the full ensemble of crystallites. It would be rather interesting for many investigations of microstructure if $g(D)$ could be deduced from a line profile. However, a careful distinction must be made between the size distribution function $g(D)$ and the full-sample column height distribution function $p(L)$ that are related by

$$p(L) = \int P(L,D)g(D)dD \qquad (3.37)$$

The combined influence of both $g(D)$ and $p(L)$ on the reflection profile can be evaluated by firstly considering the interference function due to the shape according to Bertaut (Eq. (3.29)) and secondly by integrating over the set of all sizes. This recipe translates into the formula

$$\Im(Q) = \int g(D)\left(\iint \frac{\sin^2(aP(L,D)Q/2)}{\sin^2(aQ/2)}dxdy \Big/ \iint dxdy\right) dD \qquad (3.38)$$

Equation (3.29) is seen to be recovered as the limiting case of Eq. (3.38) for $g(D) = \delta(D - D_0)$. The approach of Eq. (3.38) may principally yield the full interference function, albeit rather complicated integrals may be achieved for different realistic models of crystallite size distributions. The problem might either be solved numerically or it might generally be avoided by deriving only some characteristic parameters for the description of $g(D)$ and $p(L)$ instead of solving for the full distributions. This justifies a closer look at $p(L)$ on its own.

A useful approach in the analysis of statistical distributions generally consists in considering their statistical moments. These are obtained by integrating over an appropriate power of the probability variable and weighting with the probability distribution. The nth-order moment $\langle L^n \rangle$ of the column height distribution function $p(L)$ is defined by ($p(L)$ normalized)

$$\langle L^n \rangle = \int L^n p(L)dL \qquad (3.39)$$

Some important quantities of distributions may be formulated concisely with the concept of statistical moments. For instance, the average mean value is the first moment $\langle L^1 \rangle$, while the variance is the difference between the second moment and the square of the first moment, $\sigma^2 = \langle L^2 \rangle - \langle L^1 \rangle^2$.

When applying the Fourier analysis to the reflex profile of a size-broadened peak, Bertaut derived two relations between the Fourier coefficients and the moments of the distribution function $p(L)$. The first relation links the average column height to the inverse slope of $A^S(L)$ coefficients when L approaches $L = 0$ [22, 26]. It reads

$$\langle L^1 \rangle = \frac{-A^S(0)}{\lim_{L \to 0}\left(\dfrac{dA^S(L)}{dL}\right)} \tag{3.40}$$

An application of this expression is shown in Fig. 3.6(b) for the example of the Pd nanocrystallites introduced above [22]. The $A^S(L)$ coefficients from the size-broadened peak are given as a function of the variable L. The slope $dA^S(L)/dL$ for $L \to 0$ is given by a dashed line intersecting the ordinate at $A^S(0) = 1$. From the intersection of this line with the abscissa the average column height in the Pd sample is concluded to be $\langle L^1 \rangle = 13.1$ nm.

The second relation derived by Bertaut relates the first two statistical moments of the $p(L)$ distribution and the sum or integral over all $A^S(L)$ coefficients

$$\langle L^2 \rangle = 2 \langle L^1 \rangle \int_0^\infty A^S(L) dL \tag{3.41}$$

Also the second moment can thus be deduced from the plot of $A^S(L)$ and the knowledge of the first. This has been done in the example in Fig. 3.6(b) by integrating the course of $A^S(L)$ data points (dark dots) over L. The application of Eq. (3.41) yielded $\langle L^2 \rangle / \langle L^1 \rangle = 19.6$ nm. From these data a variance of the column height distribution function of 85.2 nm^2 and a standard deviation of 9.2 nm can be derived. Although $\langle L^1 \rangle$ and $\langle L^2 \rangle$ do not account for the full column height distribution function $p(L)$, these two parameters can be regarded as its condensate. And in many cases of practical interest these two values are sufficient for subsequent operations like the optimization of the preparation process. It can be concluded that the analysis of the course of Fourier coefficients $A^S(L)$ yields important parameters characterizing the microstructure of the sample.

A further, not to be underestimated, result of Bertaut's analysis was that the meaning of the variable L was extended from a pure mathematical variable to a physical quantity: instead of representing only the Fourier-conjugated variable of momentum transfer, L turns out to account for the column height of crystallites being subjected to a distribution the statistical moments of which follow Eqs. (3.40) and (3.41). Accordingly, L is often termed a correlation length, which becomes evident from a further look at Fig. 3.6. The A_n or $A(L)$ coefficients may be considered as correlation amplitudes that increase the more pairs of unit cells exist for the particular value of n or L (both are related via Eq. (3.19)). Evidently, the largest number of unit cell pairs in one column are found for $n = 1$ (neighbors) or even $n = 0$ (autocorrelation) and the number of pairs will decrease with increasing distance L. From this argument the monotonic decrease of $A(L)$ with increasing L might be understood, which does not only hold for the example in Fig. 3.6, but is generally observed.

3.4.5
Determination of the Crystallite Size Distribution Function

Investigations of the formation process of nanogranular grain ensembles and aerosols revealed that the distribution function of crystallite sizes D could be described in many cases by the so-called logarithmic normal function or, in short, lognormal function $g_{LN}(D)$ [30]. The analytical expression accounting for the lognormal function reads

$$g_{LN}(D) = \frac{1}{\sqrt{2\pi} D \ln \gamma} \exp\left(-\frac{(\ln D - \ln D_0)^2}{2(\ln \gamma)^2}\right) \qquad (3.42)$$

Here, D_0 and $\ln \gamma$ are the lognormal median and lognormal variance. The naming of the function derives from the fact that its plot versus $\ln(D/D_0)$ has the same shape as the Gauss dump bell function, see Fig. 3.11, where $g_{LN}(D)$ is plotted versus a logarithmic (a) and a linear abscissa (b) for the example case $D_0 = 5$ nm and $\gamma = 2$.

Similarly to the column height distribution function $p(L)$ statistical moments may also be calculated for any arbitrary crystallite size distribution function $g(D)$

$$\langle D^n \rangle = \int D^n g(D) dD \qquad (3.43)$$

The application of this definition to the lognormal function yields

$$\langle D^n \rangle_{LN} = D_0^n \exp\left(n^2 (\ln \gamma)^2 / 2\right) \qquad (3.44)$$

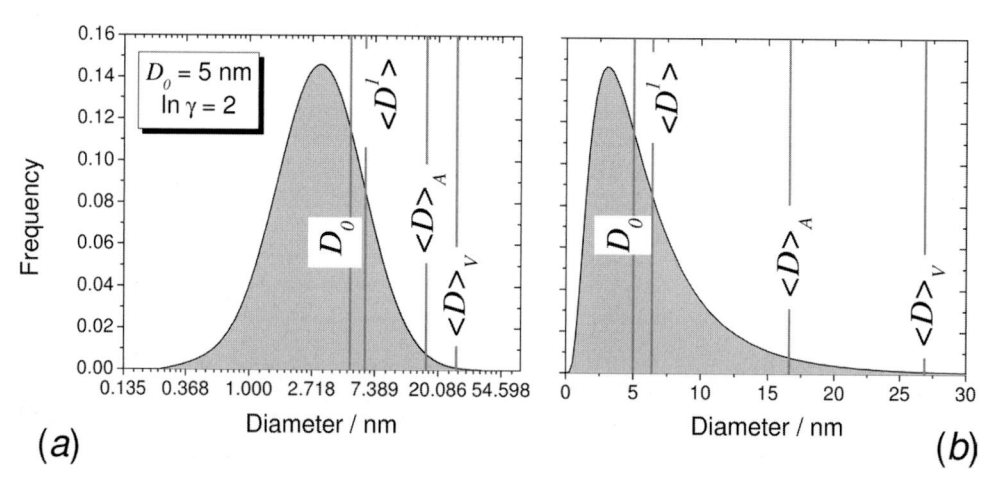

(a)

(b)

Figure 3.11 Logarithmic normal function with (a) a $\ln(D/D_0)$ abscissa and (b) a linear abscissa. The shape in (a) is of the same form as the Gauss bell function on a linear scale. The different averages of the lognormal distribution are indicated.

3.4 Reflection Broadening by Small Crystallite Size Only

This result illustrates an important property of $g_{LN}(D)$: the statistical moments of the lognormal distribution depend on both parameters D_0 and γ that describe the distribution. One consequence can be realized from Fig. 3.11, where the lognormal mean D_0 and the first moment $\langle D^1 \rangle$ are included and are seen to take different values. The mean and variance take the explicit form

$$\langle D^1 \rangle_{LN} = D_0 \exp\left((\ln\gamma)^2/2\right)$$
$$\langle D^2 \rangle_{LN} - \langle D^1 \rangle_{LN}^2 = D_0^2 \exp\left((\ln\gamma)^2\right)\left\{\exp\left((\ln\gamma)^2\right) - 1\right\} \tag{3.45}$$

As usual with distribution functions, the first moment gives an average size, which appears rather unspectacular. However, the point should seriously be considered as to what kind of averaging has been performed in obtaining $\langle D^1 \rangle$. The point is of interest, because any physical technique for the determination of crystallite sizes performs an averaging over the full grain ensemble with most of them yielding distinct average values. Determining, for instance, the first moment of the distribution function gives a so-called count-mean diameter. This would account for taking each spherical grain, measuring every diameter D_i and dividing the sum by the number of grains: $\langle D^1 \rangle = \sum D_i/N$. This is what is intuitively understood by average crystallite size, but it differs significantly from the average crystallite size obtained by microstructural investigations with XRD. In contrast, average crystallite sizes are typically determined by TEM in measuring the area A_i of individual crystallite from a micrograph like the one shown in figure 3.7 and deriving the diameter from $D_i = \sqrt{A_i/2}$. The "mean" is obtained again from $\sum D_i/N$.

It has been shown [27] by Stokes and Wilson already in 1942 that the size parameter derived from the Scherrer equation accounts for the ratio of the fourth central moment over the third moment of the crystallite size distribution function $g(D)$

$$\langle D \rangle_V = \langle D^4 \rangle / \langle D^3 \rangle \tag{3.46}$$

The average size parameter $\langle D \rangle_V$ is denoted as volume-weighted average crystallite size. This result has important consequences for the precise meaning of sizes as determined by the Scherrer equation, which is visualized in figure 3.11. It can be seen that for the lognormal distribution the crystallite size as determined by the Scherrer equation stands for a much larger value than the meadian D_0 of the distribution. This result is of general validity and means that size parameters from the Scherrer equation exceed by far what is naively thought the crystallite size would be. This explains why the crystallite size D in (3.28), (3.35) and (3.36) is characterized by the subscript "mono". In case that the crystallites exhibit a broad rather than a monodisperse distribution the size parameter in the Scherrer equation is always the volume-averaged mean.

In addition, an area-weighted average crystallite size is defined by

$$\langle D \rangle_A = \langle D^3 \rangle / \langle D^2 \rangle \tag{3.47}$$

The volume-averaged and area-averaged crystallite sizes $\langle D \rangle_V$ and $\langle D \rangle_A$ of the log-normal distribution become

$$\langle D \rangle_A = D_0 \exp\left(\frac{5}{2} \ln^2 \gamma\right)$$
$$\langle D \rangle_V = D_0 \exp\left(\frac{7}{2} \ln^2 \gamma\right) \tag{3.48}$$

These values have also been included in Fig. 3.11 and are seen to differ significantly from the first moment $\langle D^1 \rangle$ of the distribution.

These considerations have found an interesting application by Krill and Birringer [22]. The authors showed that the moments of column height distribution $p(L)$ and the area- and volume-weighted averages are related via

$$\langle L^1 \rangle = \frac{2}{3} \langle D \rangle_A \quad \text{(spheren)}$$
$$\langle L^2 \rangle / \langle L^1 \rangle = \frac{3}{4} \langle D \rangle_V \quad \text{(spheren)} \tag{3.49}$$

under the assumption that the crystallites can be assumed to be spherical [22]. One is thus able to determine the distribution function parameters D_0 and γ from the Fourier analysis of a single Bragg reflection by determining $\langle L^1 \rangle$ and $\langle L^2 \rangle$ from Eqs. (3.40) and (3.41). The application of this approach to the Pd powder shown in Fig. 3.6 yields a lognormal mean of $D_0 = 9.5 \pm 3$ nm and a lognormal variance of $\gamma = 1.71 \pm 0.15$.

3.5
Concomitant Occurrence of Size and Strain Broadening

The crystallites in a polycrystalline thin film in general exhibit defects and lattice distortions. Distortions of the periodicity in a crystal lattice can be described by displacement fields $u(r)$ that account for the displacement of atoms from their equilibrium position. These displacements vary in accordance with the precise nature of the distortion and they depend on the position r within the crystal. The displacement field can be expressed in the crystal reference frame $\{c_i\}$ by their orthogonal components

$$u(r) = u_1(r)c_1 + u_2(r)c_2 + u_3(r)c_3 \tag{3.50}$$

The adaptation of the displacement field to the lattice distortion under consideration is often the starting point for modeling the broadening of Bragg peaks by lattice strains. In some other cases it is sufficient to consider average values of strain fields, which may also be typical for certain crystalline defects.

3.5 Concomitant Occurrence of Size and Strain Broadening

Some different types of lattice distortion are introduced in Structure Box 4. The crystal lattice is distorted in the vicinity of a defect causing a slight variation of atomic bond lengths and interplanar spacings d. The larger the distance of an atom towards the defect the less its bonding surrounding will be distorted and vice versa. However, the dependency on distance r that describes the propagation of the distortion field around the defect varies with different defects. Planar lattice defects like stacking faults, for instance, cause a variation of lattice plane spacings $d - d_0$ that declines as $1/r^2$. The distortion thus decays comparatively fast. The situation is different for dislocations. They also cause a distortion of atomic coordination in their surrounding, but the decay scales only with $1/r$. Dislocations may thus affect large spatial areas in a crystalline grain and their effect on the diffraction pattern in general is essentially larger than that of two-dimensional lattice faults. It has become common practice to regard this as a hierarchy of defects in which dislocations occupy the top position. In the following lattice imperfections are always associated with dislocations leading to a distortion broadening. This is justified by the fact that most LPA techniques have been developed for dislocations.

A decisive assumption in the analysis procedure is related to the spatial distribution or the distribution function of the distortion. It is assumed that expansions, $d > d_0$, and contractions, $d < d_0$, occur in the strain field with equal probability. If a Bragg reflection of an undistorted lattice is observed at d_0 the distortion causes a part of the reflection intensity to shift to smaller d values and the same amount of intensity to shift to larger d values. The fact that interplanar spacings equally become expanded and contracted is neither easy understood nor trivial. It turns out, however, as a basic empirical fact during the investigation of material microstructure. As an example, Fig. 3.12 displays the normalized 220 reflections of an iron powder before and after being subjected to ball milling [31]. The peak is seen to be significantly broadened, which is caused by the decrease of crystallite size, but also by the introduction of lattice distortions as will be seen in the following.

The quantitative description of the variation of interplanar spacing Δd is performed by using the dimensionless quantity ε, the strain that is related via $\varepsilon = \Delta d/d_0$ to the undistorted spacing d_0. The relation between strain and line broadening might be obtained by differentiating Bragg's equation. For this purpose, approximate use is made of the difference ratio $\Delta d/\Delta(2\theta)$ rather than of the differential ratio. The differentiation of the Bragg equation with respect to 2θ results in

$$\frac{\Delta d}{\Delta(2\theta)} = \frac{\lambda \cos\theta}{4\sin^2\theta} = \frac{d_0}{2}\cot\theta \qquad (3.51)$$

This equation is interpreted such that the variation in $\Delta(2\theta)$ is identified with the integral breadth β, which is broadened due to the variation in interplanar spacing Δd. The ratio of $\Delta d/d_0$ is considered as the product of the root mean square strain ε_{rms} and a scaling factor K_D that depends on the nature of the lattice distortions and the underlying model of microstrain. It then follows for the relation between reflex broadening and strain

$$\beta_D = 2K_D \varepsilon_{rms} \tan\theta \qquad (3.52)$$

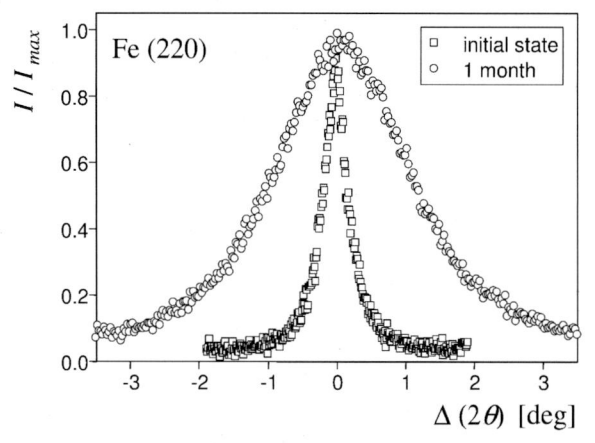

Figure 3.12 Normalized 220 reflections of an unprocessed and ball-milled iron powder (compare to Ref. [31]).

The strain appears here as root mean square strain

$$\varepsilon_{rms} = \langle \varepsilon^2 \rangle^{1/2} \tag{3.53}$$

which appears because of the probing of large sample volumes by the x-ray beam. A remarkable property of Eq. (3.52) is the dependency on the diffraction angle θ. This does not follow a $1/\cos\theta$ dependency as in the Scherrer equation, but varies with $\tan\theta$. This fundamental difference allows for a separation of reflection broadening when both microstructural causes – small crystallite size and microstrain – occur together.

The different approaches presented in the following commonly assume that size and strain broadening add to the total integral breadth of a Bragg peak

$$\beta = \beta_S + \beta_D \tag{3.54}$$

After reliable expressions were achieved for the size-broadening integral breadth β_S in the previous section we are left with the task of solving for the distortion-broadening integral breadth β_D.

3.5.1
Analysis According to Williamson and Hall

The distinct θ dependencies of both effects laid the basis for the separation of size and strain broadening in the analysis of Williamson and Hall [32]. Addition of the Scherrer equation and Eq. (3.52) results in

$$\beta_{2\theta} = \frac{K_S \lambda}{\langle D \rangle_V \cos\theta} + 2K_D \varepsilon_{rms} \tan\theta \tag{3.55}$$

3.5 Concomitant Occurrence of Size and Strain Broadening

where K_S is the Scherrer constant. One may now perform a numerical regression for the integral breadth β as a function of scattering angle 2θ and fit the course of $\beta_{2\theta}$ to the model function of Eq. (3.55) thereby obtaining $\langle D \rangle_V$ and ε_{rms} from the fit parameters. At the time when Williamson and Hall introduced their scheme, however, they proposed a graphical evaluation scheme. Multiplying both sides of Eq (3.55) by $\cos\theta/\lambda$ yields

$$\beta_{2\theta} \frac{\cos\theta}{\lambda} = \frac{K_S}{\langle D \rangle_V} + K_D \varepsilon_{rms} \frac{2\sin\theta}{\lambda} \qquad (3.56)$$

This equation shows that the plot of $\beta\cos\theta$ versus $2\sin\theta/\lambda$ should give a straight line that allows for the average crystallite size $\langle D \rangle_V$ and the r.m.s. strain $K_D \varepsilon_{rms}$ to be determined from the intercept and slope. This graph is generally termed the Williamson–Hall plot. The abscissa in this plot is either given by $2\sin\theta/\lambda$ or equally by the momentum transfers Q. As an example Fig. 3.13(a) displays the Williamson–Hall plot of a nanogranular iron powder that has been prepared by ball milling [31].

As for most of the techniques presented in this chapter the Williamson–Hall plot was developed for the investigation of powders and other voluminous samples rather than for thin films. A plot of integral breadth according to Eq. (3.56) should be interpreted with caution due to the various anisotropies that may be associated

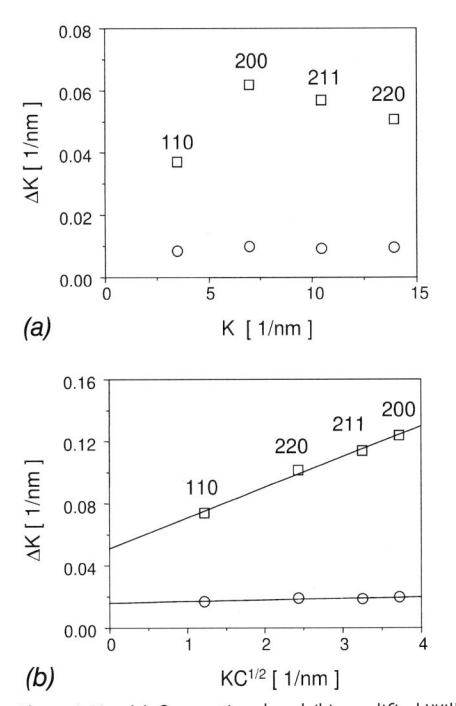

Figure 3.13 (a) Conventional and (b) modified Williamson–Hall plots of nanogranular iron powder (from Ref. [33]). Instead of Q and ΔQ the axes are denoted by K and ΔK (with $Q = 2\pi K$ being valid).

with the microstructure of the sample and that can cause considerable scatter of the data around an ideal Williamson–Hall line. Consider, for instance, the case of an anisotropic growth velocity as is often observed in thin-film growth. For a certain crystallographic direction $\langle hkl \rangle$ the grains will grow faster than others yielding an hkl-dependent crystallite size in the direction of the substrate normal s_3. The growth anisotropy would modulate the average crystallite size $\langle D \rangle_V$ as a function of θ or hkl. Endeavoring to fit such data to a straight line accounts for a further averaging process, i.e. the microstructure is forced into a model of one single average size where the averaging is also over crystallographic directions $\langle hkl \rangle$. Such a procedure might be considered as a systematic failure to model correctly the sample microstructure. Depending on the analytical question under investigation, however, such an approach might be justified. In any case the analyst should be aware of the averaging to which the data are subjected.

A further concern with the Williamson–Hall plot relates to the precise meaning of ε_{rms}. This average strain was introduced in a rather *ad hoc* fashion from the derivative of Bragg's law. This is a practical approach and might be useful for first estimates, but does not represent a firm theoretical basis. Due to the imprecise introduction the actual meaning of ε_{rms} and the values obtained from Eqs. (3.52) and (3.56) are subject to some uncertainty.

A well-founded model has been introduced, however, for the description of dislocation-induced microstrain by the so-called dislocation contrast [34–36]. The basic idea of dislocation contrast becomes evident by a closer inspection of Fig. s4.1 in Structure Box 4. The edge dislocation shown there is seen to displace the surrounding atoms in the plane of the paper. No displacements u and strains ε would, however, arise in the perpendicular direction. Therefore, a strain-induced broadening would be invisible for a relative orientation of scattering vector Q and Burgers vector b having $Qb = 0$, while the maximum displacements are observed for $Qb = Qb$. It can be concluded that the dislocation strain-induced broadening sensitively depends on the relative orientation of Q and b. This is the basis for the dislocation contrast in x-ray line broadening, where the contrast is mainly determined by the orientation of the scattering vector with respect to the dislocation.

It is evident that a kind of averaging procedure has to be performed when the dislocation contrast of a set of dislocations is to be worked out for a polycrystalline sample. Moreover, a thorough analysis of dislocation contrast must include the precise nature (edge or screw dislocation), the dislocation density ρ_d and the magnitude of the displacement fields u the latter of which will depend on the elastic constants of the material. In order to deal with these complex requirements so-called dislocation contrast factors C_{hkl} were introduced that depend on the Miller indices hkl of the reflection, the scattering vector Q and Burgers vector b, the elastic modulus E and the Poisson ratio v of the material under investigation. In a powder pattern all dislocations from the set of equivalent slip systems are mapped into the same Bragg reflection. In order to deal with this fact an averaging over the different C_{hkl} (for the same hkl) has to be performed yielding so-called average dislocation contrast factors $\bar{C}_{hkl}$. The effect of the $\bar{C}_{hkl}$ is to cause a modulation of peak broadening for various orders of reflection hkl. It should be noted that screw and edge dislocation have different average contrast factors. In the case of cubic materials the $\bar{C}_{hkl}$ were shown to be of the form

$$\bar{C}_{hkl} = \bar{C}_{h00}(1 - q\Gamma_{hkl}^2) \tag{3.57}$$

where $\bar{C}_{h00}$ is the average dislocation contrast factor of a $h00$ reflection, q depends on the elastic constants of the material and Γ_{hkl} is the ratio of the two permutation invariantes of fourth order polynomials of Miller indices hkl

$$\Gamma_{hkl} = \frac{h^2k^2 + h^2l^2 + k^2l^2}{(h^2 + k^2 + l^2)^2} \tag{3.58}$$

Table 3.4 Dislocation contrast factors for (110) slip planes in bcc-structured α-Fe [39].

Reflection	$\bar{C}_{hkl}^{edge}$	$\bar{C}_{hkl}^{screw}$
110, 220, 211	0.1781	0.1040
200	0.2648	0.3055
310	0.2335	0.2330
222	0.1492	0.03684

It has been shown by Ungár and coworkers that the dislocation contrast factors may be applied to formulate a modified Williamson–Hall equation, which then incorporates the anisotropy of dislocation-induced peak broadening [31, 33, 36]. The contrast factors enter as square-rooted prefactors to the second term in Eq. (3.55) that accounts for strain broadening and the modified Williamson–Hall equation reads

$$\beta_{2\theta} \frac{\cos\theta}{\lambda} = \frac{K_S}{\langle D \rangle_V} + \varepsilon\sqrt{\bar{C}_{hkl}} \frac{2\sin\theta}{\lambda} \tag{3.59}$$

The average microstrain now takes a defined meaning

$$\varepsilon = b\sqrt{\pi A \rho_d / 2} \tag{3.60}$$

where A depends on the elastic constants and amounts to about 10 [31]. Sometimes, an additional quadratic term is made use of in Eq. (3.59), which directly scales with $\bar{C}_{hkl}$ instead of $\bar{C}_{hkl}^{1/2}$, but this will not be considered here.

The modified Williamson–Hall plot technique has been applied to the nanogranular iron powder sample already presented in Fig. 3.13(a). Equal densities of screw and edge dislocations were assumed in the calculation of the contrast factors [31], which were inserted into an evaluation scheme according to Eq. (3.59). The result is shown in Fig. 3.13(b), where the abscissa is given by $\bar{C}_{hkl}^{1/2} Q$ instead of Q causing the data points to shift in the horizontal direction. It is seen from a comparison of Fig. 3.13(a) and (b) that the inclusion of dislocation contrast leads to a considerable improvement in the regression of the data points. Hence, the modeling of the microstructure of the nanogranular sample by the dislocation contrast approach yields much more reliable results. An average crystallite size of 18 nm and a dislocation density ρ_d of more than 10^{16} m^{-2} were derived from this study. The latter value translates to an average dislocation distance of between 4 and 8 nm. This result shows that the concept of lattice distortions by dislocations is still valid in very small crystalline entities.

It has been shown for various materials like Cu and Si_3N_4 that Eq. (3.59) can better account for the observed line broadening in nanocrystalline powders than Eq. (3.56) [40]. These investigations were accompanied by TEM studies in order to compare the average crystallite sizes obtained with both methods. Based on these works the modification of the original Williamson–Hall equation by contrast factors appears to be a useful extension for the description of line broadening from Bragg reflections. An important presupposition that has to be fulfilled for the application of Eq. (3.59) is, however, that the grains are randomly oriented. This assumption enters into the calculation of contrast factors. If preferred orientation of grains does occur in the sample the dislocation contrast factors C_{hkl} should be calculated individually for each reflection [41].

3.5.2
Method of Warren and Averbach

The Warren–Averbach method makes use of the Fourier coefficients of Eq. (3.23) of at least two harmonic reflections like hkl and $2h2k2l$. Size and strain parameters are derived by expanding the two line profiles into Fourier sums with coefficients $A(L)$ that depend on the Q-conjugated variable L [21]. Of course, the Fourier coefficients have to be corrected for instrumental broadening effects before doing so. The full derivation of the formulae is given in Ref. [7] of chapter 1. Here, only the basic ideas and the resulting equations are presented.

The sample is assumed to consist of small crystallites of arbitrary shape. The modeling of size and strain broadening in the Warren–Averbach scheme starts with decomposing the displacement field $u(r)$ into its orthogonal components as given in Eq. (3.50). The displacement field is inserted into the interference function $\Im(Q)$ and the diffracted intensity is expanded into a Fourier sum

$$\Im(Q) = \frac{1}{N} \sum_L A(L)\cos(QL) \tag{3.61}$$

All the structural information is contained in the $A(L)$ coefficients, whereas sin coefficients $B(L)$ have been neglected. From the full derivation of Warren and Averbach the $A(L)$ coefficients are found to depend on the product of two factors

$$A(L) = A^S(L)A^D(L) \tag{3.62}$$

3.5 Concomitant Occurrence of Size and Strain Broadening

that are related to the size of the columns, $A^S(L)$, or to displacements of atoms within one column, $A^D(L)$. The distortion-related coefficients $A^D(L)$ depend on the square of the microstrain $\langle \varepsilon_L^2 \rangle$ averaged over the length L within individual columns and take the explicit form

$$A^D(L) = \exp(-2\pi^2 L^2 h^2 \langle \varepsilon_L^2 \rangle / a^2) \tag{3.63}$$

In this expression $h^2 = h^2 + k^2 + l^2$ holds for the Miller indices of the reflection under consideration. The size coefficients $A^S(L)$ obey Eqs. (3.40) and (3.41). They are thus related to the column height distribution function $p(L)$ and their first and second statistical moments $\langle L^1 \rangle$ and $\langle L^2 \rangle$ may accordingly be derived by evaluating these equations. Before this can be done, size and strain coefficients A^S and A^D have to be separated.

The separation can be performed either by a fitting or by a plotting procedure that rely on the h^2 dependence of the A^D term. This is the part of the analysis where use has to be made of the availability of two or more reflections, h and $2h$. Therefore, more than one set of $A(L)$ coefficients from two or more analyzed reflections have to be available. Forming the logarithm of $A(L)$ coefficients yields a sum of two terms, one independent of h, the other one depending on h^2

$$\ln A(L) \approx \ln A^S(L) - 2\pi^2 L^2 h^2 \langle \varepsilon_L^2 \rangle / a^2 \tag{3.64}$$

which holds for small $Lh\langle \varepsilon_L^2 \rangle$. If the $\ln(A(L))$ values are plotted versus the square of reflection order h^2 and parameterized by the variable L a set of curves is obtained, from which the slope $-2\pi^2 L^2 \langle \varepsilon_L^2 \rangle / a^2$ may be determined. The slope of the curves directly gives the averaged microstrain $\langle \varepsilon_L^2 \rangle$. In addition, the size-related $A^S(L)$ coefficients may be derived from the intercept of the curves at $h^2 = 0$. A plot of $A^S(L)$ values has already been given in the example in section 3.3, see figure 3.6. The course of $A^S(L)$ coefficients may be inserted into (3.40) and (3.41), from which the first and second moments $\langle L^1 \rangle$ and $\langle L^2 \rangle$ are obtained. One might stop at this point of the analysis, since important parameters have been achieved that may already represent sufficient information on the sample's microstructure.

Further information may only be derived by introducing assumptions and thereby interpreting the results. It may be assumed, for instance, that the sample is composed of spherical grains exhibiting a lognormal diameter distribution. In this case the moments $\langle L^1 \rangle$ and $\langle L^2 \rangle$ may be inserted into Eqs. (3.48) and (3.49) that relate the L and D moments for a lognormal distribution and the parameters D_0 and γ of the diameter distribution are obtained.

The problem with the Warren–Averbach technique is related to the systematic errors that might be introduced during the different data processing steps of Fourier transforming, deconvolution, strain–size separation via fitting of Eqs. (3.63) or (3.64) and the one- and twofold differentiation of $A^S(L)$ in order to derive $\langle L^1 \rangle$ and $\langle L^2 \rangle$ [42]. Also the choice of the "true" background function, an imprecise standard profile and peak overlap may all introduce artifacts into the numerical analysis like the hook effect already mentioned. The few harmonic reflections that are typically available impose a further restriction. In metals seldom more than two reflections might be inserted into the analysis. Various modifications and improvements have been developed for the Warren–Averbach technique: hints can be found in Refs. [22, 43].

Also the Warren–Averbach method has been extended by Ungár and coworkers by introducing dislocation contrast factors $\bar{C}_{hkl}$ into the analysis scheme [36]. In this approach the microstrain is modeled as arising from screw and edge dislocations of equally populated glide systems. The method has already successfully been applied to nanogranular powders of Cu, Fe, Si$_3$N$_4$, WC and others [36, 40, 44, 45]. Most of the formalism proceeds as outlined above, except for the distortion-related coefficient $A^D(L)$, which is now given by the modified Warren-Averbach equation.

$$\ln A(L) = \ln A^S(L) - \rho_d B \frac{L^2}{4\pi^2} \ln(R_e/L)(Q^2\bar{C}) \tag{3.65}$$

where R_e is the effective outer cut-off radius of dislocations and the other symbols have their meanings as introduced above. The application of this technique to a nanodisperse Si$_3$N$_4$ powder is displayed in Fig. 3.14 [44].

Also in the Warren–Averbach approach and extensions it is assumed that the column height distribution $p(L)$ is isotropic and independent of any selected sample direction. This point deserves considerable attention in thin films, where out-of-plane structural properties may differ significantly from in-planes ones. In these cases Eq. (3.62) solely accounts for the components of size and strain along which the scattering vector Q was oriented during the measurement.

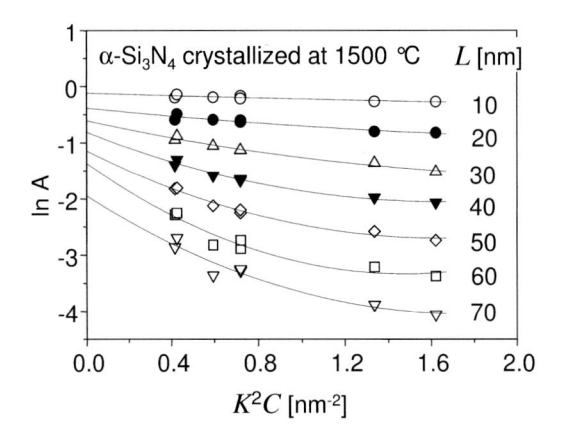

Figure 3.14 Logarithm of Fourier coefficients A(L) as determined from nanodisperse Si$_3$N$_4$ powder (from [44]). Instead of $Q^2\bar{C}_{hkl}$ the abscissa is given by $K^2\bar{C}_{hkl}$ (having $Q = 2\pi K$).

3.5.3
Single-Line Analysis

The Williamson–Hall plot makes use of as many reflections from which an integral breadth can be determined, while the Warren–Averbach scheme relies on at least two harmonic reflections. One may also endeavor to determine size and strain information from the line profile of only one Bragg peak. It has already been mentioned that experimental results implied crystallite size to cause a Cauchy-type broadening while strain broadening was associated with a Gauss profile. Based on these observations the Cauchy and Gauss fractions of a single reflection can be interpreted to signal the individual effects of size and strain broadening [42, 46, 47]. In the case of a Voigt-type line profile the Cauchy and Gauss parts h_C and h_G of the measured line profile $h(2\theta)$ are individually obtained from the convolution of the Cauchy and Gauss fraction of the instrumental and the sample profile

$$h_C = f_C \otimes g_C$$
$$h_G = f_G \otimes g_G \qquad (3.66)$$

In this case, the integral breadths of the sample-broadened profile β_G^f and β_C^f may be obtained from the integral breadth of the measured and the instrumental profile according to Eq. (3.16). Both β_G^f and β_C^f may then be assumed to be either due to size broadening or strain broadening and would translate according to

$$\langle D \rangle_V = K_S \lambda / (\beta_C^f \cos\theta) \qquad (3.67)$$
$$\tilde{\varepsilon} = \frac{1}{4} \beta_G^f \cot\theta$$

into an average crystallite size $\langle D \rangle_V$ and a weighted-average strain $\tilde{\varepsilon}$. For a Gauss-like strain distribution the latter quantity is related via

$$\tilde{\varepsilon} = \frac{1}{2}(2\pi)^{1/2} \langle \varepsilon^2 \rangle^{1/2} \qquad (3.68)$$

with the r.m.s. strain ε_{rms} introduced above.

The single-line approach was applied by Keijser et al. to various metallurgical samples, among them an electrodeposited Ni layer on a brass substrate [46]. The authors made use of a Scherrer constant $K_S = 1$, implicitly assuming cube-shaped grains, but the procedure may undoubtedly be generalized to other crystallite shapes by inserting appropriate values for K_S. The investigations revealed significantly different crystallite sizes for (hhh) and $(h00)$ oriented grains and also distinct microstrains within both grain ensembles [46] (see Table 3.5). Obviously, in this deposition process (hhh) grain growth in the direction of the surface normal leads to larger grains and reduced microstrain than growth of $(h00)$ oriented grains. The effect can be understood from anisotropic growth, i.e. by differences in grain growth for distinct crystallographic directions, which has been observed in various thin-film studies. The example clearly demonstrates the wealth of information that may be revealed by this type of investigation for elucidating information relevant to the microstructure and the growth process of thin films.

Table 3.5 Microstructural parameters obtained for a 40 μm Ni layer on brass from different Bragg reflections hkl (from Ref. [46]).

hkl	$\langle D \rangle_V$ (nm)	$\tilde{\varepsilon}$ (10^{-3})
111	52.0 ± 3.0	3.6 ± 0.1
222	42.0 ± 6.0	3.5 ± 0.1
200	21.9 ± 1.0	4.3 ± 0.2
400	22.0 ± 9.0	4.6 ± 0.9

Initially, the procedure was devised to be applied to Voigt profiles only, for which Eqs. (3.16) and (3.66) are obeyed, but the approach has been extended to Pearson and pseudo-Voigt model functions. For this purpose the P and pV profiles have to be approximated by Voigt functions [47]. A numerical analysis of this problem yields the following relations between the Cauchy and Gauss components of a corresponding Voigt profile for the Pearson function

$$\beta_C = \left(0.750445/m + 0.247681/m^2\right)\beta$$
$$\beta_G = \left(1.092228 - 1.163332/m + 1.316944/m^2 - 1.131115/m^3\right)\beta \quad (3.69)$$

and for the pseudo-Voigt function

$$\beta_C = \left(0.017475 + 1.500484\eta - 0.534156\eta^2\right)\beta$$
$$\beta_G = \left(0.184446 + 0.812692\sqrt{1 - 0.998497\eta} - 0.659603\eta + 0.445542\eta^2\right)\beta \quad (3.70)$$

Here, β denotes the integral breadth of either the P or pV profile, while m and η stand for the shape-forming parameters as introduced in Section 3.1. The errors in approximating the integral breadths from these equations increase towards the extrema of pure Cauchy or Gauss profiles [47]. The numerical relations allow the evaluation of the β_G^f and β_C^f components of P and pV profiles that can be inserted in Eq. (3.67) in order to derive the average crystallite size and mean strain. Of course, the experimental profiles have to be corrected for instrumental broadening.

A compilation of further single-line approaches was given by Delhez et al. [42].

3.5.4
Techniques of Whole-Pattern Fitting

With peak profile model functions and modern computer technology at hand a whole $\theta/2\theta$ diffraction pattern may be simulated by a superposition of various peaks at appropriate positions $2\theta_0$ and a background function $B(2\theta)$. The intensity as a function of scattering angle is then denoted by $y(2\theta)$ and expressed by the sum

$$y(2\theta) = B(2\theta) + \sum_i I_{0,i} f(2\theta - 2\theta_{0,i}, \beta_i, m_i, k_i, \eta_i ...) \quad (3.71)$$

3.5 Concomitant Occurrence of Size and Strain Broadening

where the refinement is performed in order to elucidate the peak profile parameters like $2\theta_0$ and β, m, k or η. The sum has to be extended over all peaks indexed by i that affect the intensity in the refined 2θ range. In this variant the whole-pattern fitting (WPF) is termed pattern decomposition and the peak parameters obtained may be inserted into the microstructure analysis schemes as introduced above.

A comparable approach was introduced by Rietveld in 1967 for the modeling of neutron diffraction $\theta/2\theta$ patterns [48, 49] and became widely known as Rietveld refinement. However, the fit parameters in the Rietveld technique are the atomic positions in the unit cell r_n and the lattice parameters and are thus different from the line profile parameters in pattern decomposition. Rietveld techniques were developed for the solution of crystal structures from powder patterns. In neutron diffraction Bragg peaks may be accounted for by Gauss profiles and the basic approach of Rietveld was to start with the model function

$$y(2\theta) = B(2\theta) + \sum_h I_{0,h} \exp(-\pi^2 (2\theta - 2\theta_{0,h})^2 / \beta_h^2) \qquad (3.72)$$

for the fitting of the measured pattern over the full range of scattering angles 2θ. Here, $I_{0,h}$ and β_h are the maximum intensity and integral breadth of reflection $h = hkl$. The approach aimed at the structural refinement of unit cell parameters from the integral intensities $I_{0,h}\beta_h$, i.e. the fitting of the structure factor F_h

$$|F_h(r_n)|^2 \propto I_{\text{int},h} = I_{0,h}\beta_h \qquad (3.73)$$

in order to elucidate positional parameters of atoms r_n in the unit cell (compare Eq. (1.43)). In general, structure refinements might be complicated by the overlapping of peaks that increases for low-symmetry structures and might impede the precise determination of $I_{0,h}$ and β_h from single reflections. The fitting of the complete pattern according to eq. (3.72) enables an improved refinement of the full set of shape parameters and consequently the structure factors of individual reflections.

After its introduction the Rietveld technique was also applied to x-ray powder diffraction, where it was used and still is used with great success for the solution of unknown crystal structures. Owing to the great dissemination of the method in refining complete 2θ powder patterns many other WPF techniques were also termed Rietveld refinement, which sometimes appears inadequate. The different approaches in WPF are shown in Fig. 3.15. For thin-film work or powder studies of one-elemental metals the focus generally is not on the refinement of the crystallographic structure, but on microstructure. Microstructural parameters might directly be obtained from WPF procedures by refining the crystallite size and lattice fault parameters in accordance with one of the models presented above. A variety of such procedures has been developed for determining microstructure [50–54] or a crystallized fraction [55].

A recent example is demonstrated in Fig. 3.16 [53]. In this work Scardi and Leoni investigated the effect of ball milling on a Ni powder sample. Figure 3.16(a) displays the $\theta/2\theta$ pattern of the sample after a milling time of 96 h leading to severely

Pattern decomposition	Rietveld refinement	Fundamental parameter approach
Emphasize of fit on peak profile parameters $2w$, β, m, η, k .. and background function $B(2\theta)$	Emphasize of fit on positional parameters of atoms in the unit cell r_j and cell parameters $a, b, c, \alpha, \beta, \gamma$	Emphasize of fit on crystallite size $\langle D \rangle_V$ and microstrain ε
Input: choice of peak profile model function	Input: choice of peak profile model function	Input: quantitative description of diffractometer configuration
Computer codes: DRXWin, ProfileFit, XFIt, ...	Computer codes: ARIT, Fullprof, GSAS, Koalariet, MAUD, ..	Computer codes: XFit, Topas, ..

Figure 3.15 Different approaches in whole-pattern fitting.

broadened Bragg peaks. The evolution of crystallite size and lattice faults with milling time was analyzed by the authors in the framework of a whole powder pattern modeling scheme that differs from the usual pattern fitting by dispensing with analytical profile functions. Lattice parameter, dislocation density, outer cut-off radius, contrast factor, twin and deformation fault probabilities could be refined in combination with the parameters of the grain size distribution (see Fig. 3.16(b)–(e), where some of these parameters are given). Important microstructural information is seen to derive from this analysis technique.

Recent extensions of the Rietveld approach aim at establishing a combined modeling of crystallographic structure, texture, microstructure and residual stress from various $\theta/2\theta$ patterns of one sample. Chateigner and Lutterotti have promoted this approach and further details are given in Chapter 5. For an appropriate choice of computer code for the microstructural question under investigation the reader is referred to the cited literature and the CCP14 suite (see the hints in Instrumental Box 4). For thin-film work the WPF program of choice should preferably allow for the refinement of the attenuation coefficient μ, the thickness t or the μt product of one layer or a set of these parameters for a multilayer system. This requirement is fulfilled by the MAUD program of Lutterotti [56].

A built-in approach for the deconvolution of the instrumental profile has been devised by Cheary and Coelho for x-ray powder patterns and has been termed the fundamental parameter approach [13, 57, 58]. The authors made use of analytical expressions of all instrumental aberration functions due to axial and in-plane divergence, finite length of x-ray source, sample and receiving slit, flat-specimen error and sample absorption. The instrumental line profile is the convolution of all beam-shaping effects

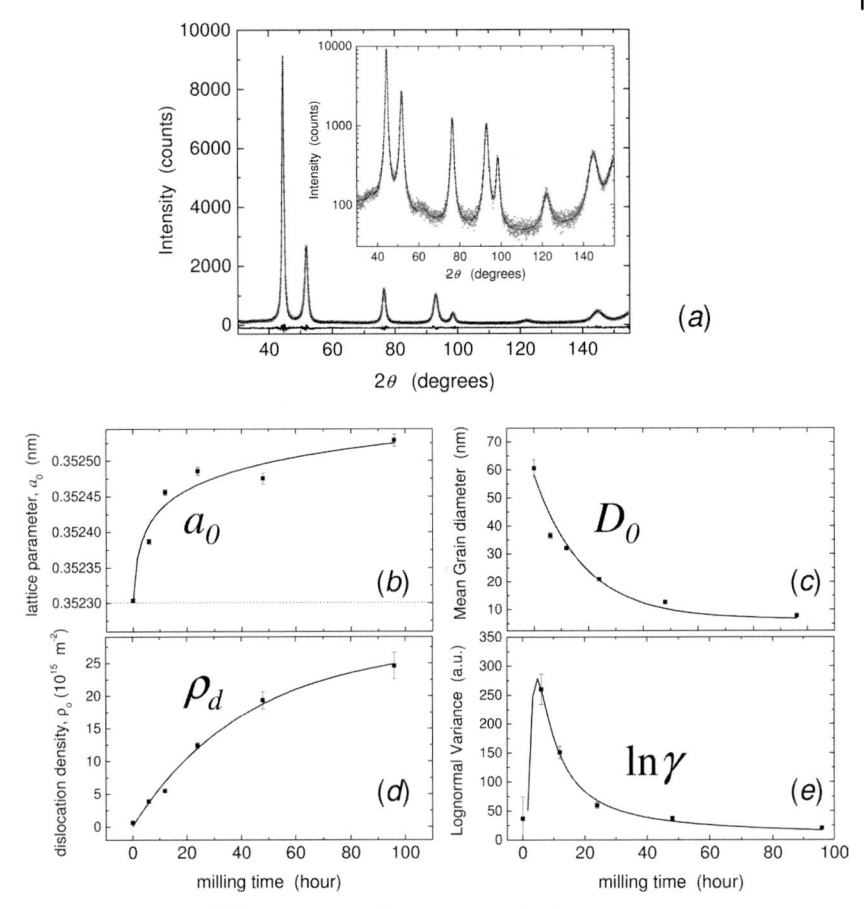

3.5 Concomitant Occurrence of Size and Strain Broadening | 133

Figure 3.16 (a) $\theta/2\theta$ diffraction pattern of a ball-milled Ni sample after 96 h and whole powder pattern modeling (straight line); the inset shows the same on a logarithmic intensity scale. (b–e) Microstructural parameters a_0, D_0, ρ_d and $\ln\gamma$ as a function of milling time (from Ref. [53]).

$$g(2\theta) = g_1(2\theta) \otimes g_3(2\theta) \otimes g_3(2\theta) \otimes \ldots \otimes g_N(2\theta) \qquad (3.74)$$

This line profile is directly deconvoluted from the measured profiles according to Eq. (3.10), which are accounted for by expressions including microstructural parameters, like Eq. (3.67). The instrumental line profile is thus not obtained from the measurement of a standard, but – fundamentally – from the diffractometer setting.

3.6
Applications

Microstructural investigations by x-ray LPA techniques have been performed for thin films of various material systems. In a recent study Santra et al. elucidated the microstructure of thin cuprous oxide films prepared by electrodeposition [59]. Cu_2O Bragg reflections from symmetric $\theta/2\theta$ powder patterns of 4 to 13 µm thick films were analyzed in the framework of different LPA techniques. It was shown in this study that crystallite sizes and microstrains obtained by the Warren–Averbach method and the integral breadth approach were comparable, while the parameters derived from the Williamson–Hall plot yielded larger values for both quantities. Another interesting example is given by the investigation of hard nitride coatings and metal multilayer coatings, for which it was endeavored to estimate the mechanical hardness from the x-ray diffraction line profile [60, 61]. The density of both dipole dislocations and dislocation pile-ups has been determined from the Warren–Averbach analysis of TiN or multilayer Bragg peaks. The hardness was derived from the linear relation with the maximum shear stress at the head of the pile-ups. This nondestructive technique would be very relevant for thin-film work, since the conventional determination of mechanical hardness by indentation techniques suffers from damage of the inspected area or in some cases from interference effects with the substrate. It would be helpful to have available more comparative studies, in which the hardness of the same layer systems has been determined by both techniques.

One problem with LPA procedures concerns the often unmentioned or unemphasized presuppositions and assumptions that enter into the formulas relating line shape and microstructure. The assumptions are related to the crystallite size distribution function (monodisperse or lognormal), the crystallite shape and its distribution, in some cases to the absence of microstrain or the restriction to dislocation-induced microstrain only and many others. It is generally recommended to specify carefully the implicit assumptions of the underlying microstructural model whenever one of the above evaluation schemes is applied. Regarding the application of these techniques to thin solid films a severe restriction is that ideally texture and residual stress should be absent. The formulation of principles as to when and how this condition might be circumvented has begun [41, 56], but remains an important issue that will have to be dealt with. From a practical point of view the thin-film grower will often be interested in relative values only, for instance, if crystallite sizes in a polycrystalline film should be increased. In these cases a set of consistent data for samples prepared under various deposition conditions may enable the identification of the optimum preparation recipe and might thus be sufficient, since no absolute microstructural data are required.

Nanocomposite layers have also been prepared from metallic particles embedded in an amorphous matrix, with metal-doped derivatives of a-C:H representing an important example. Hydrogenated amorphous carbon layers, a-C:H, are also termed diamond-like carbon (DLC) and can be compared to semiconducting a-Si:H layers introduced in Chapter 2. Thin DLC films were originally developed as razor edge coatings in the early 1980s, but were also identified as exhibiting remarkable tribological properties. In a metal-doped variant, Me-C:H layers are used for coatings of injection devices in automotive applications enabling high injection pressures in the 2000 bar range and paving the way for diesel engine technologies with high torques in the low rotational speed regime [62]. Figure 3.17 presents the $\theta/2\theta$ diffraction patterns of thin W-C:H films with different W contents from a study on metal-doped C:H layers of Schiffmann et al. [63]. Broad Bragg peaks from cubic WC_{1-x}, $x = 0.5$–0.625, particles become clearly visible with increasing W content. Also small-angle x-ray scattering (SAXS) was applied in this study and it was concluded from the combination of different structure-evaluating techniques that nanocluster radii and intercluster distances monotonically increased with increasing metal content. For low metal concentrations of less than about 5 at% cluster radii of only 0.5–1 nm could be detected corresponding to nanoclusters of only 10–100 atoms [63].

This chapter has focused on one-dimensional lattice faults and reference was made to the existence of two-dimensional distortions like twinning and stacking faults. These phenomena can have complicated effects on Bragg reflection, which are asymmetric peak broadening and peak shift [64, 65]. In two recent studies by Boulle et al. the line profile broadening due to stacking faults in an epitaxial Sr-$Bi_2Nb_2O_9$ (SBNO) layer on $SrTiO_3$ has been investigated [66] and compared with resuts obtained for powder samples of the same material [67]. SBNO belongs to the group of Aurivillius phases with $SrNb_2O_3$ units in perovskite structure stacked upon Bi_2O_3 slabs, which may give rise to stacking faults along the crystallographic c axis. 00l Bragg peaks were fitted by Voigt functions and from the variation of the Cauchy integral breadth β_C the stacking fault density was shown to diminish with annealing time.

The increasing ability of materials scientists to prepare nanocrystalline samples in the form of both bulk and thin films has brought the question of crystallite surfaces onto the agenda of LPA. The distorted coordination of surface atoms may af-

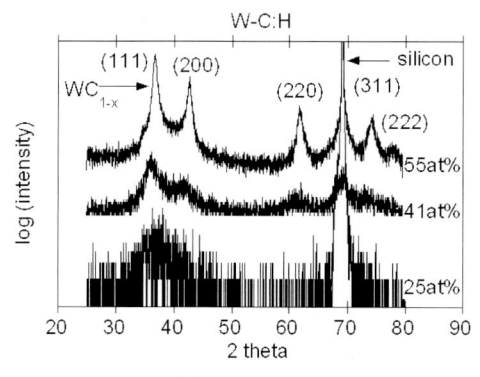

Figure 3.17 $\theta/2\theta$ diffraction patterns with a logarithmic intensity scale from thin W-C:H films of different W content (from Ref. [63]).

fect the line shape of Bragg peaks when their volume fraction reaches the few percent range as achievable in nanocrystalline materials. In their studies of anomalous magnetic surfaces of ferrites Nunes and Lin observed decidedly asymmetric tails in powder diffraction peaks from these materials. The authors proposed a model of surface relaxation that assumed a gradually increasing expansion of the interplanar spacing on the outermost layers of $CoFe_2O_4$ nanoparticles. The model predicted a shift of scattering power from the main 440 reflection to the low-scattering-angle range and yielded the observed asymmetric peak profiles in accordance with their observation [68].

The field of LPA of polycrystals in general and of thin films in particular is still in an active phase of development, which is motivated by the large number of microstructural models, model parameters and their complex nature. Interestingly, some open questions even relate to the fundamentals. For instance, in their thorough fundamental parameter approach studies on a LaB_6 standard reference sample, Cheary and Coelho found that size and microstrain broadening could both be accounted for by Cauchy-type functions [57], which contradicts the usual separation into a Gauss and a Cauchy function. A further example is given by the investigation of the natural profile of the Cu Kα emission line, where Deutsch et al. pointed to the fact that the process of convolution yields only a reliable instrumental line for either a spatially or spectrally very narrow line [12]. These conditions are only met in high-precision work.

The last remarks underline again that the results of microstructural parameters determined by LPA should always be regarded with skepticism. For a first investigation of a sample's microstructure the newcomer to the field is recommended to start by deconvoluting the sample and instrumental line profile and applying the single-line analysis approach (Section 3.5.3). This may yield relevant results that may be compared for differently prepared samples. On this basis more thorough techniques may subsequently be applied and more anisotropies may be allowed for in modeling and refining the data. It is obvious that the sophistication of the microstructural model applied to a set of samples will critically depend on the amount of time available in a thin-film R&D project.

Exercises

3.1 Nickel crystallizes in the fcc structure, having $a = 0.3524$ nm. Assume that a polycrystalline Ni sample comprises monodisperse cube-shaped grains. Calculate the number of atoms N_{sur} that reside on the surface versus the total number N_{tot} in the cube for a grain edge of $D = 2$ µm. Depict N_{sur}/N_{tot} for D between 2 nm and 2 µm.

3.2 Compare different Pearson functions with varying exponent m, but constant parameters I_0 and $2w$. How will the tails of a peak evolve for increasing m?

3.3 List the experimental causes that may contribute to the background in a $\theta/2\theta$ pattern.

3.4 Asymmetric profiles. Identify the range of FWHU 2w and μt products for which the absorption factor $A_{\theta 2\theta}$ may account for line profile asymmetries above the 1, 5 and 10% levels.

3.5 Sketch the course of a computer program to perform the Rachinger correction for a set of measured $(2\theta_i, I_i)$ data. How would you handle the intensity correction if the 2θ coordinate according to Eq. (3.12) does not belong to the set of measured data?

3.6 Show that the integral breadth Eq. (3.25) converges versus $0.903N$ when the integration limits are set to the first minima $\pm 1/N$ and the number of unit cell N diverges to infinity. Would it make sense to define the Scherrer constant K_S = 0.903 for cube-shaped crystallites?

3.7 Assume that a Bragg peak is to be analyzed that was measured at 33° with λ = 0.154 nm. Calculate the resolution Δ_L in L space when the data window ranged over (a) 32–34° and (b) 30–36°. What starting and end values $2\theta_S$ and $2\theta_E$ have to be chosen if a resolution of 0.23 nm is desired in L space?

3.8 What are the discrete Fourier transforms $H(L) = \widetilde{\mathfrak{F}}(h(Q))$ when $h(Q)$ is identified with the Gauss, Cauchy and the Voigt functions? Make use of the integral breadth representation of the model functions.

3.9 A polycrystalline specimen comprises spherical grains and exhibits a monomodal distribution with a diameter of 10 nm. How many crystallites cover a surface of 1 cm² when the film thickness is 1 µm? Assume the orientation of crystallites is isotropic and calculate how many grains would contribute to a Bragg reflection when the divergence slit is set to 0.25° and 1°? How are the results affected when the diameter expands to 50 and 100 nm?

3.10 Derive the interference function for cube-shaped crystallites (Eq. (3.25)) from the general-shape interference function (Eq. (3.29)).

3.11 A polycrystalline thin film may be composed from a set of elliptically shaped crystallites with major axes a, b and c. The variation of each axis may be accounted for by a lognormal distribution. Calculate the interference function $\widetilde{\mathfrak{F}}(Q)$ for a Bragg reflection hkl.

3.12 A polycrystalline thin film may be composed from a set of rectangular shaped crystallites of edges a, b and c. The variation of each crystallite may be accounted for by a lognormal distribution. Calculate the interference function $\widetilde{\mathfrak{F}}(Q)$ for a Bragg reflection hkl.

3.13 Demonstrate the convergence of the lognormal distribution $g(D)$ towards the delta function $\delta(D - D_0)$ in the limit of $\ln \gamma \to 0$.

3.14 Assume you dispose of a (rather abnormal) sample comprising 2×10^9 cubes with one half having D_{cub} = 5 nm and the other half 10 nm, when the cubic unit cell edges measure 0.5 nm. Give expression for $p(L)$, $g(D)$ and calculate $\langle L^n \rangle$ and $\langle D^n \rangle$ for n = 1, 2, 3, 4.

3.15 In the early days of LPA x-ray diffraction patterns were obtained as strip charts from an x–t' plotter rather than as $(2\theta_i, I_i)$ digital data sets. Integral intensities were obtained by sizing and weighting the Bragg reflections, while $2w$ and I_0 parameters were derived from length measurements. Give some reasoning on the precision of I_{int} and β obtained by this procedure.

References

Monographs and reviews relevant to this chapter

E. J. Mittemeijer, P. Scardi (eds.), *Diffraction Analysis of the Microstructure of Materials*, Springer, Berlin, 2004.

J. I. Langford, D. Louer, Powder diffraction, *Rep. Prog. Phys.* **59** (1996) 131.

M. A. Krivoglaz, *X-ray and Neutron Diffraction in Nonideal Crystals*, Springer, Berlin, 1996; *Theory of X-ray and Thermal-Neutron Scattering by Real Crystals*, Springer, Berlin, 1969.

A. J. C. Wilson, *X-Ray Optics: The Diffraction of X-rays by Finite and Imperfect Crystals*, Methuen, London, 1949.

Special papers

[1] M. von Laue, Lorentz-Faktor und Intensitätsverteilung in Debye-Scherrer-Ringen, *Z. Kristallogr.* **LXIV** (1926) 115.

[2] K. Ishii, T. Ohba, T. Hara, A new process for producing a granular material, *Mater. Sci. Eng.* **A217/218** (1996) 232.

[3] M. Birkholz, U. Albers, T. Jung, Nanocomposite layers of ceramic oxides and metals prepared by reactive gas-flow sputtering, *Surf. Coat. Technol.* **179** (2004) 279.

[4] M. M. Hall, V. G. Veeraraghavan, H. Rubin, P. G. Winchell, The approximation of symmetric X-ray peaks by Pearson type VII distribution, *J. Appl. Cryst.* **10** (1977) 66.

[5] A. Brown, J. W. Edmonds, The Fitting of Powder Diffraction Profiles to an Analytical Expression and the Influence of Line Broadening Factors, *Adv. X-ray Anal.* **23** (1980) 361.

[6] W. Voigt, *Sitzungsber K. Bayer. Akad. Wiss.* **42** (1912) 365.

[7] J. I. Langford, A rapid method for analysing the breadths of diffraction and spectral lines using the Voigt function, *J. Appl. Cryst.* **11** (1978) 10.

[8] Abramowitz, I., *Handbook of Mathematical Functions*, Dover, New York, 1965.

[9] G. K. Wertheim, M. A. Butler, K. W. West, D. N. E. Buchanan, Determination of the Gaussian and Lorentzian content of experimental line shapes, *Rev. Sci. Instrum.* **45** (1974) 1369.

[10] P. Thompson, D. E. Cox, J. B. Hastings, Rietveld Refinement of Debye-Scherrer Synchrotron X-ray Data from Al_2O_3, *J. Appl. Cryst.* **20** (1987) 79.

[11] M. Ahtee, L. Unonius, M. Nurmela, P. Suortti, A Voigtian as profile shape function in Rietveld refinement, *J. Appl. Cryst.* **17** (1984) 352.

[12] M. Deutsch, G. Hölzer, J. Härtwig, J. Wolf, M. Fritsch, E. Förster, $K\alpha$ and $K\beta$ x-ray emission spectra of copper, *Phys. Rev. A* **51** (1995) 283.

[13] R. W. Cheary, A. Coelho, A fundamental parameters approach to x-ray line-profile fitting, *J. Appl. Cryst.* **25** (1992) 109.

[14] A. Kern, A. A. Coelho, R. W. Cheary, Analysis of the full diffraction pattern, in *Diffraction Analysis of the Microstructure of Materials*, ed. E. J. Mittemeijer, P. Scardi, Springer, Berlin, 2004, p. 17.

[15] M. Deutsch, G. Hölzer, J. Härtwig, J. Wolf, M. Fritsch, E. Förster, $K\alpha$ and $K\alpha$ x-ray emission spectra of copper, *Phys. Rev. A* **51** (1995) 283 and G. Hölzer, M. Fritsch, M. Deutsch, J. Härtwig, E. Förster, $K\alpha_{1,2}$ and $K\beta_{1,3}$ x-ray emission lines of the 3d transition metals, *Phys. Rev. A* **56** (1997) 4554.

[16] W. A. Rachinger, A Correction for the $\alpha_1\alpha_2$ Doublet in the Measurement of Widths of X-ray Diffraction Lines, *J. Sc. Instrum.* **25** (1948) 254.

[17] A. J. C. Wilson, *Mathematical Theory of X-Ray Powder Diffractometry*, Philips Technical Library, Eindhoven, 1963.

[18] G. Caglioti, A. Paoletti, F. P. Ricci, Choice of Collimators for a Crystal Spectrometer for Neutron Diffraction, *Nucl. Instr. Meth.* **3** (1958) 223.

[19] J. I. Langford, D. Louer, Powder diffraction, *Rep. Prog. Phys.* **59** (1996) 131.
[20] A. R. Stokes, A Numerical Fourier-analysis Method for the Corrections of Widths and Shapes of Lines on X-ray Powder Photographs, *Proc. Phys. Soc. London* **61** (1948) 382.
[21] B. E. Warren, B. L. Averbach, The separation of cold-work distortion and particle size broadening in X-ray patterns, *J. Appl. Phys.* **23** (1952) 497.
[22] C. E. Krill, R. Birringer, *Estimating grain-size distributions in nanocrystalline materials from X-ray diffraction profile analysis*, Philos. Mag. A **77** (1998) 621, and C. E. Krill, *Personal communication*, 2005.
[23] L. E. Alexander, H. P. Klug, *X-ray Diffraction Procedures for Polycrystalline and Amorphous Materials*, Wiley-Interscience, New York, 1974.
[24] S. Sun, C. B. Murray, Synthesis of monodisperse cobalt nanocrystals and their assembly into magnetic superlattices, *J. Appl. Phys.* **85** (1999) 4325.
[25] P. Scherrer, Bestimmung der Größe und der inneren Struktur von Kolloidteilchen mittels Röntgenstrahlen, *Nachr. Königl. Gesell. Wiss. Göttingen* (1918) 98.
[26] E. F. Bertaut, Raies de Debye-Scherrer et répartition des dimensions des domaines de Bragg dans les poudres polycristallines, *Acta Crystallogr.* **3** (1950) 14.
[27] A. R. Stokes, A. J. C. Wilson, A method of calculating the integral breadths of Debye–Scherrer lines, *Proc. Cambr. Philos. Soc.* **38** (1942) 313.
[28] J. I. Langford, A. J. C. Wilson, Scherrer after sixty years: a survey and some new results in the determination of crystallite size, *J. Appl. Cryst.* **11** (1978) 102.
[29] J. I. Langford, D. Louër, Diffraction Line Profiles and Scherrer Constants for Materials with Cylindrical Crystallites, *J. Appl. Cryst.* **15** (1982) 20.
[30] W. C. Hinds, *Aerosol Technology*, Wiley, New York, 1999.
[31] Á. Révész, T. Ungár, A. Borbély, J. Lendvai, Dislocations and grain size in ball-milled iron powder, *Nanostruct. Mater.* **7** (1996) 779.
[32] G. K. Williamson, W. H. Hall, X-ray line broadening from filed aluminium and wolfram, *Acta Metall.* **1** (1953) 22.
[33] T. Ungár, G. Tichy, The effect of dislocation contrast on X-ray line profiles in untextured polycrystals, *phys. stat. sol. (a)* **171** (1999) 425.
[34] M. A. Krivoglaz, *X-ray and Neutron Diffraction in Nonideal Crystals*, Springer, Berlin, 1996; *Theory of X-ray and Thermal-Neutron Scattering by Real Crystals*, Springer, Berlin, 1969.
[35] M. Wilkens, The determination of density and distribution of dislocations in deformed single crystals from broadened X-ray diffraction profiles, *phys. stat. sol. (a)* **2** (1970) 359.
[36] T. Ungár, A. Borbély, The effect of dislocation contrast on x-ray line broadening: a new approach to line profile analysis, *Appl. Phys. Lett.* **69** (1996) 3173.
[37] P. Klimanek, R. Kuzel, X-ray diffraction line broadening due to dislocations in non-cubic materials: I. General considerations and the case of elastic isotropy applied to hexagonal crystals, *J. Appl. Cryst.* **21** (1988) 59.
[38] R. Kuzel, P. Klimanek, X-ray diffraction line broadening due to dislocations in non-cubic materials: II. The case of elastic anisotropy applied to hexagonal crystals, *J. Appl. Cryst.* **21** (198) 363.
[39] N. Armstrong, P. Lynch, Determination of the Dislocation Contrast Factor for X-ray Line Profile Analysis, in:*Diffraction Analysis of the Microstructure of Materials*, p. 249.
[40] T. Ungár, J. Gubicza, G. Ribárik, A. Borbély, Crystallite size distribution and dislocation structure determined by diffraction profile analysis: principles and practical application to cubic and hexagonal crystals, *J. Appl. Cryst.* **34** (2001) 298.
[41] A. Borbély, J. H. Driver, T. Ungár, An X-ray method for the determination of stored energies in texture components of deformed metals: application to cold worked ultra high purity iron, *Acta Mater.* **48** (2000) 2005.
[42] R. Delhez, T. H. de Keijser, E. J. Mittemeijer, Determination of crystallite size and lattice distortions through X-ray diffraction line profile analysis, *Fresenius. Z. Anal. Chem.* **312** (1982) 1.
[43] I. Lucks, P. Lamparter, E. J. Mittemeijer, An evaluation of methods of diffraction-line broadening analysis applied to ball-milled molybdenum, *J. Appl. Cryst.* **37** (2004) 300.

[44] J. Gubicza, J. Szépvölgyi, I. Mohai, G. Ribárik, T. Ungár, The effect of heat-treatment on the grain-size of nanodisperse plasmathermal silicon nitride powder, *J. Mater. Sci.* **35** (2000) 3711.

[45] T. Ungár, A. Borbély, G. R. Goren-Muginstein, S. Berger, A. R. Rosen, Particle-size, size distribution and dislocations in nanocrystalline tungsten-carbide, *Nanostruct. Mater.* **11** (1999) 103.

[46] T. H. de Keijser, J. I. Langford, E. J. Mittemeijer, A. B. P. Vogels, Use of the Voigt function in a single-line method for the analysis of X-ray diffraction line broadening, *J. Appl. Cryst.* **15** (1982) 308.

[47] T. H. de Keijser, E. J. Mittemeijer, H. C. F. Rozendaal, The determination of crystallite-size and lattice-strain parameters in conjunction with the profile-refinement method for the determination of crystal structures, *J. Appl. Cryst.* **16** (1983) 309.

[48] H. M. Rietveld, Line profiles of neutron powder-diffraction peaks for structure refinement, *Acta Crystallogr.* **22** (1967) 151.

[49] H. M. Rietveld, A profile refinement method for nuclear and magnetic structures, *J. Appl. Cryst.* **2** (1969) 65.

[50] M. Ferrari, L. Lutterotti, Method for the simultaneous determination of anisotropic residual stresses and texture by x-ray diffraction, *J. Appl. Phys.* **76** (1994) 7246.

[51] G. Ribárik, T. Ungár, J. Gubicza, MWP-fit: a program for multiple whole profile fitting of diffraction peak profiles by ab initio theoretical functions, *J. Appl. Cryst.* **34** (2001) 669.

[52] P. Scardi, M. Leoni, Diffraction line profiles from polydisperse crystalline systems, *Acta Crystallogr.* **A57** (2001) 604.

[53] P. Scardi, M. Leoni, Whole powder pattern modelling, *Acta Crystallogr.* **A58** (2002) 190.

[54] M. Leoni, P. Scardi, Nanocrystalline domain size distributions from powder diffraction data, *J. Appl. Cryst.* **37** (2004) 629.

[55] X. Orlhac, C. Fillet, P. Deniard, A. M. Dulac, R. Brec, Determination of the crystallized fractions of a largely amorphous multiphase material by the Rietveld method, *J. Appl. Cryst.* **34** (2001) 114.

[56] L. Lutterotti, D. Chateigner, S. Ferrari, J. Ricote, Texture, residual stress and structural analysis of thin films using a combined X-ray analysis, *Thin Sol. Films* **450** (2004) 34.

[57] R. W. Cheary, A. A. Coelho, Axial divergence in a conventional X-ray powder diffractometer: II. Realization and evaluation in a fundamental-parameter profile fitting procedure, *J. Appl. Cryst.* **31** (1998) 862.

[58] R. W. Cheary, A. A. Coelho, Axial divergence in a conventional X-ray powder diffractometer: I. Theoretical foundations, *J. Appl. Cryst.* **31** (1998) 851.

[59] K. Santra, P. Chatterjee, S. P. Sen Gupta, Powder profile studies in electrodeposited cuprous oxide films, *Sol. Energy Mater. Sol. Cells* **57** (1999) 345.

[60] N. Zhang, W. Zhai, Y. Wang, A. Wagendristel, X-ray studies of Ti–N coatings, *Vacuum* **44** (1993) 51.

[61] F. L. Shan, Z. M. Gao, Y. M. Wang, Microhardness evaluation of Cu–Ni multilayered films by X-ray diffraction line profile analysis, *Thin Solid Films* **324** (1998) 162.

[62] W.-D. Haecker, Harte Schichten für hohe Beanspruchungen, *Bosch Research Info*, 3/2001.

[63] K. I. Schiffmann, M. Fryda, G. Goerigk, R. Lauer, P. Hinze, A. Bulack, Sizes and distances of metal clusters in Au-, Pt-, W- and Fe-containing diamond-like carbon hard coatings: a comparative study by small angle X-ray scattering, wide angle X-ray diffraction, transmission electron microscopy and scanning tunneling microscopy, *Thin Solid Films* **347** (1999) 60.

[64] L. Velterop, R. Delhez, T. H. de Keijser, E. J. Mittemeijer, D. Reefman, X-ray diffraction analysis of stacking and twin faults in f.c.c. metals: a revision and allowance for texture and non-uniform fault probabilities, *J. Appl. Cryst.* **33** (2000) 296.

[65] M. Leoni, Y. H. Dong, P. Scardi, Strain-Texture Correlation in r.f. Magnetron Sputtered Thin Films, *Mater. Sci. Forum* **321-324** (2000) 439.

[66] A. Boulle, C. Legrand, R. Guinebretière, J. P. Mercurio, A. Dauger, X-ray diffraction line broadening by stacking faults in $SrBi_2Nb_2O_9/SrTiO_3$ epitaxial films, *Thin Solid Films* **391** (2001) 42.

[67] A. Boulle, C. Legrand, R. Guinebretière, J. P. Mercurio, A. Dauger, Planar faults in layered Bi-containing perovskites studied by X-ray diffraction line profile analysis, *J. Appl. Cryst.* **34** (2001) 699.

[68] A. C. Nunes, D. Lin, Effects of surface relaxation on powder diffraction patterns of very fine particles, *J. Appl. Cryst.* **28** (1995) 274.

4
Grazing Incidence Configurations

The x-ray attenuation coefficient μ has been introduced in previous chapters and shown to have a value of about 10^4–10^5 m^{-1}. The penetration depths of x-rays, i.e. the corresponding $1/\mu$ values, are accordingly found in the 10–100 µm range. In most thin-film investigations the thickness is substantially less causing a large fraction of the diffractogram as measured in the symmetric $\theta/2\theta$ configuration to stem from the substrate. In the case of thin films with thicknesses in the nanometer range or less only negligible structural information can be gained in this measuring configuration. The path traveled by the x-ray in the sample, which is of the order of $\sin\theta/\mu$, is too short for typical Bragg angles to deliver x-ray reflections of sufficient peak-to-noise ratio.

Especially for the analysis of thin films, x-ray diffraction techniques have been developed for which the primary beam enters the sample at very small angles of incidence. In its simplest variant this configuration is called GIXRD: grazing incidence x-ray diffraction. The small entrance angle causes the path traveled by the x-rays to increase significantly and the structural information contained in the diffractogram to stem primarily from the thin film. The parafocusing geometry as applied in conventional powder diffractometry cannot be used in these experiments and new beam-shaping devices for parallel x-ray optics have had to be developed. These and further details of the measurement geometry are described in Section 4.1 and in Instrumental Box 5. Moreover, the GIXRD technique allows one to obtain depth-resolved structural information by measuring diffractograms under different angles of incidence. This is evaluated in Section 4.2. Up to this point the chapter will be concerned with diffraction phenomena only. In addition to the information given in the previous chapters the peculiarities are developed that arise only under very small angles of incidence.

Presentation is made of an x-ray technique that is named x-ray reflectometry or x-ray reflectivity (XRR), and which does not rely on the diffraction of x-rays. Rather it makes use of reflection and refraction of radiation and similarly as in optical ellipsometry the real and imaginary parts of the complex index of refraction of a thin film can be determined. The decisive distinction between both methods is the wavelength of the radiation, which differs by about three orders of magnitude. For this reason the associated phenomena may be observed only under very small incidence angles in the case when x-rays are applied. A further characteristic of XRR

Thin Film Analysis by X-Ray Scattering. M. Birkholz
Copyright © 2006 WILEY-VCH Verlag GmbH & Co. KGaA, Weinheim
ISBN: 3-527-31052-5

Instrumental Box 5:

Parallel Beam Optics

Two distinct techniques for x-ray beam optics are in use and may be named beam-selective and beam-conditioning techniques. With the selective methods those parts of the beam are cut off that are not intended to reach the detector. To this group belong all kinds of slits like divergence slits and Soller slits. Also crystal monochromators like the secondary HOPG monochromator introduced in Chapter 2 belong to this group, since they simply scatter the undesired intensity into a direction away from the detector. Beam conditioners, however, are intended to direct the full beam intensity either to the sample or to the detector by varying the beam divergence δ. Such devices became available only recently and are not as easily fabricated as optical lenses, which is a direct consequence from the small deviation of the x-ray refractive index n from unity. Beam-conditioning devices accordingly operate under very small incidence angles for the received x-ray beam.

In the case of the GIXRD configuration shown in Fig. 4.2(a) only beam-selective elements are applied. A parallel plate collimator (PPC) is inserted on the secondary side. The suppression of nonparallel parts of the beam is the greater for a longer the slit assembly. If the length and distance are designated by L and S, the full width of maximum peak width introduced by this device amounts to $\arctan(S/L)$. The effect of the parallel plate collimator may be compared with the influence of the Soller slits. However, in the latter the plates are oriented parallel to the scattering plane in order to reduce δ_a, while the PPC diminishes δ_p by plates that are oriented perpendicular to the scattering plane. Commercially available PPCs are accommodated to the appropriate diffractometer geometry to yield divergences δ_p of a few 0.1°. The measurement mode also relies on the increase of radius of the focusing circle to very large values for grazing beam incidences. The bundle of reflected x-rays then becomes almost parallel and can be collected by the PPC.

Behind this collimator the x-ray beam irradiates a flat single-crystal monochromator, which is most often made from LiF. This crystal monochromator is fabricated as a flat plate in contrast to the HOPG secondary monochromator introduced in Instrumental Box 2, which most often is bent with a bending radius defined by the distance to the detector. The LiF monochromator is a single crystal with the (100) surface exposed to the beam. For Cu Kα radiation only a small fraction of the incoming intensity is absorbed by the crystal, while the main part is reflected as the 200 peak under a Bragg angle $\theta_B = 22.499°$ into the detector. One should be aware of the necessary correction of the Lorentz polarisation factor when the detector is to be used for the determination of integral intensities.

The small angle of the incident beam in GIXRD experiments causes the x-ray footprint on the sample to take large values (see Eq. (i3.1)). Divergent slits of small opening angles have accordingly to be used to avoid the illumination of the sample holder, which might lead to the occurrence of unexpected Bragg reflections in the diffraction pattern. Numerous spurious peaks are frequently encountered in thin-film GIXRD that may even appear to migrate through the diffraction patterns when the incidence angle α or the incident beam divergence are varied. Different sources were identified for anomalous peaks in GIXRD [1], and for wide-angle diffraction those of single-crystalline substrates turn out to be the most relevant ones (see Exercise 4.3). A practical approach for the solution of such phenomena is to measure the uncoated substrate or the substrate holder separately in the same geometry.

In the second configuration used in GIXRD (see Fig. 4.2(b)), the divergence of the beam is reduced by a laterally graded multilayer mirror (LGMM). This device is often constructed as a parabola segment with the x-ray emitting tube spot situated in the parabola focus (Fig. i5.1). The mirror surface is a stack of alternating low-density and high-density layers made from Mo or W and Si or C, for instance. The layer thickness is between 1.5 and 10 nm and increases with increasing distance r from the x-ray tube. The layer period $D(r)$ is laterally adapted to the incidence angle on the mirror surface $\theta_m(r)$ such that the Bragg condition $2D(r)\sin\theta_m(r) = \lambda$ is equally fulfilled for all positions r along the mirror surface, $D(r) = D(r_0)/2\sin\theta_m(r)$. The divergence of the beam may be reduced by such a mirror down to values of a few 0.01°, while the peak intensity loss may be less than 30%. The insertion of a β filter becomes unnecessary if a multilayer mirror is used, since it also suppresses the β component. LGMMs for x-ray diffraction applications are typically fabricated from 20 to 100 layer pairs that are deposited by sputter techniques and are also known as Göbel mirrors [2, 3]. The use of parallel-beam conditioners like LGMMs is also advantageous if diffraction patterns of curved surfaces or less absorbing samples have to be measured. Moreover, the peak shift due to a height misalignment of the sample as observed in the parafocusing mode vanishes in the parallel-beam configuration.

Parallel-beam optics are also appropriate to be used in x-ray reflectometry since the measurement is performed in the $\theta/2\theta$ mode. Comparative investigations revealed that higher counting rates and better signal-to-noise ratios may be obtained if the radiation

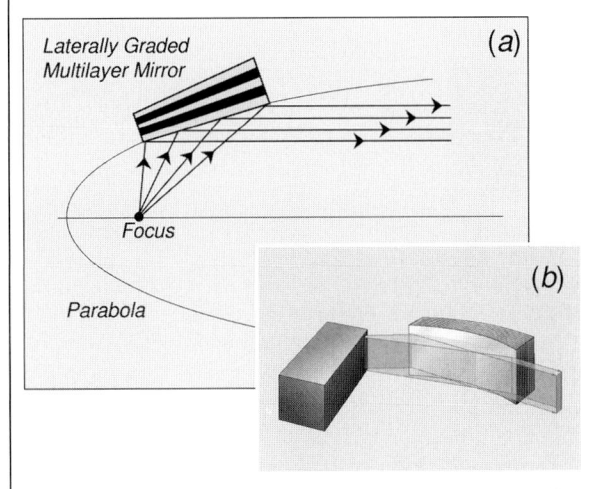

Figure i5.1 Laterally graded multilayer mirror: (a) principle of operation in top view and (b) schematic of combination with x-ray tube in side view (from Ref. [2]).

is rendered parallel by a LGMM [3]. The use of a beam attenuator for small scattering angles 2θ is a must in XRR experiments in order to manage the intensity variation over many orders of magnitude. Especially, if a semiconducting detector is used, direct irradiation by the x-ray tube for $\theta < \theta_c$ may cause irreversible damage. The plateau in reflected intensity as mentioned in Section 4.6.1, $I(2\theta) \approx$ constant for $\theta < \theta_c$, may not be observed as flat as expected. The effect is due to the self-shadowing of the sample for incidence angles close to zero that diminishes with increasing θ. The increase of intensity in this region follows a $f(B_p/h)$ behavior, where B_p is the length of the sample in direction of the beam and h is the height or diameter of the x-ray beam. If the material parameters β and δ of the sample are to be determined with high precision, B_p and h have carefully to be adjusted with respect to θ_c. The determination of β or μ sensitively depends on a reliable regression of the model curve to the measured reflectivity. The length of the probing x-ray beam on the sample's surface may be adjusted by a beam knife. By lowering the height of the beam knife above the sample the beam length B_p can be contracted sufficiently, which is unavoidable when curved samples are to be measured.

The sample adjustment prior to the XRR experiment should be performed with great care. First, the sample height should be adjusted by a height scan, while setting the diffractometer to $\theta = 2\theta = 0$ (beam attenuator!). The correct height is identified in the $I(z)$ curve from $I = I_{max}/2$. Subsequently, the θ circle has to be adjusted and again a height alignment should be carried out. A ψ adjustment should also be performed if the diffractometer has this degree of freedom. As general in x-ray diffraction, the measurement of a standard before the measurement of the sample should be carried out in order to check for instrumental artifacts.

The x-ray tube is generally operated in the line focus mode for GIXRD and XRR experiments. For GID measurements, however, a change to the point focus mode becomes necessary. The tilting of the sample holder with respect to the (K_0, s_3) plane must be performed with respect to the center of mass of the beam spot on the sample surface. The tilting is performed by virtue of a so-called Euler cradle, which is described in Chapter 5.

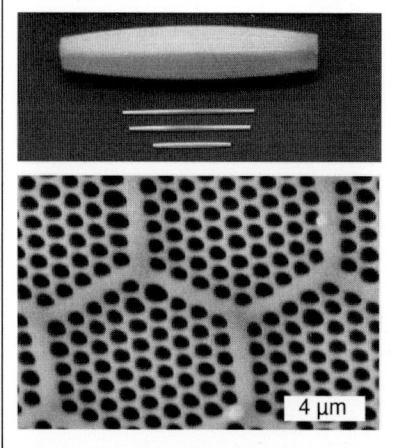

Figure i5.2 Polycapillary lenses (top) configured for use in focusing applications and (bottom) cross-section through the set of hollow fibers (kindly provided by IfG Institute for Scientific Instruments GmbH, Berlin).

> In the case where the point focus mode is used the parallel beam shaping is ideally not performed by a multilayer mirror, but by a polycapillary. This kind of beam conditioner consists of a bundle of about 10^4–10^6 hollow glass fibers as shown in Fig. i5.2 [4]. The x-rays are guided within the fibers by total reflection at the internal air–glass boundary. The fibers are tapered and exhibit an increase of curvature with increasing distance from the polycapillary's central axis. The design rule to be fulfilled for beam guidance reads $R/\varnothing_i > 2\alpha_c^2$, where R is the radius of fiber curvature, $\varnothing_i$ the internal fiber diameter and α_c the critical angle. The beam leaving the polycapillary may exhibit a reduced divergence of about 0.15° or more. Large intensity gains of some 10^3 become possible when a polycapillary is used instead of a pinhole collimator.
>
> **References**
>
> **Monographs relevant to this box**
>
> A. Erko, V. Aristov, B. Vidal, *Diffraction X-Ray Optics*, Institute of Physics, Bristol, 1996.
>
> **Special papers**
>
> [1] R. A. Vaia, M. S. Weathers, W. A. Bassett, Anomalous peaks in grazing incidence thin film X-ray diffraction, *Powder Diffract.* **9** (1994) 44.
> [2] C. Michaelsen, J. Wiesmann, C. Hoffmann, K. Wulf, L. Brügemann, A. Storm, *Recent Developments of Multilayer Mirror Optics for Laboratory X-Ray Instrumentation*, SPIE, Seattle, 2002.
> [3] M. Schuster, H. Göbel, Application of graded multilayer optics in X-ray diffraction, *Adv. X-Ray Anal.* **39** (1996) 1.
> [4] A. Bjeoumikhov, N. Langhoff, R. Wedell, V. Beloglazov, N. Lebed'ev, N. Skibina, New generation of polycapillary lenses: manufacture and applications, *X-Ray Spectr.* **32** (2003) 172.

derives from the fact that the index of refraction n is smaller than unity for x-rays. Therefore, by passing from air into the film the beam enters an optically less dense medium, which is the opposite situation to that encountered in optical ellipsometry. Section 4.4 is devoted to this interesting phenomenon.

XRR is not a diffraction technique, but belongs to the larger group of scattering methods. Its use for the investigation of thin films has steadily increased over the last few years. This development may arise from the fact that diffractometers with their goniometer stages may equally be used for this type of measurement, although they should be named reflectometers in this case as is frequently done. In addition, the XRR technique yields important information on the structural quantities of thin films, like density, chemical composition, thickness and surface roughness. And because the technique does not rely on diffraction phenomena the investigated sample may be amorphous and does not have to consist of crystalline matter.

Finally, a third method has been developed that makes use of both a grazing incidence and grazing exit angle and which is generally denoted by GID. This technique again relies on the diffraction of x-rays by crystalline lattice planes – but this time the measured lattice planes are vertically tilted with respect to the substrate plane. By virtue of this technique structural information on layers with submonolayer thickness may be obtained. This method is described in Section 4.7.

Thin films and multilayers from the group of oxides will serve as illustrative examples and some of their applications in optical and semiconductor technology are given. In the latter field, the development of high-k dielectrics is an urgent issue for further miniaturization of transistor devices, while in optics thin oxide layers have found wide application for optical lenses in cameras, spectacles and many other devices that extend to experimental laboratory setups for purposes like wavelength filtering, reduction of reflection losses, etc. Additionally, optical coatings are increasingly applied to architectural glass, projectors, display panels, solar cells, low-heat transmission windows and many other applications. A basic design rule for optical coatings is the combination of layers with as different as possible refractive indices (one should not confuse the refractive index in the visible spectrum with that for x-ray wavelengths). Titanium dioxide in the rutile form is a transparent oxide with a large refractive index in the visible region. The rutile structure is discussed in Structure Box 5.

4.1
Grazing Incidence X-ray Diffraction (GIXRD)

The measuring configuration in the GIXRD experiment is shown schematically in Fig. 4.1. The angle between the incoming beam and the sample surface is very small and amounts to only a few degrees or even less. This angle is different from half the scattering angle and will be denoted by α. The x-rays travel a maximum path of $l = t/\sin\alpha$ within the layer, which might be a high multiple of the thickness t when the entrance angle is chosen very flat. For the investigation of a substance with attenuation coefficient μ a first criterion for a reliable choice of α is that the average path in the layer should compare with the inverse attenuation coefficient, i.e. $l \approx 1/\mu$. This condition translates to the recommendation that $\alpha \approx \arcsin(\mu t)$ should hold between the incidence angle and the μt product for an effective measurement of the layer (α in radians). It will be seen in the following that further restrictions have to be considered for a reliable choice of α, which arise when α be-

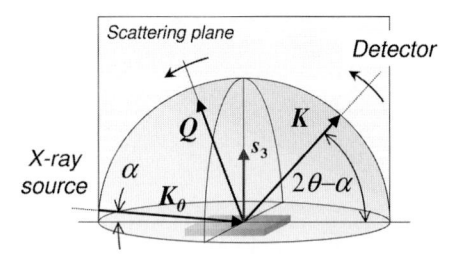

Figure 4.1 The geometry in grazing incidence diffraction is characterized by a small incidence angle α that is kept constant during the measurement.

Structure Box 5:

Rutile Structure

The rutile structure falls into the symmetry class of tetragonal crystals. The naming derives from one modification of the prototype compound titanium dioxide, TiO_2, but the structure type is also known as cassiterite. Various other oxides and fluorides occur in this structure. The rutile unit cell comprises two molecular units AB_2, where A represents the less polarizable atoms/ions compared to B. Figure s5.1 displays the crystallographic unit cell. In TiO_2 rutile and other compounds taking this structure the c lattice parameter is typically smaller than a. A selection of a and c values is given in Table s5.1. The A species reside on centrosymmetric (000) and (½½½) positions, while the B species occupy positions along the diagonal in the basal plane and at height $z = ½$. The x and y fractional coordinates are symmetry related in the rutile structure, and the B positions read $(\bar{u}\bar{u}0)$, $(uu0)$, $(½+u, ½-u, ½)$ and $(½-u, ½+u, ½)$. The A atoms can be seen from Fig. s5.1 to be surrounded by an octahedron of B atoms and the rutile structure may thus be imagined to be comprised of AB_6 octahedrons that are interconnected via shared B atoms.

The structure factor F_{rutile} divides into two parts F_A and F_B due to scattering at A and B sublattices. F_A is the structure factor of the body-centered tetragonal lattice that is adopted, for instance, by one-elemental metals In and Sn. The contribution of F_B is more complex due to the positional parameter u. The calculation of the temperature independent factor F_{rutile} according to Eq. (1.23) yields

$$F_{rutile}(hkl) = \begin{cases} 2f_A + 4f_B \cos(2\pi hu)\cos(2\pi ku) & h+k+l = \text{even} \\ 4f_B \sin(2\pi hu)\sin(2\pi ku) & h+k+l = \text{odd} \end{cases} \quad (s5.1)$$

The extinction conditions to be derived from this expression are that Bragg reflections would not be observed if the sum $h + k + l$ becomes an odd number and $h = 0$ or $k = 0$ or both, which holds, for instance, for 001, 003,...100, 012,... . Moreover, the rutile structure exhibits the interesting property that $h + k + l = $ odd reflections are fully caused by the sublattice of B atoms/ions.

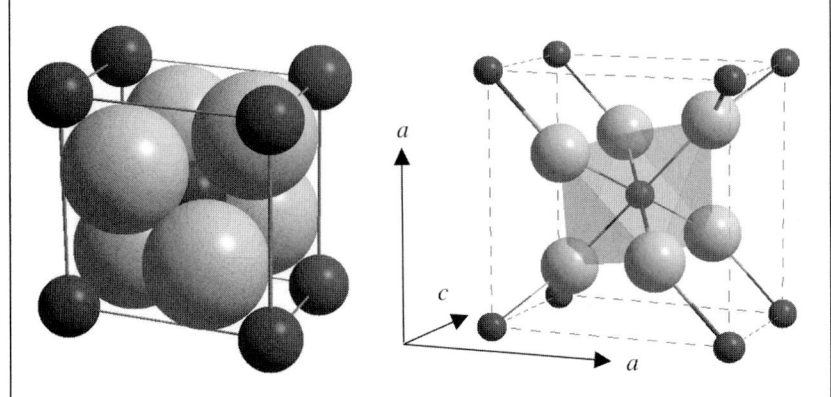

Figure s5.1 Two crystallographic unit cell representations of rutile TiO_2.

Table s5.1 Notations and symmetries in the rutile structure (from Landolt-Börnstein).

Name (Struktur-bericht designation)	Space group (number)	Examples	Lattice parameters a and c (nm)	Positional parameter u
Rutile structure (C4)	$P4_2/mnm$ (136)	SnO_2	0.4738; 0.3187	0.307
		TiO_2	0.4593; 0.2959	0.305
		CoF_2	0.4695; 0.3182	0.304
		MnF_2	0.4874; 0.3310	0.305
		NiF_2	0.4650; 0.3084	0.304

It should be noted that titanium dioxide TiO_2 may take different crystallographic structures than only the rutile structure. Well known are, for instance, the anatase or brookite modifications. Especially, when stoichiometry deviations like an oxygen-deficiency in TiO_{2-x} occurs during thin film growth an additional variety of crystallographic structures become possible, by which an observed $\theta/2\theta$ powder diffraction pattern may be explained. Titanium dioxide may also be deposited as an amorphous layer for too low deposition temperatures or too little ion irradiation. Therefore, the interpretation of thin film diffraction pattern often is not straightforward, when intentionally TiO_2 should have been grown. Comparable statements are valid for other compounds crystallizing in the rutile structure.

comes so small that it approaches the so-called critical angle. As in the symmetric $\theta/2\theta$ configuration the wavevectors K_0, K and the sample normal s_3 all lie in the same plane in GIXRD and it thus belongs to the so-called coplanar configurations.

The measurement of the diffractogram is performed such that the angle α is kept constant, while the detector is moved along the 2θ circle. This is the decisive distinction compared to the symmetric configuration where the entrance angle θ is also changed during the measurement. However, the scattering angle 2θ again denotes the angle between the outgoing beam and the elongation of the incoming beam. If the collected intensity is then plotted versus 2θ, Bragg reflections are found at comparable positions as in a symmetrically measured pattern. In contrast to the latter, the angle between outgoing beam and sample surface is $2\theta - \alpha$ in the GIXRD experiment instead of θ as in the symmetric scan.

The different measurement geometry has important consequences for the orientation of the scattering vector Q and the beam optics. In the symmetric $\theta/2\theta$ configuration the scattering vector Q is always parallel with the surface normal s_3 having $s_3 \parallel Q$. The crystallographic lattice planes causing an observed Bragg reflection are all oriented in parallel with the substrate plane (see Fig. 1.9). In the grazing incidence configuration, however, Q is neither parallel to s_3 nor does its orientation remain constant. Rather the orientation of Q changes in the course of the measurement, during which the scattering vector is tilted from the start position close to s_3 towards an end position ever closer to the incoming beam K_0. The angles formed by the different vectors and the sample surface are given in Table 4.1 for the symmetric and asymmetric cases. In the GIXRD configuration reflections with

Table 4.1 Angles between sample surface and beam directions K_0 and K and the scattering vector Q for the symmetric and asymmetric diffractometer configurations.

	Symmetric configuration		Asymmetric configuration	
	Angle with surface	Variable?	Angle with surface	Variable?
K_0	θ	Yes	α	No
K	θ	Yes	$2\theta - \alpha$	Yes
Q	$\pi/2$	No	Exercise 2	Yes

distinct Bragg angles θ_{hkl} are caused by lattice planes that are neither parallel with the substrate surface nor with each other. In the case of a random orientation of crystallites, as in a powder, a comparable pattern will be measured in both the symmetric and asymmetric configurations. For thin layers, however, where structural parameters may sensitively depend on film thickness and orientation, the different tilt of grain ensembles has to be considered for the interpretation of diffractograms measured in GIXRD geometry.

The necessity for modified beam optics follows from the fact that the conditions of the parafocusing geometry are not obeyed in the asymmetric configuration. Because the incidence angle is fixed the measuring circle and the focusing circle do not intersect throughout the whole scan, but only for a single point. Without taking special measures, the divergent beam leaving the x-ray tube and impinging on the sample would enter the detector completely unfocused. The solution to this problem is to run the diffractometer in the parallel beam mode. Figure 4.2 displays the two different configurations that are in use when the tube operates in the line focus mode. In the first setup (see Fig. 4.2(a)) the parallel fraction of the beam is collected on the secondary side by virtue of long collimator slits. Any remaining unparallel beam fractions are subsequently cut off by a flat LiF crystal monochromator. The second, now widely used, parallel beam setup shown in Fig. 4.2(b) makes use of shaping the primary beam into a parallel bundle just after it leaves the x-ray source. This technique can be performed by laterally graded multilayer mirrors (LGMM), which are discussed in more detail in Instrumental Box 5. Sample rotation around the substrate normal or ϕ axis should be avoided in most GIXRD experiments, since this may cause an extra broadening of reflections if the sample is not very precisely positioned. An averaging over many crystallites, which is the reason for ϕ rotation in symmetric x-ray diffraction, is already obtained in GIXRD through the illumination of many crystallites caused by the flat entrance angle.

In consequence of the distinct beam course in GIXRD the absorption factor A has a different form from that given in Eq. (1.31). The diffraction of the beam in a depth z is considered in Fig. 4.3 in order to derive the new absorption factor. A configuration factor k is now introduced to account for the beam path in the experiment. The k factor allows a general formulation of the absorption factor for any measurement configuration as will become clear in the following chapters. For the GIXRD experiment it takes the form

4 Grazing Incidence Configurations

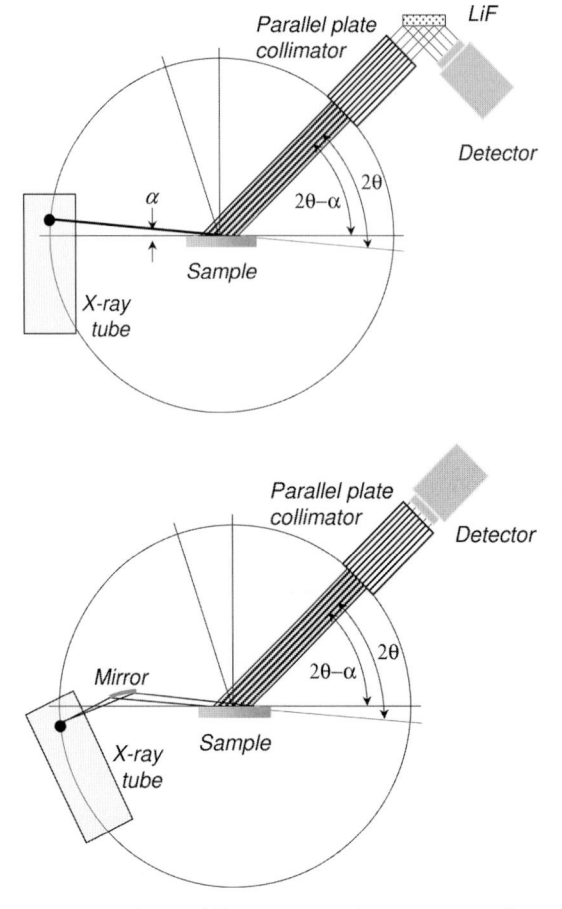

Figure 4.2 The two diffractometer configurations most frequently applied in GIXRD. The x-ray tube is operated in the line focus mode. The plane of the figure is the scattering plane.

$$k_\alpha = \frac{1}{\sin\alpha} + \frac{1}{\sin(2\theta - \alpha)} \qquad (4.1)$$

The index α is used as subscript of k in order to characterize it as the appropriate GIXRD factor. Other indices will be used for other configurations. The intensity scattered at depth z scales with $\exp(-\mu z/\sin\alpha)$ and $I_0/\sin\alpha$, since the incoming beam is attenuated and inclined. After being scattered the intensity is attenuated proportional to $\exp[-\mu z/\sin(2\theta - \alpha)]$

$$dP(z) = \frac{SCF \cdot I_0}{\sin\alpha} \exp\left(-\mu z\left[\frac{1}{\sin\alpha} + \frac{1}{\sin(2\theta - \alpha)}\right]\right) dz \qquad (4.2)$$

4.1 Grazing Incidence X-ray Diffraction (GIXRD) | **153**

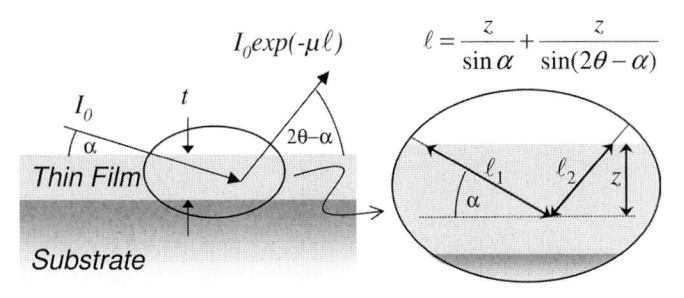

Figure 4.3 Schematic representation of the illuminated sample depth for the derivation of the absorption factor A_α in GIXRD.

Here, SCF is a constant that includes terms of the integral intensity not considered in the following. The total scattered intensity follows from the integration over all depths

$$P = \frac{SCF \cdot I_0}{\sin\alpha} \int_0^t \exp(-\mu z k_\alpha) dz = \frac{SCF \cdot I_0}{\mu} \frac{(1-\exp(-\mu t k_\alpha))}{k_\alpha \sin\alpha} \tag{4.3}$$

If P_∞ is used to symbolize the integral intensity of a Bragg reflection of an infinitely thick sample, $P_\infty = SCF \cdot I_0/(\mu k_\alpha \sin\alpha)$ is obtained. As in Chapter 1, the thin-film absorption factor A is obtained from Eq. (4.3) by division by P_∞. The GIXRD absorption factor A_α can accordingly be written

$$A_\alpha = \left(1 - \exp(-\mu t k_\alpha)\right) \tag{4.4}$$

It is by this amount that a Bragg reflection from a thin-film sample is reduced in GIXRD geometry when compared to the reflex intensity from an infinitely thick sample measured under the same conditions. The absorption factor A_α depends on the angles α and 2θ as well as on the μt product of the layer. Therefore, intensity corrections may only be performed when the latter is known. The μt product may be determined from the measurement of a substrate reflection before and after the deposition as outlined in Chapter 2, or it may be calculated when the thickness t, chemical composition and density of the layer are known. In principle, the same arguments as given in Section 2.3 hold with respect to the determination of μ, t or μt and the propagation of errors. In Fig. 4.4 a set of absorption factors A_α is shown for various incidence angles α and μt products. It can be seen that for all chosen combinations of α and μt the reduction of intensity is rather constant over the 2θ range from 20° to 60°. This contrasts remarkably to the situation encountered in symmetric $\theta/2\theta$ diffractometry, for which the absorption coefficient $A_{\theta 2\theta}$ was realized to deacrease significantly with increasing scattering angle 2θ (see Fig. 1.16). This distinct behavior of A_α and $A_{\theta 2\theta}$ is caused by the different configuration factors k_α and $k_{\theta 2\theta}$ and demonstrates the versatility of GIXRD for the investigation of sample with low μt products like thin films.

Figure 4.5(a) displays the GIXRD pattern of a thin TiO_2 film prepared on a Ti-coated steel substrate measured with Cu Kα radiation and an incidence angle of α

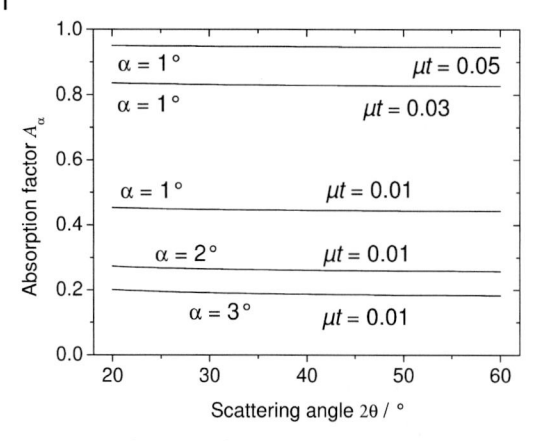

Figure 4.4 Absorption factor A_α in GIXRD geometry for various incidence angles α and μt products.

= 1.8° for 2θ between 22° and 60° [1]. Titanium dioxide occurs in a variety of phases, of which tetragonal rutile and anatase are the most prominent ones. In addition, thin TiO_2 films are often prepared fully amorphous and are either transformed to crystalline phases via subsequent annealing steps or they are used in the amorphous state when allowed for by the intended application. The optical refractive index of TiO_2 phases ranges from 2.2 to 2.8 and is found to approach the largest values for rutile films, the preparation of which either requires a minimum deposition temperature or a high ion bombardment during film growth [2]. Clearly, the observed peaks in Fig. 4.5(a) can be assigned to the rutile phase (ICDD card 00-021-1276), albeit the measured reflections are shifted to somewhat smaller 2θ angles. It turns out that the formation of the rutile phase at comparatively low temperatures is assisted by the high ion bombardment during the deposition process, which is also responsible for the intrinsic film stress indicated by the shift of Bragg reflections (see Chapter 6). For comparison the $\theta/2\theta$ pattern of the same film is shown in Fig. 4.5(b). The symmetrically measured pattern is mainly characterized by the Bragg peaks of the steel substrate and the Ti interlayer, as is seen from the superimposed ICDD cards 00-044-1294 for hcp Ti and 00-006-0696 for bcc Fe. Moreover, the $\theta/2\theta$ pattern of the same sample exhibits a pronounced background and a worse signal-to-noise ratio due to x-ray fluorescence of the iron substrate excited by the Cu Kα radiation. For all but the leftmost peak it is hard to decide whether they stem from the substrate, the interlayer or the TiO_2 top layer. Also, many details of the rutile pattern obtained from the GIXRD measurement are completely buried under substrate and bottom layer peaks in the symmetric $\theta/2\theta$ scan. This example illustrates convincingly the extra information available from grazing incidence diffraction of a thin-film sample.

The GIXRD technique outlined above is also known by various terms in the literature, like grazing incidence asymmetric Bragg diffraction (GIABD) and others.

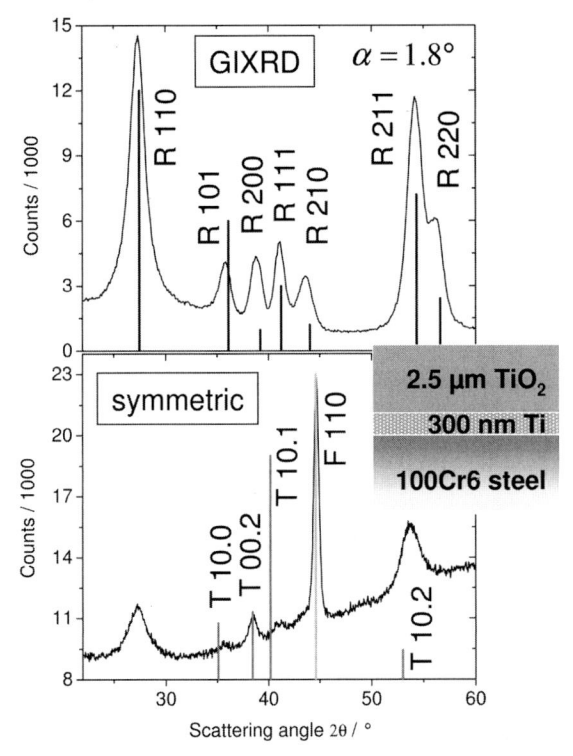

Figure 4.5 (a) GIXRD, $\alpha = 1.8°$, and (b) $\theta/2\theta$ diffraction pattern from the same TiO$_2$/Ti layer system on steel. Vertical lines indicate the positions of Bragg reflections hkl from rutile (R), titanium (T) and ferrite (F) according to appropriate ICDD cards.

4.2
Penetration Depth and Information Depth

The growth of thin films is often associated with thickness-dependent evolutional processes. For instance, the grain size may be observed to increase with increasing thickness or the composition of distinct crystallographic phases may change either continuously or abruptly at a critical thickness. GIXRD investigations may deliver valuable experimental data on such property gradients by measuring a set of diffraction pattern $I(\alpha, 2\theta)$ under varying incidence angles α. However, the transformation of the measured structural information into a uniquely defined depth coordinate is not straightforward. The complexity of the problem is demonstrated by the existence of three concepts for the penetration depths τ of x-rays. These are introduced first.

The first concept of intensity damping considers the depth $\tau_{1/e}$ as the characteristic quantity for which the x-ray beam is attenuated to $1/e$ of its initial value. This yields in the case of the GIXRD geometry

$$\tau_{1/e} = \frac{\sin\alpha}{\mu} \qquad (4.5)$$

By virtue of the 1/e depth it is already possible to understand qualitatively distinct diffraction patterns obtained under different grazing angles α. Especially in the case of a thickness-dependent variation of crystalline phases a plot of the relative intensities (corrected by dividing by A_α) versus $\tau_{1/e}$ may yield interesting results.

In the second approach, a penetration depth τ_{63} is defined by the condition that the exponent in the absorption factor of Eq. (4.4) equals unity, $\mu\tau_{63}k_\alpha = 1$. The meaning behind this definition is that for a powder sample of infinite thickness this depth specifies the thickness of the upper layer that accounts for $(1 - 1/e) = 63\%$ of the measured intensity. Applying this definition to the GIXRD configuration yields

$$\tau_{63} = \frac{1}{\mu k_\alpha} = \frac{\sin\alpha \sin(2\theta - \alpha)}{\mu(\sin\alpha + \sin(2\theta - \alpha))} \qquad (4.6)$$

In contrast to the $\tau_{1/e}$ approach this concept is related to the integral intensity of a Bragg reflection. A clear advantage of this second approach over the first is that both the incident and exiting beam are included, which appears more adequate for describing the information content of a diffracted beam received by the detector.

The disadvantage of both concepts is that the quantities $\tau_{1/e}$ and τ_{63} may exceed the thickness t of a thin layer and should therefore be used with caution in thin-film work. Obviously, any useful depth definition for the ordering of structural information from different layer depths will have to include the film thickness t or will yield the film thickness as limiting value. These conditions are obeyed by the definition of the so-called average information depth $\bar\tau$. This quantity is calculated from the weighted sum of all beams that differ by the amount of damping they undergo by being diffracted in various depths. The exponential term, $\exp(-\mu z k)$, has the meaning of a weight function while the integration is over all depths from the film surface to the interface with the substrate. The average information depth is of the general form

$$\bar\tau = \frac{\int_0^t z\exp(-\mu z k)dz}{\int_0^t \exp(-\mu z k)dz} \qquad (4.7)$$

Normalization is achieved by dividing by the integral over the weight function. This expression is valid for all measuring configurations presented so far and may be adopted to the special configuration in use by inserting the appropriate configuration factor k. In the case of GIXRD an information depth $\bar\tau_\alpha$ is calculated from Eq. (4.7) by inserting k_α and it is obtained as

$$\bar\tau_\alpha = \frac{1}{\mu k_\alpha} + \frac{t}{1 - \exp(\mu t k_\alpha)} \qquad (4.8)$$

For an infinitely thick layer the average information depth approaches the penetration depth, $\lim \bar\tau \xrightarrow{t\to\infty} \tau_{63}$.

The appearance of the symbol τ in the context of a penetration depth is sometimes confusing in the literature. In some calculations it may be found as a term inserted for $1/\mu k$, where it has the meaning of an abbreviation. In addition to this, τ is frequently used when the utilization of $\bar{\tau}$ would be more appropriate. It is helpful to always consider thoroughly the actual meaning of τ whenever the quantity is found in calculations or diagrams.

Information and penetration depths $\bar{\tau}$ and $\tau_{1/e}$ are given in Table 4.2 for a 500 nm thin TiO$_2$ rutile film with linear absorption coefficient of $\mu = 0.0528$ μm^{-1} for Cu Kα radiation. Both depths are seen to vary only slightly for constant incidence angles α and scattering angles 2θ between 22° and 60°. Any variation of α, however, causes substantial variations of $\bar{\tau}$ and $\tau_{1/e}$. In the limit of large incidence angles or small absorption, the average information depth $\bar{\tau}$ approaches half the film thickness $t/2$ (Exercise 6)

$$\lim_{\mu k \to 0} \bar{\tau} = \frac{t}{2} \tag{4.9}$$

This limit corresponds to the situation that the incident x-ray intensity is hardly damped over the whole thickness as is shown schematically in Fig. 4.6. In this situation each partial layer of the film is almost equally illuminated and the measured diffraction information arises on the average from the middle of the film. Bragg reflections measured under large α values would contain information from all depths.

Table 4.2 also shows that the penetration depth $\tau_{1/e}$ increases much more than $\bar{\tau}$ when the incidence angle α is increased. For $\alpha > 1.5°$ the penetration depth $\tau_{1/e}$ is greater than the film thickness. The use of the quantity in these cases is doubtful to say the least and a modification of μ to a quantity also containing the substrate absorption would be more adequate. This example demonstrates that the quantity $\tau_{1/e}$ is the best used as an abbreviation for $1/\mu k$, but its use should be avoided for $\tau_{1/e} > t$. In addition, the meaning of an information depth should be reserved for $\bar{\tau}$ as defined in Eq. (4.7). It should be noted that Eq. (4.9) is fully general for any arbitrary measuring geometry, which enters into $\bar{\tau}$ by use of an appropriate configuration factor k.

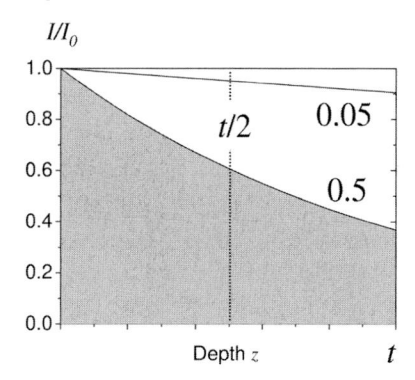

Figure 4.6 Course of x-ray intensity reaching the depth z of a thin film of thickness t for $\mu tk = 0.05$ and 0.5.

Table 4.2 Average information depth $\bar{\tau}$ and 1/e penetration depth $\tau_{1/e}$ in the scattering angle 2θ range 22–60° for different angles of incidence α. Values are calculated for a $t = 500$ nm thin TiO_2 rutile film, $\mu = 0.0528$ μm^{-1}, and Cu Kα radiation.

α (°)	$\bar{\tau}$ (nm)	$\tau_{1/e}$ (nm)
0.9	180–182	298
1.8	212–214	595
2.7	224–226	892

4.3
Depth-Dependent Properties

The formalism just derived for the average information depth may be developed for any other depth-dependent layer property. This property may be, for instance, the concentration of a chemical phase, the amorphous fraction, the crystallite size, microstrain or macrostrain or any other. Let us denote this depth-dependent property quite generally by $P(z)$. It is realized from the above considerations that the determination of $P(z)$ by analyzing a Bragg reflection yields a weighted average of this property $\bar{P} = \langle P(z) \rangle$ with the weight given by the attenuation of x-ray intensity. On the basis of the considerations given above one may define an attenuation length τ that allows one to express the depth-averaged property derived from a Bragg reflection by

$$\bar{P} = \frac{\int_0^t P(z)\exp(-z/\tau)dz}{\int_0^t \exp(-z/\tau)dz} \tag{4.10}$$

In principle, this formula may be applied using any of the different length scales introduced in the previous section. Most investigations performed so far have been carried out by inserting the 1/e penetration depth $\tau_{1/e}$. The denominator again serves for the purpose of normalization and can be abbreviated by N_τ. Let us now assume that a set of $\bar{P}_\alpha$ values has been determined from the measurement of a Bragg peak under various incidence angles α. The analytical form in the numerator of $\bar{P}$ in Eq. (4.10) is that of a Laplace transform $\mathcal{L}$ or to be precise that of a Fredholm integral equation, because the integral does not extend to infinity, but to the maximum depth t, which is the thickness of the layer. Assuming for the moment that the integral would extend to infinity the property-depth Eq. (4.10) might be brought into the condensed form

$$\bar{P}_\alpha = \frac{1}{N_\tau}\mathcal{L}\big(P(z), 1/\tau\big) \tag{4.11}$$

Use has been made of the notation $\mathcal{L}(f(x), 1/y)$ where $f(x)$ gives the function to be subjected to transformation and y being the new coordinate of the transformed function. Equation (4.11) represents an important result, because the task of solving for P(z) is now realized as a problem from the theory of Laplace transformation. The straightforward approach in obtaining the property function accordingly is to invert the equation and calculate P(z) from the inverse Laplace transformation $\mathcal{L}^{-1}$ of the set of $N_\tau \overline{P}_\alpha$ values

$$P(z) = \mathcal{L}^{-1}\left(N_\tau \overline{P}_\alpha(\tau), z\right) \qquad (4.12)$$

This formula allows for the determination of property profiles from thin films and surface layers by x-ray diffraction. Its result relies on the availability and reliability of $\overline{P}_\alpha$ data. The latter are usefully determined by GIXRD procedures, because only this technique allows for the measurement of $\overline{P}_\alpha$ data that are reasonable spaced on the τ coordinate.

Equation (4.12) seems to supply a simple recipe to derive the gradient function P(z). In practice, the procedure of inverse Laplace transformation may be complicated by noisy and improperly spaced data in τ space. These two effects can cause instabilities in the numerical solution of Eq. (4.12) and may lead to physically meaningless results [3]. A reliable approach for circumventing these problems is to approximate P(z) by a model function like a polynomial and perform the transformation of Eq. (4.10) to the model function [4]. A set of model parameters $P_0, P_1, P_2...$ may then serve as fitting parameters in τ space, from which the property function is also composed in real space. Consider, for instance, the expansion of P(z) in a polynomial

$$P(z) = P_0 + P_1 z + P_2 z^2 + ... \qquad (4.13)$$

Inserting this expression in Eq. (4.10) may easily be solved analytically. The evaluation of the integrals in the numerator and denominator, $N_\tau = \tau[1 - \exp(-t/\tau)]$, yields for the gradient function in τ space

$$\overline{P}_\alpha = P_0 + P_1\left(\tau + \frac{t}{1+\exp(t/\tau)}\right) + P_2\left(2\tau^2 + \frac{t(t+2\tau)}{1+\exp(t/\tau)}\right) + ... \qquad (4.14)$$

The course of the measured data $\overline{P}_\alpha$ can be fitted to this model function to give the polynomial coefficients P_i that account for the property gradient in z space. Most applications of this concept appear to have been performed in the field of residual stress analysis in thin films and we will return to this formalism in Chapter 6.

A second remark concerns the validity of this formalism for very low incidence angles. It is seen in the following that the concept of penetration depth has to be detailed for very small angles of incidence that approach the so-called critical angle α_c. This angle typically is less than 0.5° in the case when Cu Kα radiation is employed. The formalism given above then remains valid, albeit only by inserting the adequate expression for τ. Before this can be presented, the refractive index of x-rays and the effect of total external reflections of x-rays have to be introduced.

4.4
Refractive Index for X-rays

In the case where the angle of incidence shrinks to ever-smaller values effects are observed, which derive from the refractive index n of x-rays being less than unity. The deviation of n from unity can be understood from the refractive index of an ensemble of damped harmonic oscillators like electrons of mass m that are oscillating in the electric field of the x-ray beam. This reads

$$n = 1 + \frac{\rho_e}{2\pi m} \frac{e^2}{4\pi\varepsilon_0} \sum_i \frac{s_i}{v_i^2 - v^2} \tag{4.15}$$

where ρ_e is the density of electrons in the specimen, while s_i and v_i account for their oscillator strengths and resonance frequencies; v is the frequency of the exciting electromagnetic wave. The oscillator strengths are usually normalized by the condition $\sum s_i = 1$. An extended derivation of Eq. (4.15) can be found in textbooks on optics and on the theory of electromagnetic fields [5]. In the case of x-rays the stimulating frequency exceeds by far any resonance frequency of the solid, $v \gg v_i$, which lie in the infrared, visible or ultraviolet part of the spectrum. Evaluating the mathematical limit yields

$$n = 1 - \rho_e \frac{\lambda^2 r_e}{2\pi} \tag{4.16}$$

Again, r_e is the classical radius of the electron, $r_e = 2.82 \times 10^{-6}$ nm. The refractive index is accordingly smaller than unity, which will have far-reaching consequences for x-ray diffraction under small angles of incidence. However, it has to be stated that the deviation from unity is rather small, which is seen by inserting electron densities ρ_e of condensed matter into Eq. (4.16) that extend over about $1\frac{1}{2}$ orders of magnitude from 139 (lithium) to 4438 nm^{-3} (uranium). A typical value is taken by rutile TiO_2, having $\rho_e = 1217$ nm^{-3}. The small deviation is denoted by δ and n can thus be expanded according to

$$n = 1 - \delta \tag{4.17}$$

The expansion is justified, since radiation with wavelengths $0.1 < \lambda < 0.2$ nm causes deviations of n from unity of only some 10^{-6}. The combination of both equations results in an expression for a monatomic solid having ρ_a atoms per volume and a nuclear charge Z

$$\delta = \rho_a(Z + f') \frac{\lambda^2 r_e}{2\pi} \tag{4.18}$$

The scattering power of the Z electrons per atom has to be modified by adding the dispersion correction f'. Often, the real part susceptibility δ is expressed by the mass density ρ instead of electron density ρ_e or atomic density ρ_a. For the general case of a solid composed of different component atoms having nuclear charges Z_i and atomic weights M_i one obtains

$$\delta = \rho N_L \frac{\sum_i (Z_i + f_i') \lambda^2 r_e}{\sum_i M_i \quad 2\pi} \qquad (4.19)$$

where the sums have to be extended over all atoms in the unit cell and N_L is Loschmidt's number.

We now have a formula that relates the real part of the x-ray refractive index of a solid with its density and chemical composition. In addition, the imaginary part β of the dielectric susceptibility is needed to account fully for the effects observed by x-ray diffraction under very small angles of incidence. β is related to the attenuation coefficient μ via

$$\beta = \frac{\lambda}{4\pi} \mu \qquad (4.20)$$

The absorption index can be given in a form comparable to Eq. (4.19) by writing

$$\beta = \rho N_L \frac{\sum_i f_i'' \lambda^2 r_e}{\sum_i M_i \quad 2\pi} \qquad (4.21)$$

where f'' is the imaginary anomalous dispersion factor introduced in Chapter 1.

Summarizing, the complex refractive index for x-rays is given by

$$\hat{n} = 1 - \delta - i\beta \qquad (4.22)$$

with the real and imaginary part of the dielectric susceptibility given by the specimen's atomic density and appropriate atomic form factors. (The use of δ and β for denoting the real and imaginary deviation of the refractive index from unity is widely used in the literature. Other symbols would have been preferred, since both have been used with different meanings previously. In this chapter, however, δ and β will be used with the meaning expressed by Eq. (4.22).) All interesting x-ray optical effects as they occur under grazing incidence configurations derive from this special form of $\hat{n}$ or equally from $\delta, \beta \ll 1$.

4.5
Total External Reflection and Critical Angle

If refraction and reflection of x-rays under small incidence angles are considered in the formalism of geometrical optics, it turns out that below a critical angle the effect of total reflection will occur. The critical angle is denoted by α_c. The effect is similar to geometrical optics, where a light beam changes from an optically dense medium into a less dense one and which can be applied to confine a light beam in a glass fiber. An expression for α_c as a function of δ and β is now derived with the help of Fig. 4.7. Wave vectors are indicated in the following by upper- and lower-case letters K and k, respectively, according to whether the electromagnetic wave propagates in air/vacuum or within the thin film.

According to Snell's law of refraction, the sine of the incoming and outgoing angles are proportional to the refractive indices of the two media at the interface of which the transition occurs. Since in x-ray terminology the beam angles are measured with respect to the sample surface and relative to the substrate normal in geometrical optics, the sine functions in Snell's law have to be substituted by cosines and one arrives at

$$\frac{1}{\hat{n}} = \frac{\cos\alpha_t}{\cos\alpha} \tag{4.23}$$

where $\hat{n}$ is the refractive index of the thin film or the substrate, and α_t and α are the angles of transmitted and incident beam, respectively. Both angles may be assumed very small such that the cosine functions may be approximated, $\cos x \approx (1 - x^2/2)$ for $x \approx 0$.

If absorption is firstly neglected and products of small quantities are assumed to vanish, like $\delta\alpha_t \approx 0$, if follows for the angle of the plane wave transmitted into the medium

$$\alpha_t = \sqrt{\alpha^2 - 2\delta} \text{ for } \beta \approx 0 \tag{4.24}$$

Total reflection, i.e. $\alpha_t = 0$, will be observed for all angles of incidence for which the condition $\alpha^2 \leq 2\delta$ is fulfilled. The critical angle α_c is thus defined from this condition and it holds

$$\alpha_c = \sqrt{2\delta} \tag{4.25}$$

It may be estimated from the electron densities cited above that α_c takes values between 0.1° and 0.4° if Cu Kα radiation is used.

In a second approach, absorption is taken into account and one starts from

$$\alpha_t = \sqrt{\alpha^2 - \alpha_c^2 - 2i\beta} \text{ for } \beta > 0 \tag{4.26}$$

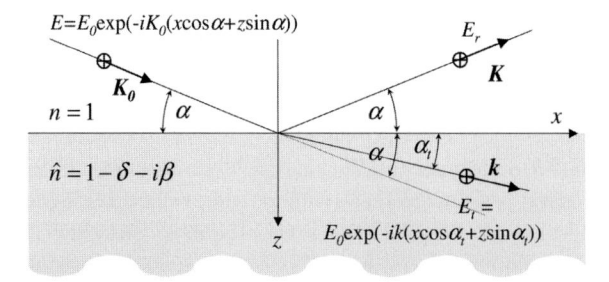

Figure 4.7 Electric field vectors E and wave vectors K_0, K and k of electromagnetic plane waves at the surface of an illuminated sample (σ polarization).

4.5 Total External Reflection and Critical Angle

where the same approximations are made as above and α_c^2 has been inserted instead of 2δ. In Fig. 4.7 the angle of the transmitted wave with respect to the surface is denoted by α_t. This angle can be developed into a sum of a real and an imaginary part

$$\alpha_t = \text{Re}(\alpha_t) + \text{Im}(\alpha_t) = B_+ - iB_- \tag{4.27}$$

where the B_+ and B_- terms may be represented by the real and imaginary deviation of n from unity

$$B_\pm^2 = \frac{1}{2}\left\{\sqrt{(\alpha^2 - \alpha_c^2)^2 + 4\beta^2} \pm (\alpha^2 - \alpha_c^2)\right\} \tag{4.28}$$

The electric field entering the material $E_t = E_0 \exp(-ik(x\cos\alpha_t - z\sin\alpha_t))$ is simplified by making use of $k = \hat{n}K_0 = \hat{n}K$ and Snell's law to arrive at the phase factor $-iK(x\cos\alpha - \hat{n}z\sin\alpha_t)$. Inserting the approximation $\hat{n}\sin\alpha_t \approx \alpha_t$ and the Euler representation (4.27) yields

$$E_t = E_0 \exp\left(-iK(x\cos\alpha - zB_+)\right)\exp(-zKB_-) \tag{4.29}$$

For $\alpha < \alpha_c$, this expression describes a plane wave propagating along x, but that is exponentially damped in the z direction, i.e. an evanescent wave at the surface. It is the imaginary part of α_t that accounts for the penetration depth of the incoming wave. The penetration depth introduced in Section 4.3 may be written for very small incidence angles close to the critical angle

$$\tau_{1/e} = \frac{\lambda}{4\pi B_-} = \frac{\sqrt{2}\lambda}{4\pi}\left\{\sqrt{(\alpha^2 - \alpha_c^2)^2 + 4\beta^2} - (\alpha^2 - \alpha_c^2)\right\}^{-1/2} \tag{4.30}$$

Within this depth the intensity of the impinging beam is decreased to $1/e$ of its incident value. The phenomenon of total reflection of x-rays therefore enables the investigation of the topmost layer of a thin film or a substrate by setting $\alpha < \alpha_c$. The depth from which structural information is monitored then amounts to only a few nanometers making highly sensitive surface analyses possible.

The penetration depth $z_{1/e}$ is shown in Fig. 4.8 as function of the incident angle α on a double-logarithmic scale. Numerical values for the TiO_2 example were used and the application of Cu Kα radiation was assumed. It can be recognized that the penetration depth divides into three ranges: it is nearly constant significantly beneath the critical angle α_c, where $z_{1/e} \approx \lambda/4\pi\alpha_c$ approximately holds. Above α_c the penetration depth $z_{1/e}$ approaches the limit $\mu/\sin\theta$ as has already been given in Eq. (4.5) and as represented by the dashed line the plot. A transition is observed to occur in the vicinity of the critical angle between both limiting cases. Obviously the sample may be monitored with respect to depths that differ by orders of magnitude by varying the incidence angle around α_c. For measurement settings of $\alpha < \alpha_c$, however, only structural information from the uppermost layer is collected.

The transition from Eq. (4.5) to Eq. (4.30) describes the effect of taking into account total reflection and absorption at very low incidence angles. The penetration depth given in Eq. (4.30) thus represents the required generalization with which Section 4.3 ended. Equation (4.30) has to be inserted into the Laplace transformation and its inverse in Eqs. (4.10) to (4.12) when depth-dependent properties are to be evaluated with incidence angles close to the critical angle α_c.

In the GIXRD configuration the refraction of x-rays also causes a shift of reflection centroids from their Bragg peak position θ_B. This shift increases as the angle of incidence α approaches the critical angle α_c. The effect is shown schematically in Fig. 4.9 from which it may be derived that the shift $\Delta(2\theta)$ is

$$\Delta(2\theta) = (2\theta_0 - 2\theta_B) =$$
$$\alpha - \text{Re}(\alpha_t) = \alpha - B_+ \xrightarrow{\beta=0} \alpha - \sqrt{\alpha^2 - \alpha_c^2} \qquad (4.31)$$

The shift thus is always in the direction to higher 2θ angles compared to $2\theta_B$. This fact is understood from the refraction of the beam away from the surface normal, which causes the Bragg angle to be reached later than in the hypothetical case that no refraction occurs. The shift ultimately has to be considered for the interpretation of GIXRD pattern that have been recorded with an incidence angle close to the critical angle, $\alpha \approx \alpha_c$. This should in particular be done when conclusions are to be drawn from the shift of reflection position with respect to residual stresses within the film (Chapter 6). Stress might only be deceived by the shift of reflection positions due to refraction. For $\alpha < \alpha_c$ and $\alpha \gg \alpha_c$ the shift $\Delta(2\theta)$ becomes 2α and $\delta_{\cot}\alpha$, respectively [15].

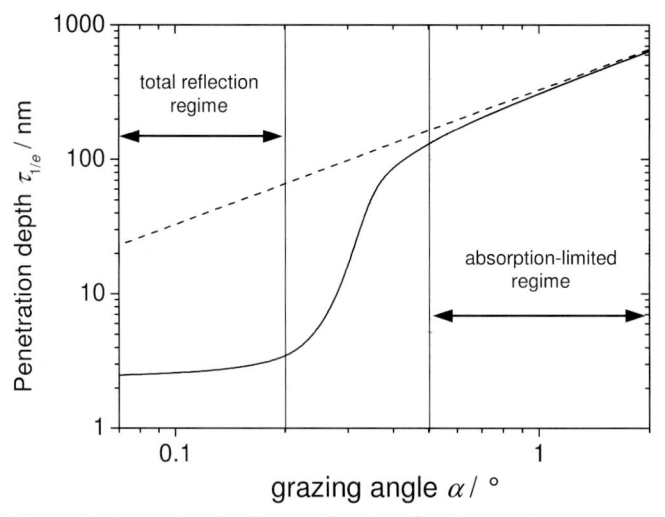

Figure 4.8 Penetration depth $\tau_{1/e}$ as function of incident angle α for TiO_2 and Cu Kα radiation according to Eq. (4.30) (solid curve) and Eq. (4.5) (dashed line).

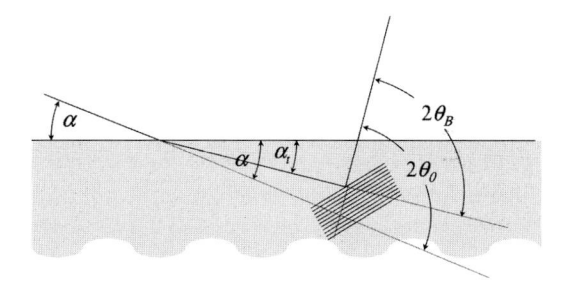

Figure 4.9 Schematic drawing to derive the Bragg peak position shift in GIXRD due to refraction of the incoming beam.

In passing we note that grazing incidence experiments are sometimes termed glancing angle measurement when the incidence angle very closely approaches the critical angle α_c. This notation is justified by the fact that practically the total intensity impinging on the sample becomes reflected. However, this notation is not used here.

4.6 X-ray Reflectivity (XRR)

It is evident from the discussion above that the deviation of the refractive index from unity $\delta + i\beta$ contains interesting information about the specimen, like density and chemical composition. Measurements of the x-ray refractive index under grazing angles moreover allow these properties to be determined in the close vicinity to a surface, which is ideally suited for the investigation and characterization of surface layers and thin films. One way of evaluating $\delta + i\beta$ is by the technique of x-ray reflectometry or reflectivity (XRR). In this method the diffractometer is operated in the symmetric $\theta/2\theta$ configuration, but with much smaller θ angles than those in $\theta/2\theta$ diffraction. Figure 4.10 visualizes the XRR setup in the usual representation of the sample reference frame. The XRR technique is introduced in the following by presenting simulated and experimental examples for a substrate and

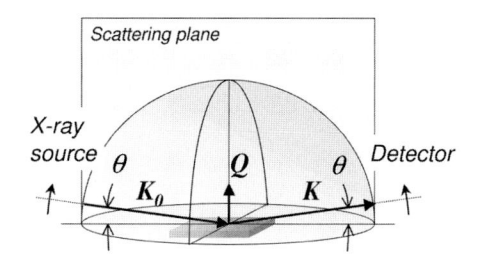

Figure 4.10 XRR experiment from the viewpoint of the sample reference frame.

various thin-films systems. Simulations and regressions to experimental curves were performed using the computer code for rocking curve and reflectivity simulations (RCRefSim) [7]. In the following, the incidence angle is denoted by θ instead of α, since the operation mode is that of a $\theta/2\theta$ scan. In contrast to the wide-angle diffraction scan, the XRR experiment has to be performed in the parallel beam configuration, i.e. the in-plane beam divergence should not exceed 0.05° and ideally would be even lower. It will be seen that XRR does not only allow for the determination of δ and β, but also for the surface and interface roughness and – most importantly – the thickness of thin films and multilayers.

4.6.1
Reflectivity of a Substrate

Figure 4.11(a) displays the x-ray reflectivity of a quartz glass substrate simulated for Cu Kα radiation. Quartz glass is an amorphous modification of SiO_2 and is frequently used as a substrate for the optical characterization of thin films deposited on it. The material is typically characterized by a high chemical purity and, therefore, well-defined properties, but due to high costs it is mainly applied for scientific rather than industrial purposes. A reflectogram is usually presented as an intensity plot versus either diffraction angle θ or scattering angle 2θ, while the ordinate is given on a logarithmic scale, since the reflected intensity may vary by many orders of magnitude.

The XRR curve is seen to separate into three different regions: region I of increasing intensity, an intensity plateau II and region III of steeply decreasing intensity. A quantitative description of the reflectivity curve $I(2\theta)$ can be derived from the Fresnel coefficients [8]. For this purpose one makes use of the continuity of the tangential electrical field component of the incident, refracted and outgoing beam and arrives at the ratio of reflected over incident intensity

$$\frac{I_R}{I_0} = \left| \frac{\theta - \sqrt{\theta^2 - \theta_c^2 - 2i\beta}}{\theta + \sqrt{\theta^2 - \theta_c^2 - 2i\beta}} \right|^2 \tag{4.32}$$

A fitting of the curve using Eq. (4.32) reveals that regions II and III are adequately accounted for and are separated by the critical angle $\theta_c = (2\delta)^{1/2}$.

The plateau that occurs for scattering angles smaller than the critical angle, $\theta < \theta_c$, of nearly constant reflectivity is caused by the total reflection of the primary beam, which spreads as an evanescent wave on the sample surface and enters the material exponentially damped. In the region above the critical angle, $\theta > \theta_c$, the reflected intensity is seen to decrease steeply over several orders of magnitude. For an ideally smooth surface the decrease scales with the inverse fourth power of the wave vector transfer, $1/Q^4 = (4\pi\lambda/\sin\theta)^4$, which is called the Porod slope [9, 10]. In the region around the critical angle, $\theta \approx \theta_c$, a transition occurs from the total reflection regime to the Porod regime. According to Eq. (4.32) the reflectivity curve is directly related to the materials parameters δ and β or ρ_e and μ, respectively, that both may be derived by fitting the curve with the intensity ratio given in Eq. (4.32).

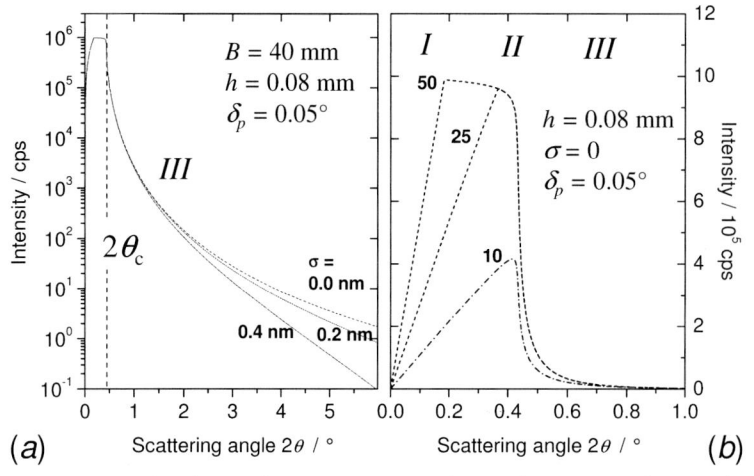

Figure 4.11 (a) Simulation of an XRR scan from a quartz glass substrate of various surface roughness values σ probed by Cu Kα radiation. (b) XRR scan of quartz glass substrate for different substrate lengths B of 10, 25 and 50 mm (linear ordinate).

The electron density ρ_e may be converted to the mass density ρ according to $\rho_e = N_L \rho Z/M$ or by calculating the sum over all atoms in the unit cell (compare Eq. (4.19)).

Various instrumental effects have to be considered in an XRR scan. First, in the plateau region II almost all the intensity emitted from the x-ray source may be reflected into the detector and precautions have to be taken to avoid any damage. Different practices are in use: (a) cutting of the full 2θ scan range into separate ranges that are measured with different powers of the x-ray tube or (b) insertion of an automatic beam attenuator into the beam path that opens after the intensity at the detector has fallen below a certain threshold value. Second, the first region of the reflectivity curve is sensitively affected by the geometrical extension of the sample. This is understood from the fact that both affect the footprint of the beam on the sample and therefore the amount of reflected intensity (Exercise 4.15). The effect of this factor on the measurable intensity in region I and II is shown in Fig. 4.11(b) for a set of various substrate lengths that were simulated for the same sample as in Fig. 4.11(a). It should be noted that the intensity ordinate is now plotted on a linear scale in order to emphasize the observed variations. It can be seen that the precision in the determination of the critical angle θ_c can be severely affected by too short a sample.

Moreover, the absorption coefficient μ affects the shape of the transition region. It has already been shown by Parrat that this effect is caused by the β/δ ratio, leaving the plateau of region I flatter as the ratio becomes smaller. There are various computer programs available to perform a regression of the measured reflectivity curves to the intensity ratio given in Eq. (4.32). Computer codes have even been developed that avoid the approximations made in deriving Eq. (4.32), but fit the full expression without any approximating assumptions [7].

So far only the reflectivity of ideally smooth surfaces has been considered. How will the reflectivity vary if real surfaces are measured that are rough, i.e. show a variation of height? Experimentally it is found that the steep decrease in region III is even stronger for rough surfaces and that the slope can be modeled *ad hoc* by multiplying Eq. (4.32) with an exp($-\sigma^2 Q^2$) factor that resembles closely a Debye–Waller factor. Here, σ accounts for the r.m.s. roughness (root-mean square) of the surface, which is given by

$$\sigma = \int \sqrt{z(x,y)^2 - \bar{z}^2}\, dxdy \bigg/ \int dxdy \qquad (4.33)$$

where $\bar{z}$ is the average height. Figure 4.11(a) demonstrates the effect of surface roughness on the measured pattern by displaying reflectivity curves of a perfectly smooth quartz glass and those with a small r.m.s. roughness of 0.2 and 0.4 nm. The general effect of surface roughness is seen to enhance the steep intensity decrease.

It turns out that the surface roughness may exhibit a diffuse intensity component observed in the reflectivity curve. The phenomena are summarized by considering the reflectivity to be composed of a specular and a diffuse component

$$I(2\theta) = I_{\text{specular}} + I_{\text{diffuse}} \qquad (4.34)$$

The term specular derives from the Latin word for mirror (speculum). For a strong surface roughness observed at high scattering angles the diffuse component may become of the same order of magnitude as the specular one and then also has to be taken into account. In many thin-film applications, however, it suffices to evaluate the XRR curve as being purely of the specular type and to include the effects of surface roughness by modeling the intensity ratio in combination with modified Debye–Waller factors.

4.6.2
Reflectivity of a Single Layer

Figure 4.12 shows an XRR pattern of a single Ta_2O_5 layer of 25 nm thickness on a quartz glass substrate simulated for Cu Kα radiation. The critical angle θ_c has shifted to larger values in comparison to Fig. 4.11, which is explained by the larger electron density ρ_e of Ta_2O_5 compared to quartz glass (see Table 4.3). The sequence of critical angles depends on the electron density of substrate and thin film: the larger the number of electronic scatterers the higher the θ_c value (see Eq. (4.25)). The most remarkable feature in the $I(2\theta)$ plot is the occurrence of intensity oscillations, which are called Kiessig oscillations or Kiessig fringes after H. Kiessig who published on this phenomenon in 1931 [8]. The appearance of intensity oscillations is understood from the same formalism developed for thin-film optics, i.e. maximum intensities are observed whenever the phase difference Δ between the reflected and refracted beam is a multiple of the wavelength λ.

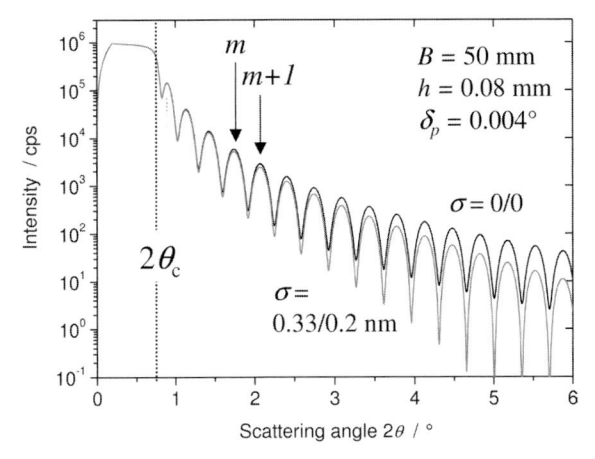

Figure 4.12 XRR of a Ta_2O_5 thin film, $t = 25$ nm, on a quartz glass substrate simulated for Cu Kα radiation. Upper curve for smooth surface and interface (black), lower curve for 0.2 nm interface roughness and 0.33 nm surface roughness (gray).

Figure 4.13 displays a schematic beam trace comparable to the figure in Kiessig's original publication. He argued that the phase difference Δ between the two beams is given by the relation

$$\Delta = (AB + BC)n - AD \quad (4.35)$$

between the individual path lengths of both beams, where n is the refractive index of the layer. A closer inspection of the figure reveals (Exercise 4.17) that the phase shift is related via

$$\Delta = 2t \sin \theta_t \quad \text{or} \quad \Delta \approx 2t\theta_t = 2t\sqrt{\theta^2 - \theta_c^2} \quad (4.36)$$

Table 4.3 Mass densities ρ, according to Ref. [6], electron densities ρ_e and critical angles θ_c of some oxide compounds used as optical coatings. Critical angles are given for Cu Kα radiation.

Compound	ρ (g cm^{-3})	ρ_e (nm^{-3})	θ_c (°)
Quartz glass SiO_2	2.20	662	0.216
Quartz SiO_2	2.64–2.66	794–800	0.237–0.238
Anatase TiO_2	3.84	1100	0.279
Rutile TiO_2	4.26	1217	0.294
Nb_2O_5	4.47	1236	0.294
ZnO	5.61	1576	0.326
Amorphous Pr_2O_3	7.07	2040	0.352
SnO_2	6.95	1833	0.359
Ta_2O_5	8.28	2099	0.372
HfO_2	9.68	2437	0.399

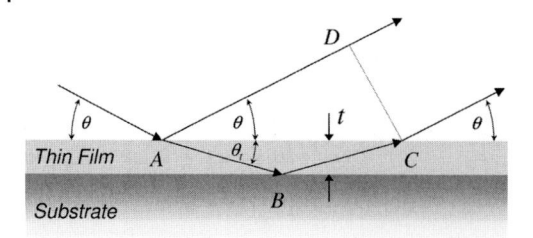

Figure 4.13 Schematic representation of reflected and refracted beams for the derivation of their phase difference Δ.

to the film thickness and the critical angle. Interestingly, this phase difference relation (left-hand equation) compares to the Bragg equation, with the film thickness t substituting for the interplanar spacing d. A maximum is observed whenever the phase difference is a multiple of the wavelength, $\Delta = m\lambda$. The evaluation proceeds by assigning an index m to each oscillation maximum and forming the square

$$\theta_m^2 = \theta_c^2 + \left(\frac{\lambda}{2t}\right)^2 m^2 \tag{4.37}$$

Similarly to ellipsometric investigations of thin films the θ_m^2 values can be plotted versus m^2 to obtain the film thickness from the slope s of the linear fitting curve

$$t = \frac{\lambda}{2\sqrt{s}} \tag{4.38}$$

In modern XRR evaluation programs the determination of t is performed automatically by fitting the reflectivity curve with the thickness t as a refinable parameter. It is evident from Eq. (4.37) that the distance between two adjacent fringes for $2\theta_{m+1}$ and $2\theta_m$ significantly above θ_c scales like

$$2\theta_{m+1} - 2\theta_m \approx \frac{\lambda}{t} \tag{4.39}$$

and is thus proportional to $1/t$. Hence, the maxima will become more closely spaced with increasing layer thickness. For $\lambda = 0.154$ nm and $t = 154$ nm the difference in scattering angle 2θ is only of the order of 0.06°. Because the resolution in 2θ cannot be made arbitrarily small, the measurable thickness of thin films by XRR is limited to a maximum value. Although the fringe separation increases for higher orders m, it might become difficult to determine too thick films by XRR. The precise value of the upper limit will depend on the instrumental performance and the properties of the sample like the beam divergence, possible bowing of the substrate, thickness inhomogeneities, surface and interface roughness and other factors that contribute to the washing out of sharp fringes. As a rule of thumb it may be remembered that thickness values of single layers in excess of some 100 nm are hardly measurable when Cu Kα radiation is employed. This statement is unaffected by the precision in the x-ray wavelength λ which is known to less than 1 ppm (see Table i2.1).

It has to be emphasized that the analysis of XRR fringing directly yields the film thickness t, which is uncoupled from the linear absorption coefficient μ as in the μt product. Besides the chemical composition the thickness t probably is the most important physical quantity characterizing a thin film. However, the film thickness is generally not easily determined with a relative error of less than a few percent. The analysis of Kiessig oscillations enables its determination as a pure geometric quantity, i.e. the distance between (average) interface plane and film surface. The thickness value obtained is practically unaffected by the refractive index n, which contrasts to optical or infrared techniques, where often the product of both is determined from measurements of optical path differences. It is a direct consequence of the small deviation of n from unity for electromagnetic radiation in the x-ray regime that highly precise thickness values may be determined by XRR.

Figure 4.12 also displays the simulation of the Ta_2O_5/SiO_2 system with the inclusion of roughness effects. An interface roughness of 0.2 nm and a surface roughness of 0.33 nm were assumed for the second simulation. As in the case of a single substrate the roughness is seen to cause a decrease of intensity received by the detector. Also the scattering-plane beam divergence δ_p may have a pronounced effect on the measured reflectivity pattern. Fine structural details like the pronounced intensity minima are washed out by larger δ_p values. The analyst is recommended to perform a set of simulations of the layer/substrate system under investigation prior to the measurement in order to verify whether the intended structural properties may be derived from the measurement. These simulations may also be helpful in choosing the correct instrumental setup and may be performed with most reflectometry computer programs.

4.6.3
Reflectivity of Multilayers and Superlattices

In XRR measurement of a multilayer system the incident beam intensity is reduced by the two effects of reflection at interfaces and absorption in individual layers. The calculation of the reflected beam intensity is based on the continuity condition of the tangential components of the electric fields and the conservation of tangential components of the wave vectors at the different interfaces. Parrat solved this task in 1954 during an investigation of surface oxidation effects of metallic layers [11]. In fact, the surface sensitivity of the XRR technique allows for the determination of thicknesses in the few nanometers range of oxide layers, as they typically occur on technologically relevant surfaces.

A schematic of a multilayer system is depicted in Fig. 4.14. The layers are numbered in order from the top to the bottom. In the matrix technique approach [12] the reflectivity of the multilayer system is described by virtue of so-called reflection coefficients $r_{j,j+1}$

$$r_{j,j+1} = \frac{k_{z,j} - k_{z,j+1}}{k_{z,j} + k_{z,j+1}} \qquad (4.40)$$

that contain the z wave vector component $k_{z,j}$ and $k_{z,j+1}$ in the two neighboring media and account for the reflection of the incoming and the exiting beam at the $j, j + 1$ interface. A set of 2×2 matrices is introduced for each interface $j, j + 1$ and for each layer j. These are the refraction and translation matrices $R_{j,j+1}$ and T_j, respectively. The product of all of them establishes the transfer matrix M, with the ratio of M_{12} and M_{22} components giving the amplitude of the multilayer reflectivity. Applying the procedure for a one-layer system one ends up with an expression for the intensity of the reflected beam

$$\frac{I_R}{I_0} = \frac{r_{0,1}^2 + r_{1,2}^2 + 2r_{0,1}r_{1,2}\cos 2k_{z,1}t}{1 + r_{0,1}^2 r_{1,2}^2 + 2r_{0,1}r_{1,2}\cos 2k_{z,1}t} \tag{4.41}$$

that depends on the reflection coefficients of the air–film interface (0,1) and the film–substrate interface (1,2) according to Eq. (4.40). The ratio also accounts for the oscillation phenomena observed in Fig. 4.13 via the cosine terms that depend on the layer thickness t. This analytical expression has been the basis for the simulation of the Ta_2O_5 reflectivity curve. The recursion scheme of the matrix technique allows effective integration in a computer code. Moreover, surface roughness effects may also be included by multiplying the reflection coefficients with Debye–Waller factor-like damping terms that contain the r.m.s. roughness of all individual $j, j + 1$ interfaces.

Analytical expressions for the reflectivity of multilayers with two and more layers become rather unclear for recognizing the underlying physics. Instead of presenting the generalization of eq. (4.41) to two-layer, three-layer systems, and so on, an example of a complex multilayer architecture is presented. Figure 4.15 displays a set of simulations from a thermal barrier multilayer that comprises a stacking of oxide and metal layers, as shown in the inset. Thermal barrier coatings are used, for instance, on architectural glass or car glass for the control of the heat flow into buildings, automobiles, etc. An essential part of this layer stacking is the so-called

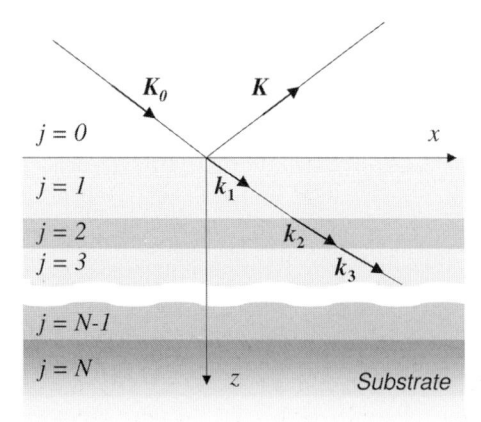

Figure 4.14 Notation for participating wave vectors in XRR of a multilayer system.

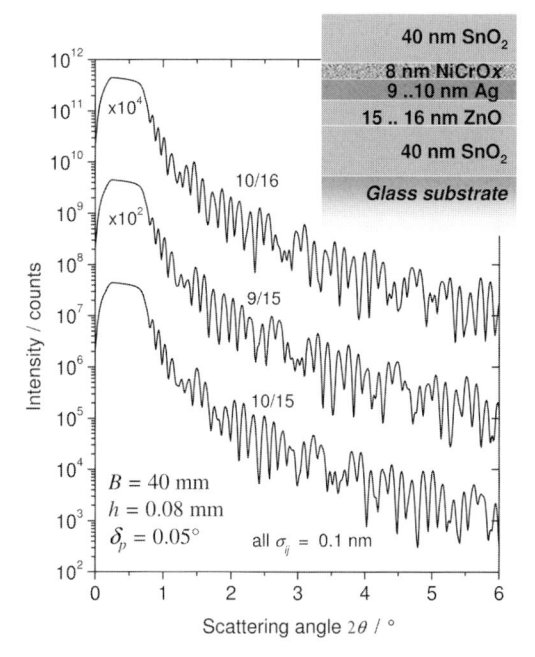

Figure 4.15 Simulated XRR patterns of thermal barrier coating SnO/NiCrO$_x$/Ag/ZnO/SnO on a glass substrate recorded with Cu Kα radiation. The numbers above each pattern indicate the supposed thickness of the Ag and ZnO layer, respectively (in nm). Other thicknesses were kept constant.

blocking layer, which blocks the infrared part of the solar or any other outdoor spectrum from being transmitted by the glass. In the layer stacking shown, the role of the blocking layer is played by the NiCrO$_x$ layer. The blocking layer thickness, interface roughness, etc., are decisive parameters for the functionality of the stack. Figure 4.15 displays various reflectivity simulations of this multilayer system, where only slight deviations are performed between the different patterns. It can be seen that small differences in the assumed architecture cause substantial changes in the XRR pattern. A wealth of structural information is contained in the many fringes that can be observed.

XRR may elucidate the inner architecture of periodic multilayers that are also termed superlattices (as a special case of multilayers). These layer structures essentially comprise a stack of N pairs of layers of type A and B, where the thicknesses of individual layers t_A and t_B add up to the superlattice period t_{per}

$$t_{per} = t_A + t_B \tag{4.42}$$

Such superlattices are fabricated for applications as so-called Bragg mirrors in light-emitting devices or as x-ray mirrors as is outlined in Instrumental Box 5. The XRR curve of such multilayers is characterized by various features that are illustrated for the example of an x-ray mirror system shown in Fig. 4.16. Also shown are the reflectograms of a $10 \times$ (6 nm TiO$_2$/6 nm C) superlattice on a Si substrate. As

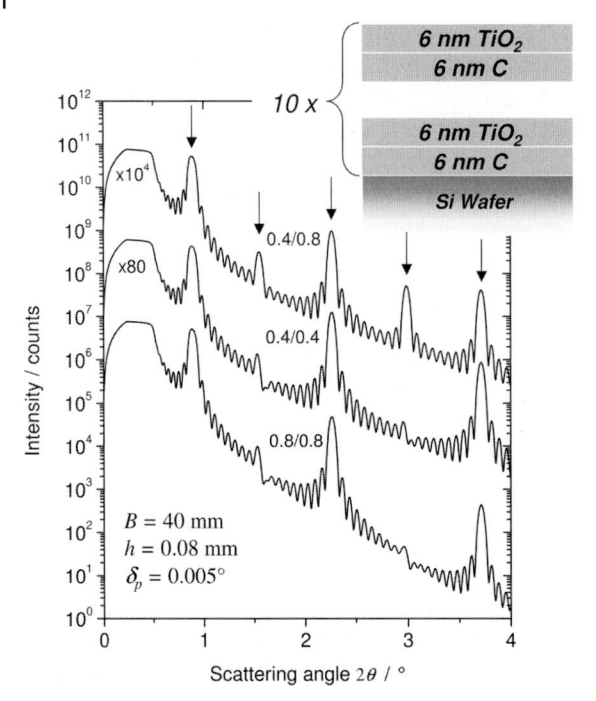

Figure 4.16 Simulation of XRR of a (TiO$_2$/C) superlattice on Si wafer. The numbers above each pattern give the roughness of TiO$_2$ and C layers, respectively (in nm). Substrate roughness was set constant to 0.4 nm. Superlattice peaks are indicated by arrows.

for the single-layer system a modified Bragg equation may be derived, Eq. (4.36), describing the occurrence of the superlattice peaks, which cause large interference maxima in the reflectivity curve. However, in the case of a superlattice the thickness has to be substituted by the superlattice period and it is obtained

$$\theta_m^2 = \langle \theta_c \rangle^2 + m^2 \left(\frac{\lambda}{2t_{per}} \right)^2 \tag{4.43}$$

Here, $\langle \theta_c \rangle$ is the average critical angle of the total superlattice. As usual, the thickness values for individual layers A and B and the interface roughness σ_{AB} and σ_{BA} may be determined by fitting with the matrix formalism. Figure 4.16 displays three different simulations with varying interface roughnesses. The difference between the large superlattice fringes is accounted for by Eq. (4.43), while the small fringing scales with the total thickness of the layer system, i.e. $1/(Nt_{per})$. Ideally, there are $N - 2$ small fringes between two adjacent superlattice peaks. It is seen from the figure that for certain combinations of σ_{AB} and σ_{BA} every second maximum may be severely reduced. This effect may be observed in superlattices having $t_A = t_B$ and is due to the destructive interference form x-ray beams reflected at the AB and BA interface.

Superlattices with large differences in electron density between individual layers have found application as x-ray mirrors. Such mirrors are in use in different analytical techniques like x-ray fluorescence for the detection of chemical elements, but also in x-ray diffraction as outlined in Instrumental Box 5. Typical material combinations are Mo/Si, V/C or La/B$_4$C and other systems with substantial ρ_e contrast. The layer system presented in the example has been investigated with respect to its reflectivity for characteristic K radiation from carbon at 277 eV for the purposes of elemental analysis [13]. In addition, the deployment of x-ray multilayer mirrors in the semiconductor production process is imminent. Due to the constraint of ongoing miniaturization in the microelectronic industry the use of extreme ultraviolet (EUV, $\lambda < 100$ nm) radiation is being introduced into lithography processesing. This technology requires appropriate beam-conditioning devices, for which multilayer mirrors represent one of the favored technical solutions.

4.7
Grazing Incidence Diffraction (GID)

The GID method is a hybrid technique that is based on diffraction and on the extreme surface sensitivity of incident x-rays close to the critical angle. The technique was introduced by Marra, Eisenberger and Cho in 1979 and was originally denoted by them as diffraction-reflection technique [14]. Various terminologies have occurred in the literature since then, but here it will be abbreviated by the most established acronym, GID, although it is in fact a grazing incidence–grazing exit diffraction technique.

In the GID setup a diffraction pattern is recorded while both the angle of the incoming and outgoing beam are kept at very small constant values. The incidence and exit angle are now denoted α_i and α_f. A schematic representation of the GID configuration is shown in Fig. 4.17. As usual the recording of the diffraction pattern proceeds by continuously increasing the scattering angle 2θ, but in this scanning mode the x-ray tube and the detector operate very close to the plane of the sample surface. Accordingly, the incident and the diffracted wave vectors K_0 and K are almost confined to the surface plane and so is the vector of momentum transfer $Q = K - K_0$ that is nearly perpendicular to the substrate normal, $Q \perp s_3$. Since the structure of the sample is always probed in direction of Q in diffraction, GID

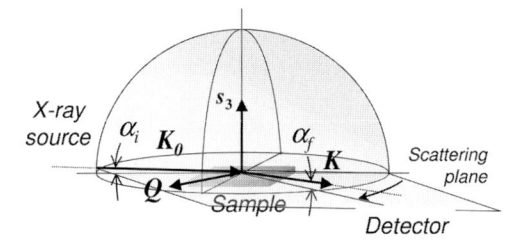

Figure 4.17 Measurement setup in (noncoplanar) GID experimentation.

probes the interplanar spacing of vertically inclined lattice planes. This is the first decisive distinction compared to the conventional $\theta/2\theta$ scan.

A further characteristic feature of the method is its extreme surface sensitivity. By choosing the incidence and exit angles α_i and α_f close to the critical angle α_c surface regions of the layer may be probed that are as close to the surface as the damping depth of the evanescent surface wave. Inserting the electron densities of real-world solids into this recipe, (see Table 4.3) it is evident that thin surface layers of only a few nanometers may be elucidated. Additionally, the choice of α_i and α_f can decide the probing depth of the experiment that may be varied from the topmost surface layers, $\alpha_i, \alpha_f < \alpha_c$, down to the measurement of the full volume of a thin film, $\alpha_i, \alpha_f \approx \arcsin(\mu t)$.

The GID experiment performed by the inventors was carried out using a laboratory setup. A rotating anode was used as the x-ray source, which also is a laboratory x-ray source, but allows for 3-5 times higher intensities than the conventional x-ray tubes presented in Instrumental Box 2. The authors already pointed to the fact that very thin layers may be studied by the technique when synchrotron radiation of much more intense and collimated x-ray beams is employed [14]. In fact, many GID studies on thin and ultrathin films have been carried out using synchrotrons, although recent improvements in laboratory equipment increasingly allows GID to be also carried out with the laboratory diffractometers. Laboratory GID experiments necessarily require measures for parallel beam configurations. As in the GIXRD configuration the transformation of peak positions $2\theta_0$ into d values may be affected by peak shifts due to refraction. For the general case of both grazing incidence and grazing exit the peak shift has recently been worked out by Genzel [15].

The extreme surface sensitivity of the method shall be demonstrated by an investigation of an ultrathin Pr_2O_3 layer that was epitaxially grown on a Si wafer of (001) orientation [16]. Various oxides and silicates like those derived from Hf, Pr and other metals are currently under development for high-k dielectrics that should operate, for instance, as gate insulators in metal–oxide–semiconductor field-effect transistors (MOSFETs). For many years the material of choice for this purpose has been SiO_2 due to the low density of surface states that it forms at the common interface with the Si substrate. Due to the ongoing miniaturization in semiconductor technology the thickness of the gate oxide will also have to be scaled down in accordance with the decreasing dimensions of the transistor gate. A further reduction by usage of SiO_2, however, is prohibited by the increase of leakage currents with too thin an oxide. Alternative gate oxides with higher dielectric constants are therefore investigated with the intention to substitute for silicon dioxide.

A highly significant issue for gate oxides relates to the formation of interface states as induced by strain relaxation due to the mismatch of the Si underlayer and the epitaxial oxide. A synchrotron GID study was performed to monitor the evolution of the Pr_2O_3 structure in the relevant thickness range between 1 and 10 nm. Thin Pr_2O_3 films of the cubic $Ia3$ form were prepared on Si (001) prior to the experiment. A beam of 0.1127 nm wavelength was applied and the incidence and exit angles were adjusted to $\alpha_i = \alpha_f = 0.3°$. This value is just below the critical angle of

Pr_2O_3 for this wavelength in order to achieve a high surface sensitivity. The result of a GID scan for a 4 nm thin oxide layer is displayed in Fig. 4.18(a). During the scan the sample was rotated by θ around the substrate normal s_3, while the detector moved by 2θ at the same time – comparable to a $\theta/2\theta$ scan. The direction of the scattering vector Q remained fixed along the $(\bar{1}10)$ direction of the silicon wafer, from which the $H = -K$ abscissa label can be understood. Apart from the Si (220) reflection only the Pr_2O_3 040 and $40\bar{4}$ reflections occurred in the pattern indicating the epitaxial growth of the oxide on the lattice of the Si substrate.

Figure 4.18(b) demonstrates the results of a closer examination of the 040 reflection for a thickness series of nominally 1, 2, 4, 6 and 10 nm thin oxide layers. It is seen that the integrated intensity is markedly low for the 1 and 2 nm thin films. Moreover, a shift of the reflection to lower 2θ angles is seen for the 1 nm thin layer. A careful analysis of this effect and further data on the size of epitaxial domains implies that the first atomic layers are affected by the growth of a silicate $(Pr_2O_3)_x(SiO)_y$ interface layer [16]. This example nicely demonstrates the high surface sensitivity and technological significance of the GID technique.

4.8
Applications

In addition to the optical films used as illustrative examples in this chapter, the scattering of x-rays under grazing incidence has extensively been applied in its different variants to magnetic thin films. In GIXRD experiments Huang investigated the depth-dependent phase composition of an intentionally γ-phase Fe_2O_3 thin film for data storage applications. By controlling the scattering depth through the variation of the incidence angle the investigations revealed a less magnetizable layer of antiferromagnetic α-phase Fe_2O_3 on the film surface. The regression of the data to an appropriate fit function of the type of Eq. (4.10) allowed a quantification of the thickness of the surface layer to 9 nm [17].

Magnetic multilayers and superlattices are routinely characterized by XRR with respect to determining individual layer thickness and interface roughness. Such layer systems are typically comprised of two distinct metallic materials with at least one of them exhibiting ferromagnetic properties in the bulk state. The magnetic properties of the whole stack are engineered by adjusting the thickness of individual layers on a nanometer scale and thereby controlling the magnetic coupling between the subset of the ferromagnetic layers. When subjected to an external magnetic field such layer stacks may show remarkable effects with respect to the vertical conduction of electrical currents through them like the giant magnetoresistance effect (GMR), from which spin valves and other spin-electronic devices can be constructed. A proper functioning of magnetic multilayers is hindered by intermixing between the different layers systems or by too large an interface roughness and many investigations are thus related to these aspects (e.g. see Refs. [18–21]).

XRR also provides a sensitive tool to study surface oxidation of metals and other materials sensitive to surface oxidation. In fact, in his seminal XRR study Parratt

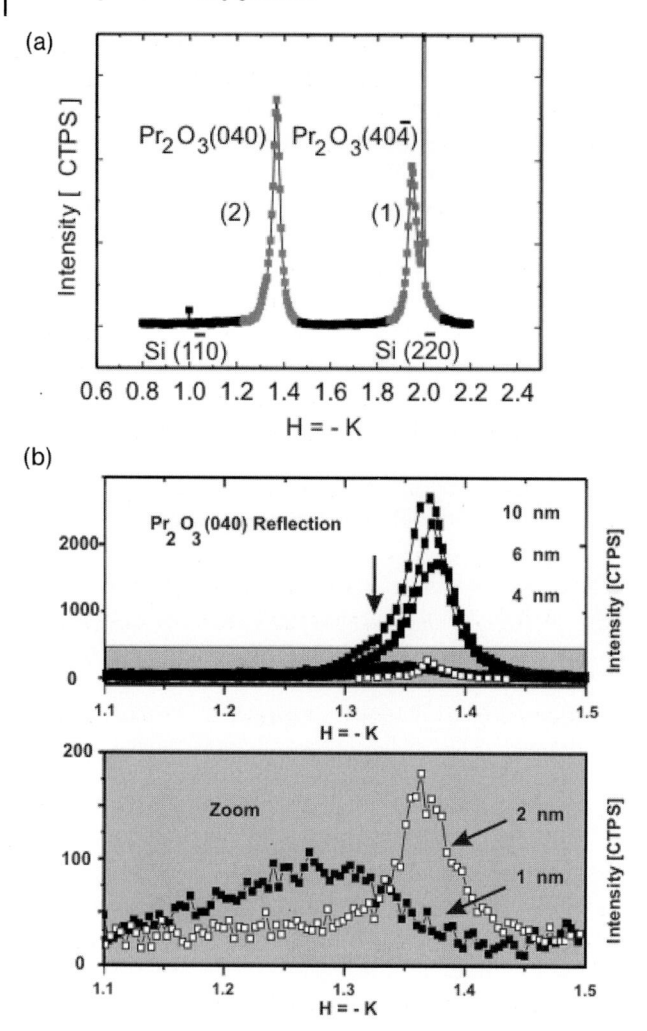

Figure 4.18 GID measurement of a thin Pr_2O_3 film: (a) $h = -k$ scan along Si $[1\bar{1}0]$; (b) 040 reflection from thickness series (from Ref. [16]).

investigated a thin Cu film on glass and could only achieve reliable agreement between the measured and the simulated reflectivity curve after introducing a 15 nm thin surface oxide layer into the model [11]. Assuming an oxide layer on the top of a thin film or multilayer system should routinely be checked in the fitting of measured reflectivity curves. Also other surface chemical reactions may be studied by XRR, GIXRD and GID, and because of the low absorption of x-rays by thin water layers the technique could be applied to *in situ* investigations of an electrochemical deposition [22].

The GID technique was initially introduced for the investigation of thin Al films on GaAs [14] and revealed the metal film to grow initially commensurate to the underlying GaAs lattice, which was associated with an in-plane extension and out-of-plane compression of the Al unit cell. With increasing thickness a stress relief was observed and for t of about 100 nm the Al lattice constant relaxed to the bulk value. Al thin-film metallization is of great interest in microelectronics where Al is used as electrical contact to semiconductors. GID investigations were thus also performed on thin Al films on oxidized and unpassivated Si wafers, in order to study strain distribution [23] and defects like dislocations [24], the control of which is of relevance for the lifetime of metal–semiconductor contacts.

The high surface sensitivity of GID qualifies the technique for the investigation of surface-induced disorder phase transitions. The first-order phase transition of Cu_3Au was studied by GID by systematically varying α_i and α_f [25, 26]. In the experiment the intensity of the (001) superlattice reflection was monitored in the vicinity of the critical temperature $T_c = 663$ K. The results indicated that the transition process starts below T_c from a disordered surface layer which grows into the ordered bulk while T approaches T_c. The transition was thus identified as an interface delocalization transition.

Finally, investigations of nanogranular thin films [27–29] and nucleation layers [30] should be mentioned that were also carried out using grazing incidence geometry. For the former investigations, shape parameters and average distances were deduced while for the latter, geometrical information on nucleation in the initial phase of thin-film growth was derived. These studies belong to the group of small-angle x-ray scattering (SAXS) experiments, which are frequently termed GISAXS when performed under grazing incidence.

Exercises

4.1 Visit colleagues who work in the optical department in their laboratories and discuss the distinctions and similarities between optical ellipsometry and x-ray reflectometry.

4.2 The scattering vector Q varies its orientation continuously during a GIXRD measurement. What is the angle between sample and surface and Q as a function of θ or 2θ?

4.3 Si wafers of (100) orientation are often used as substrates (for Si structural parameters, see Structure Box 2). What is the lowest angle between the surface normal of an Si (001) substrate and its (311) planes? Suppose you are investigating a film deposited on Si (001) by GIXRD with Cu Kα radiation and grazing angle α. The layer is only weakly absorbing such that Bragg reflections from the substrate may enter the detector. For which α does there exist an optimal probability to observe the Si 311 reflection in the diffraction pattern and what is the angle of the scattering vector Q with the surface normal s_3? Illustrate the geometry by a figure.

4.4 Continuation of preceding exercise. Switch to an incidence angle $\alpha = \alpha_{opt} + 1°$ that is larger by one degree than the optimum α to detect Si 311. For which λ (different from Cu Kα!) would Laue's condition now hold to observe Si 311?

What is the Bremsstrahlung intensity emitted from the Cu tube at this wavelength compared to the intensity of the characteristic Cu Kα line?

4.5 Derive an expression for the integral intensity of a Bragg reflection in the kinematical approach for GIXRD comparable to Eq. (1.44) of Section 1.3.

4.6 What is the analytically correct solution of information depth $\bar\tau$ in GID without making use of the approximation $\mu t < 1$?

4.7 Information depth $\bar\tau$ with $k_{2\theta}$ with approximation $\mu t < 1$?

4.8 Recall the limiting process: $\lim(\alpha \gg \alpha_c)\, z_{1/e} \to \sin\alpha/\mu$.

4.9 Show Eq. (4.9) by expanding $\exp(\mu k t)$ in a quadratic function.

4.10 Derive Eq. (4.28) from Eqs. (4.26) and (4.27).

4.11 Takayama and Matsumoto observed during their investigations of Cu/Au/Si by GIXRD the occurrence of Cu reflections not only at their usual $2\theta_{Cu}$ positions but also at adjacent positions [31]. They explained the effect by assuming that the beam entering the Cu film is first refracted ($\Delta\alpha$) and subsequently reflected at the Cu–Au interface before being diffracted. At which scattering angles did they observe the reflections according to such a beam path?

4.12 Calculate the electron density and the critical angle for Cu Kα radiation for the materials with which you have recently been working.

4.13 Derive the intensity ratio of Eq. (4.32) for XRR of an uncoated substrate of refractive index $\hat n$ from $I_r/I_0 = (\sin\theta - \hat n \sin\theta_t)^2/(\sin\theta + \hat n \sin\theta_t)^2$.

4.14 Show that the Fresnel coefficient of Eq. (4.32) approaches the Porod slope, $(\lambda/\sin\theta)^4$, in the limit of $\theta > \theta_c$.

4.15 Give the analytical dependency of the XRR signal in region I as a function of beam height h and sample length B.

4.16 Give a physical motivation for the damping of the specular component in XRR by surface roughness (hint: consider the effect of total reflection).

4.17 Derive the phase difference Δ between x-ray beams reflected at the film surface and film–substrate interface in an XRR experiment, Eq. (4.36).

4.18 Discuss the advantages and disadvantages of using of wavelengths longer than that of Cu Kα radiation in order to expand x-ray reflectograms to larger 2θ ranges and thereby improve the precision in θ_c and t to be obtained from a fit of the pattern.

4.19 Consider a single layer–substrate system with electron densities $\rho_{e,tf}$ and $\rho_{e,sub}$. Can you derive how the amplitude of Kiessig fringes depends on these two quantities?

4.20 GID configuration. Give the wave vectors $\mathbf{K_0}$ and $\mathbf{K}$ in spherical coordinates and calculate from them the angle between $\mathbf{K_0}$ and $\mathbf{K}$.

References

Monographs relevant to this chapter

J. Daillaut, A. Gibaud, *X-Ray and Neutron Reflectivity*, Springer, Berlin, 1999.
H. Dosch, *Critical Phenomena at Surfaces and Interfaces*, Springer, Berlin, 1992.
M. Born, E. Wolf, *Principles of Optics*, Pergamon Press, Oxford, 1965.
R. W. James, *Optical Principles of X-ray Diffraction*, Bell and Sons, London, 1948.

Special papers

[1] M. Birkholz, O. Bialas, T. Jung, Metalloxid-Metall Nanokompositschichten für Verschleiß- und Korrosionsschutz, *Trib. Schmierungst.* **51** (2004) 12.
[2] J. Szczyrbowski, G. Bräuer, M. Ruske, J. Bartella, J. Schroeder, A. Zmelty, Some properties of TiO_2 layers prepared by medium frequency reactive sputtering, *Surf. Coat. Technol.* **112** (1999) 261.
[3] P. Predecki, Determination of depth profiles from X-ray diffraction data, *Powder Diffract.* **8** (1993) 122.
[4] C. Genzel, X-ray stress analysis in presence of gradients and texture, *Adv. X-ray Anal.* **44** (2001) 247.
[5] M. Born, E. Wolf, *Principles of Optics*, Pergamon Press, Oxford, 1965.
[6] R. C. Weast, Physical constants of inorganic compounds, in *Handbook of Chemistry and Physics*, CRC Press, Boca Raton, FL, 1986, p. B-68.
[7] P. Zaumseil, RCRefSim (rocking curve and reflectivity simulation), Frankfurt (Oder), 2005, available from: zaumseil@ihp-microelectronics.com.
[8] H. Kiessig, Untersuchungen zur Totalreflektion von Röntgenstrahlen, and Interferenzen von Röntgenstrahlen an dünnen Schichten, *Ann. Phys.* **10** (1931) 715 and 769.
[9] G. Porod, The x-ray small angle scattering of close packed colloidol systems. I, *Kolloid-Zeitschr.* **124** (1951) 83.
[10] S. K. Sinha, E. B. Sirota, S. Garoff, H. B. Stanley, X-ray and neutron scattering from rough surfaces, *Phys. Rev. B* **38** (1988) 2297.
[11] L. G. Parratt, Surface studies of solids by total reflection of X-rays, *Phys. Rev.* **95** (1954) 359.
[12] A. Gibaud, S. Hazra, X-ray reflectivity and diffuse scattering, *Curr. Sci.* **78** (2000) 1467.
[13] C. Michaelsen, J. Wiesmann, C. Hoffmann, K. Wulf, L. Brügemann, A. Storm, *Recent Developments of Multilayer Mirror Optics for Laboratory X-Ray Instrumentation*, SPIE, Seattle, 2002.
[14] W. C. Marra, P. Eisenberger, A. Y. Cho, X-ray total-external-reflection-Bragg diffraction: a structural study of the GaAs–Al interface, *J. Appl. Phys.* **50** (1979) 6927.
[15] C. Genzel, X-ray residual stress analysis in thin films under grazing incidence – basic aspects and applications, *Mat. Sci. Technol.* **21** (2005) 10.
[16] T. Schroeder, T.-L. Lee, J. Zegenhagen, C. Wenger, P. Zaumseil, H.-J. Müssig, Structure and thickness-dependent lattice parameters of ultrathin eptaxial Pr_2O_3 films on Si(001), *Appl. Phys. Lett.* **85** (2004) 1229.
[17] T. C. Huang, Surface and ultra-thin film characterization by grazing-incidence asymmetric Bragg reflection, *Adv. X-Ray Anal.* **33** (1990) 91.
[18] M. Spasova, U. Wiedwald, R. Ramchal, M. Farle, M. Jergel, E. Majkova, S. Luby, R. Senderak, Magnetization and magnetic anisotropy of Co/W multilayers, *phys. stat. sol. (b)* **225** (2001) 449.
[19] L. Agazzi, S. Bennett, F. J. Berry, C. M., M. Rateo, G. Ruggiero, G. Turilli, Magnetic interactions and interface properties in Co/Fe multilayers, *J. Appl. Phys.* **92** (2002) 3231.

[20] Y. Li, S. X. Wang, Intermixing effects in ultrathin barrier magnetic tunneling junctions, *J. Appl. Phys.* **91** (2002) 7950.

[21] J. A. Gonz lez, J. P. Andrés, M. A. López de la Torre, J. M. Riveiro, T. P. A. Hase, B. K. Tanner, X-ray study of the interdiffusion and interfacial structure in ferrimagnetic $Gd_{1-x}Co_x$/Co multilayers, *J. Appl. Phys.* **93** (2003) 7247.

[22] M. G. Samant, M. F. Toney, G. L. Borges, L. Blum, O. R. Melroy, In-situ grazing incidence X-ray diffraction study of electrochemically deposited Pb monolayers on Ag(111), *Surf Sc.* **193** (1988) L29.

[23] M. F. Doerner, S. Brennan, Strain distribution in thin aluminium films using x-ray depth profiling, *J. Appl. Phys.* **63** (1988) 126.

[24] C. J. Shute, J. B. Cohen, Strain gradients in Al–2at% Cu thin films, *J. Appl. Phys.* **70** (1991) 2104.

[25] H. Dosch, *Critical Phenomena at Surfaces and Interfaces,* Springer, Berlin, 1992.

[26] H. Dosch, L. Mailänder, A. Lied, J. Peisl, F. Grey, R. L. Johnson, S. Krummmacher, Experimental evidence for an interphase delocalization transition in Cu_3Au, *Phys. Rev. Lett.* **60** (1988) 2382.

[27] T. Slimani, P. Goudeau, A. Naudon, G. Farges, J. L. Derep, Grazing-incidence X-ray scattering study of the microstructure of tungsten–carbon films, *J. Appl. Cryst.* **24** (1991) 638.

[28] F. D'Acapito, F. Zontone, Grazing-incidence X-ray diffraction in the study of metallic clusters buried in glass obtained by ion implantation, *J. Appl. Cryst.* **32** (1999) 234.

[29] J. Stangl, V. Holý, P. Mikulík, G. Bauer, I. Kegel, T. H. Metzger, O. G. Schmidt, C. Lange, K. Eberl, Self-assembled carbon-induced germanium quantum dots studied by small-angle x-ray scattering, *Appl. Phys. Lett.* **74** (1999) 3785.

[30] G. Renaud, R. Lazzari, C. Revenant et al., Real-time monitoring of growing nanoparticles, *Nature* **300** (2003) 1416.

[31] T. Takayama, Y. Matsumoto, Effects of Refraction and Reflection on Analysis of Thin Films by the Grazing-Incidence X-Ray Diffraction Method, *Adv. X-ray Anal.* **33** (1990) 109.

5
Texture and Preferred Orientation

The distribution of the crystallite orientations in a polycrystalline thin film is only rarely isotropic. On the contrary, in a large number of studies a certain crystallographic direction [hkl] was found to be preferentially oriented with respect to the sample reference frame. This anisotropy of crystallite orientation is named texture or preferred orientation and both terminologies will be used interchangeably for denoting the phenomenon. The effect may easily be recognized in a symmetric $\theta/2\theta$ diffraction pattern by a pronounced enhancement of certain Bragg reflections I_{hkl} – and reduction of others – when compared with a powder pattern of randomly oriented grains. Since many physical properties of polycrystals sensitively depend on the distribution of grain orientation, this structural anisotropy is directly mapped into a property anisotropy and the issue thus is a highly relevant one for technological applications.

Techniques for the investigation of textures were initially developed for bulk metallic materials like steel plates and wires that were subjected to forming processes like rolling, cold-working or deep-drawing and subsequent annealing that all have pronounced effects on the crystallite orientation distribution and the mechanical properties of macroscopic work pieces. Accordingly, many notations and methods in the field are related to their metallographic roots. The evaluation of texture in thin polycrystalline films started with retardation, but may now routinely be applied because of the development of measuring and analysis tools over the last few decades.

The full solution to the texture problem is given by the orientation distribution function (ODF). This distribution function describes the density of orientations that transform the sample reference frame $\{s_i\}$ into the reference frames of individual crystallites $\{c_i\}_j$ as shown schematically in Fig. 5.1. Most texture investigations of thin films, however, are not concerned with the determination of the ODF, but make use of simplified methods like the determination of texture factors, the measurement of pole figures of selected Bragg peaks, ω scans (rocking curves) or ϕ scans. By application of these techniques the thin-film grower has initial quantitative measures that allow for the optimization of orientation-dependent properties and these will be introduced in the following sections. In addition, the more elaborate techniques of determining the ODF are presented, since the full solution of texture is required if orientational averages of physical properties are to be calcu-

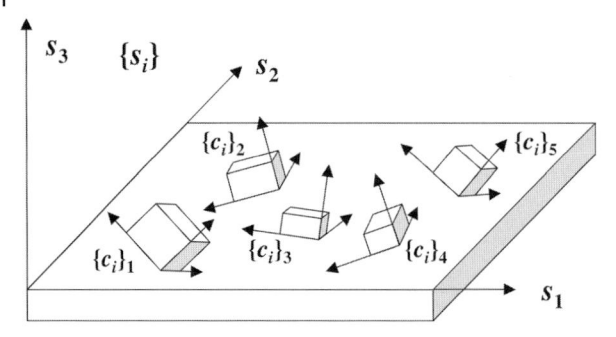

Figure 5.1 Sample reference frame $\{s_i\}$ and crystallographic reference frames $\{c_i\}_j$ of individual crystallites in a polycrystalline thin film.

lated. Instrumental Box 6 is devoted to the Euler cradle, which is a goniometer extension to the conventional $\theta/2\theta$ diffractometer and by which pole figures may be collected. It should be remembered that texture may appear as an evolutionary process in thin-film growth, i.e. preferred orientation is often observed to increase with increasing thickness. The issue of texture gradients is considered in Section 5.7.

Two kinds of textures play prominent roles in thin-film technology, with the first being the so-called fiber texture. In the case of a fiber texture the majority of crystallites is preferentially oriented with respect to a certain direction in the sample reference frame, which is denoted as the fiber axis. In an overwhelmingly large number of thin-film studies the fiber axis was identified to coincide with the substrate normal s_3. For a fiber texture the crystallites are left with one orientational degree of freedom, which is the angle of rotation around the fiber axis, and it may equivalently be denoted as a monoaxial texture. The second main case of a thin-film texture is the biaxial texture, where the orientation of crystallites is fixed along two directions in the sample reference frame. This state can be regarded as a prelude to an epitaxial film into which the polycrystal would merge for perfectly aligned crystallites. The transition between biaxial textured and epitaxial films thus is a continuous one and this relation is revisited in Chapter 7.

This chapter is illustrated by examples of fiber textures from the group of dielectric oxides that are in use and under development for piezoelectric and ferroelectric applications. In addition, the group of high-temperature superconductors (HTSC), intensively investigated since 1986, will be considered as examples of biaxial textures. The structures of the most prominent compounds, $BaTiO_3$ and $YBa_2Cu_3O_7$, are presented in Structure Box 6. For the majority of applications of dielectric and HTSC thin films the degree of preferred orientation is the decisive structural parameter for their proper function.

Structure Box 6:

Perovskite Structure and Derivatives

The prototype of this class of compounds is calcium titanate $CaTiO_3$. Other ternaries that also crystallize in the perovskite structure, like $SrTiO_3$ and $BaTiO_3$, have become increasingly important in various fields of modern technology, which also holds for the mixed perovskites like $(Pb,Zr)TiO_3$, where one of the two cation positions is randomly occupied by distinct metal atoms. An essential feature of these compounds is their tendency for structural phase transitions at certain critical temperatures, T_c, also named Curie temperatures. In the case of $BaTiO_3$, for instance, the perovskite structure is observed between 268 and 393 K and represents only one of a total of four structures the compound adopts between low and elevated temperatures.

The perovskite structure ABO_3 is characterized by crystallographic positions of non-inversional symmetry that are occupied by all atoms. At first sight, this fact might appear as a crystallographic detail of minor importance, in particular because the deviations from the centrosymmetric positions are rather small. It turns out, however, that this detail is the decisive ingredient for all physical properties on which the technological relevance of the perovskites is based. As an example, Fig. s6.1 depicts the unit cell of $BaTiO_3$ at room temperature with the central Ti atom surrounded by an oxygen bonding octahedron. The cell is recognized to be approximately cubic, but, indeed, this is only an approximation, because the c axis is slightly extended compared to the basal a axes and the structure in fact is tetragonal, $c/a = 1.0045$. The elongation of the unit cell of only 0.45% is accompanied by small shifts of Ti and O atoms in the z direction [1]. The shift of O atoms may be summarized by stating that they are all slightly shifted above the metal atom planes ($z \approx 0.02$ and 0.52 for O1 and O2, respectively). The Ti atoms are slightly displaced below the $z = 0.5$ plane. The structure factor reads

$$F_{perovskite}(hkl) = f_{Ba} + e^{\pi i(h+k)}\left[\begin{array}{l} f_{Ti}\exp\left[2\pi i\left(\frac{1}{2}+u_{Ti}\right)\right]+ \\ f_O\left\{\begin{array}{l}\exp(2\pi i l u_{O1})+ \\ \exp\left[2\pi i\left(\frac{1}{2}+u_{O2}\right)\right]\left(1+e^{\pi i(h-k)}\right)\end{array}\right\} \end{array}\right] \quad (s6.1)$$

The structure factor is simply the sum of the individual sublattices that enter almost noninteracting (no destructive interference). In consequence of the tetragonal elongation a splitting of all reflections is observed in the diffraction pattern, except for hhh-type ones (see the simulated powder pattern in Fig. s6.2). Depending on how strongly c/a deviates from unity the splitting might become difficult to resolve (see the two lowest reflections in Fig. s6.2). For $BaTiO_3$ at room temperature a splitting of $\Delta d = 4$ pm results for 100 and 001 that can be resolved by Cu Kα radiation for which the reflections are found at $2\theta = 22.04°$ and $22.26°$. The tetragonal splitting increases with increasing scattering angle, albeit in some cases the improvement might be balanced by the peak damping caused by too thin a film or by peak broadening due to microstructural defects.

On their positions in the polar unit cell all atoms undergo a polarization by nonvanishing crystal electrical fields. These fields do not cancel in the pervoskite structure as they

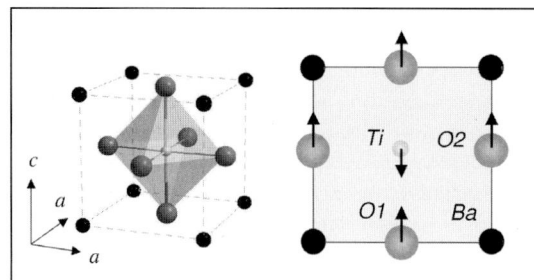

Figure s6.1 Three- and two-dimensional representation of crystallographic unit cell of perovskite-structured $BaTiO_3$. The z shift of Ti and O atoms have been indicated by arrows (right).

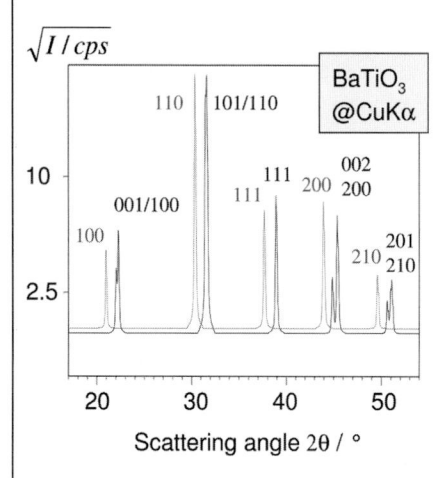

Figure s6.2 Simulated x-ray diffraction powder pattern of $BaTiO_3$ in perovskite (dark dots) and paraelectric cubic phase (gray dots). For clarity, the latter is shifted by 1° towards lower scattering angles.

do, for instance, in the sphalerite structure, which is determined by the point symmetry of atomic positions [2]. In addition, a single crystal composed of $BaTiO_3$ unit cells exhibits a measurable dipole moment, which is named pyroelectricity. In pyroelectric polycrystals the dipole moments may be aligned or even inverted by applying an external electrical field and in accordance with the notation of ferromagnetism this phenomenon is termed ferroelectricity. It is evident that by application of a mechanical load the induced strains may cause a variation of the dipole moment, which finally is called piezoelectricity. $BaTiO_3$ in the perovskite structure features all these interesting properties.

With increasing temperature the deviation of O atoms from the metal planes decreases until it fully vanishes at T_c = 393 K. At this critical temperature $BaTiO_3$ performs a phase transition from tetragonal perovskite to a cubic structure. All atoms occupy positions of inversional symmetry above T_c and neither pyro- nor ferroelectricity remains.

The cubic phase is thus denoted as paraelectric. Bragg peaks that were subjected to the tetragonal splitting in the polar phase merge for the paraelectric one (see the light gray dotted trace in Fig. s6.2).

Closely related to compounds crystallizing in the pervoskite structure are the high-temperature superconductors (HTSCs) like $YBa_2Cu_3O_7$. They also adopt crystallographic structures with oxygen and highly polarizable metal atoms on positions without inversional symmetry. In contrast to the perovskite structure materials, however, $YBa_2Cu_3O_7$ exhibits no measurable polarization, which may be understood from the fact that individual dipole moments in one unit cell compensate. Figure s6.3 shows the structure of $YBa_2Cu_3O_7$.

Some of the details of the perovskite structure are given in Table s6.1.

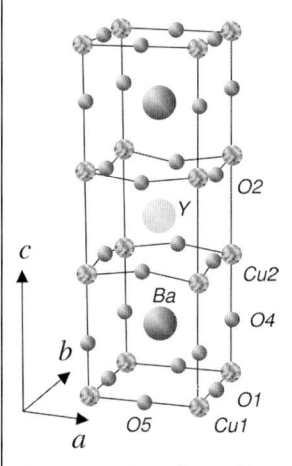

Figure s6.3 Crystallographic unit cell of orthorhombic $YBa_2Cu_3O_7$.

Table s6.1 Notations and symmetries in the perovskite structure.

Name (Struktur-bericht designation)	Space group (number)	Atomic positions		Example (unit cell edge a, c (nm))
Perovskite structure (E2$_1$)	P4mm (99)	A	0 0 0	$BaTiO_3$: 0.4000, 0.4018
		B	½ ½ ½+u_B	
		O1	½ ½ u_{O1}	
		O2	½ 0 ½+u_{O2}	

References

[1] R. H. Buttner, E. N. Maslen, Structural parameters and electron difference density in BaTiO$_3$, *Acta Crystallogr.* **B48** (1992) 764.

[2] M. Birkholz, Crystal field-induced dipole moments in heteropolar crystals: I. Concept, *Z. Phys. B* **96** (1995) 325 and II. Physical significance, *Z. Phys. B* **96** (1995) 333.

5.1
Texture Factors

Before these new structures are presented, an example is given that stems from the group of wurtzite compounds already introduced in Chapter 2. The example demonstrates the effect of texture on a $\theta/2\theta$ diffraction pattern for a thin ZnO film. Figure 5.2 displays a section of the $\theta/2\theta$ pattern with the three Bragg reflections 10.0, 00.2 and 10.1 of hexagonal ZnO. The appropriate ICDD histogram 00-036-1451 is superimposed onto the measured 2θ range and it is seen that predicted and observed intensities deviate significantly. Whereas 10.0 and 10.1 peaks are hardly recognized the 00.2 reflection contains almost all the scattering intensity in the scanned 2θ range. It can be concluded from this pattern and the vanishing intensity of reflections other than 00.l (not shown here) that most crystallites in the sample are oriented with their 00.l planes parallel to the substrate. Equivalently it may be stated that the $\langle 00.1 \rangle$ direction of most crystallites is parallel to the surface normal s_3. This 00.l-textured growth mode is typically observed for wurtzite-type thin films and is extremely pronounced for thin ZnO films in their doped and undoped variants.

The hexagonal wurtzite structure allows for piezoelectricity and enables the use of wurtzite-structured thin films in integrated electroacoustic devices. The property is based on the manipulability of the crystallites' extension by external electrical fields. The wurtzite structure allows piezoelectricity to occur in the nonbasal direction electing the c axis to become the polar axis. Electroacoustic devices are usually fabricated from single crystals of piezoelectric quartz, $LiNbO_3$ or $LiTaO_3$. Due to increasing demands for miniaturization, higher frequency devices and lightweight construction thin films have been promoted as substitutes in integrated devices as frequency filters, sensors, delay lines, actuators, etc. A schematic of a

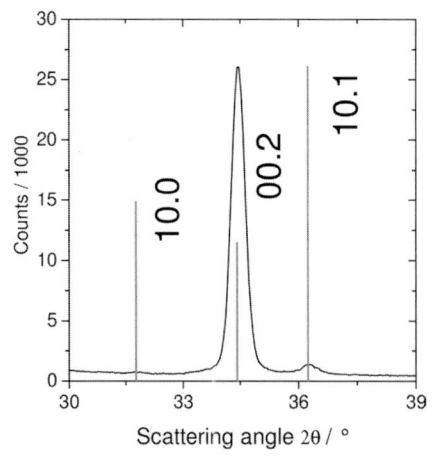

Figure 5.2 Symmetrical $\theta/2\theta$ diffraction pattern of a nominally 500 nm thin ZnO film.

surface acoustic wave (SAW) device used as frequency filter is shown in Fig. 5.3. Electrical signals are applied to one metal contact, conducted via surface elastomechanical waves along the ZnO film surface and received by the second contact. The transmissivity of the device is highly selective and is at maximum for frequencies close to the resonance frequency determined by geometrical constraints and the velocity of sound. It is obvious that the propagation losses in wurtzite-structured SAW devices are directly related to the misaligned volume fraction which should be reduced as much as possible.

It is evident from this example that the optimized preparation of piezoelectric thin films requires a technique to control the orientation of crystallites. An x-ray diffraction method for the determination of preferred orientation that has found widespread application makes use of strongly simplifying assumptions. The basic idea is to compare the intensities of a thin-film diffraction pattern measured in the symmetric $\theta/2\theta$ configuration with those from a powder sample exhibiting random orientation of individual crystallites. Since the method may be applied without tilting and rotating the sample, but simply by measuring a diffraction pattern in the symmetric configuration, the procedure has gained broad dissemination and can be found for many examples in the literature. The approach, of course, relies on the availability of integrated intensity data from a powder sample with randomly oriented grains. In many cases the intensity data from the ICDD set are accessed.

The starting point of the procedure is the quantitative expression of the integral intensity I_h^m of the measured reflection $\boldsymbol{h} = hkl$ in a $\theta/2\theta$ pattern given in Eq. (1.45), which is written here in the abbreviated form

$$I_h^m = SCF^m T_h |F_h|^2 m_h Lp A_{\theta 2\theta} \tag{5.1}$$

SCF^m is the instrument constant or scaling factor in the configuration applied to the actual measurement, while the other variables have their usual meaning. The absorption factor $A_{\theta 2\theta}$ has to be inserted in the adequate form of the symmetric $\theta/2\theta$ configuration (see Eq. (1.32)). The meaning of T_h is that of a weight factor or concentration factor that accounts for the degree of enhancement or reduction to which the hth reflex is subjected when compared with a powder sample. Therefore, $T_h = 1$ describes the weight of the Bragg reflection in a sample with random crystallite orientation. One may now express the intensity of a powder reflection in an analogous fashion by

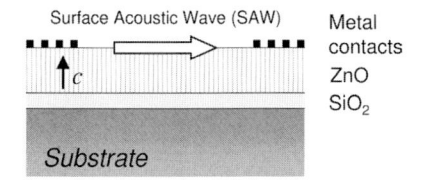

Figure 5.3 Schematic sketch of a thin-film surface acoustic wave (SAW) device.

$$I_h^{ICDD} = SCF^{ICDD}|F_h|^2 m_h LP \tag{5.2}$$

which differs from Eq. (5.1) by the absence of the absorption factor $A_{\theta 2\theta}$ since the ICDD-listed intensities all hold for samples of infinite thickness ($\mu t \gg 1$). It has to be pointed out that the absorption factor in Eq. (5.1) remains unconsidered in many thin-film studies published in the literature, which may be associated with large errors in the derived results.

In a second step the sum of integrated intensities $\sum_{h'} I_{h'}$ is calculated from all Bragg peaks observed in the 2θ range investigated. In order to get rid of the scaling factor the sum is normalized to a fixed value of 100, for instance, for both the thin-film pattern and the pattern of ideally random orientation. The ratio of $I_h/\sum_{h'} I_{h'}$ is subsequently calculated for both patterns and the texture factor T_h may be derived from the combination of both expressions to yield

$$T_h = \frac{I_h^m / A_{\theta 2\theta}(\theta_h)}{I_h^{ICDD}} \frac{\sum_{h'} I_{h'}^{ICDD}}{\sum_{h'} I_{h'}^m / A_{\theta 2\theta}(\theta_{h'})} \tag{5.3}$$

The determination of T_h factors by measuring the integral intensities and relating them to a model pattern offers a comparatively simple approach to obtain a first rough estimate of the degree of texture. It is for this purpose that texture factors have found wide dissemination in the literature. Of course, any other than ICDD intensity data I_h^m may be inserted into Eq. (5.3) provided that they account for an infinitely thick powder sample of ideally random orientation. Such data may be obtained by several methods like calculation of a powder pattern from single-crystal data according to Eq. (1.45), preparation of a powder sample or purchasing it from a commercial source and measuring it on the same diffractometer.

Because of its simplicity the texture factor technique is presented at the beginning of this chapter. However, the method should be applied with caution. On the one hand, in the day-to-day business of optimizing the textured growth of polycrystalline films, texture factors may supply helpful quantitative information on how deposition processes have to be tailored to arrive at orientationally improved samples. On the other hand, before using it the preferred orientation safely has to be verified as a fiber texture with coinciding fiber axis and substrate normal. Otherwise, completely misleading conclusions may be drawn from the texture factors. And in the case that a thorough quantitative description of the degree of texture or even the orientation distribution function is desired the procedures outlined in the following sections should be preferred.

5.2
Pole Figures

In a conventional $\theta/2\theta$ diffractogram the lattice planes contributing to a Bragg reflection I_h are all oriented parallel or nearly parallel to the thin-film surface. Only a subset of grains is monitored, for which the plane normal n_h lies in a cone of aperture angle δ around the substrate normal s_3, when δ characterizes the divergence of the x-ray bundle received by the detector. The selective detection of only a subset of grains in the $\theta/2\theta$ technique has already been visualized in Fig. 1.9. If the orientation distribution is to be determined, however, the reflections I_h of all grains in the sample must be evaluated. This is performed by scanning the reflection under various orientations, which is accounted for by the azimuth angle ϕ and the tilt angle ψ. The geometry of the experimental setup is shown in Fig. 5.4.

The angle of rotation ϕ measures the azimuth of the sample in the surface plane und has already been introduced in Chapter 2. The tilt angle ψ quantifies the amount of tilting of the sample normal s_3 with respect to the scattering vector Q or with respect to the scattering plane. For all $\psi \neq 0$ the diffractometer is out of the coplanar configuration and operates in a noncoplanar geometry. The variation of tilt angle and azimuth is performed by virtue of a so-called Euler cradle, which is described in more detail in Instrumental Box 6. The intensity of a Bragg reflection can now be measured under varying orientations yielding an intensity function I_h over the (ϕ, ψ) plane and the obtained intensity function $I_h(\phi, \psi)$ gives the numerical input for the evaluation of the orientation distribution.

There exist many possibilities for visualizing the function $I_h(\phi, \psi)$, but circular representations of the (ϕ, ψ) plane are generally used. Different ways of projecting a point P on a unit sphere with coordinates (ϕ, ψ) onto a unit circle are shown in Fig. 5.5. They all have in common that the two angles are restricted to the $0 \le \phi < 360°$ and $0 \le \psi \le 90°$ ranges. As usual, the azimuth ϕ is the circular distance from a fixed radius vector. Figure 5.5(a) shows a perpendicular projection that becomes rather close-mashed for higher ψ values, since iso-tilt lines scale with $\sin\psi$. The geometrical construction of the stereographic projection is demonstrated in Fig.

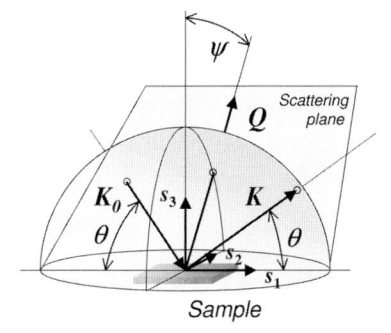

Figure 5.4 ψ tilt from the viewpoint of the sample reference frame. The scattering plane is tilted away from the surface normal s_3 towards the s_2 axis.

5.5(b). Here the tilting angle ψ is mapped into the equatorial plane by joining P with the south pole S and taking the distance from the origin to the piercing point P'. This distance turns out to scale with $\tan(\psi/2)$. In Figure 5.5(c) an equal area net is shown where the radial distance scales with $\sqrt{2}\sin(\psi/2)$. Finally, Fig. 5.5(d) displays a (ϕ, ψ) net with linear ψ scale. Except for the case of Fig. 5.5(a), the various (ϕ, ψ) nets are all in use for the visualization of pole figures. They all have distinct advantages and are used depending on which property is to be visualized for $I_h(\phi, \psi)$. It is separately mentioned in the following for each presented pole figure which net is used.

If I_h is plotted according to the stereographic projection, the $I_h(\phi, \psi)$ map is named an intensity pole figure. The naming is based on the use of the stereo-

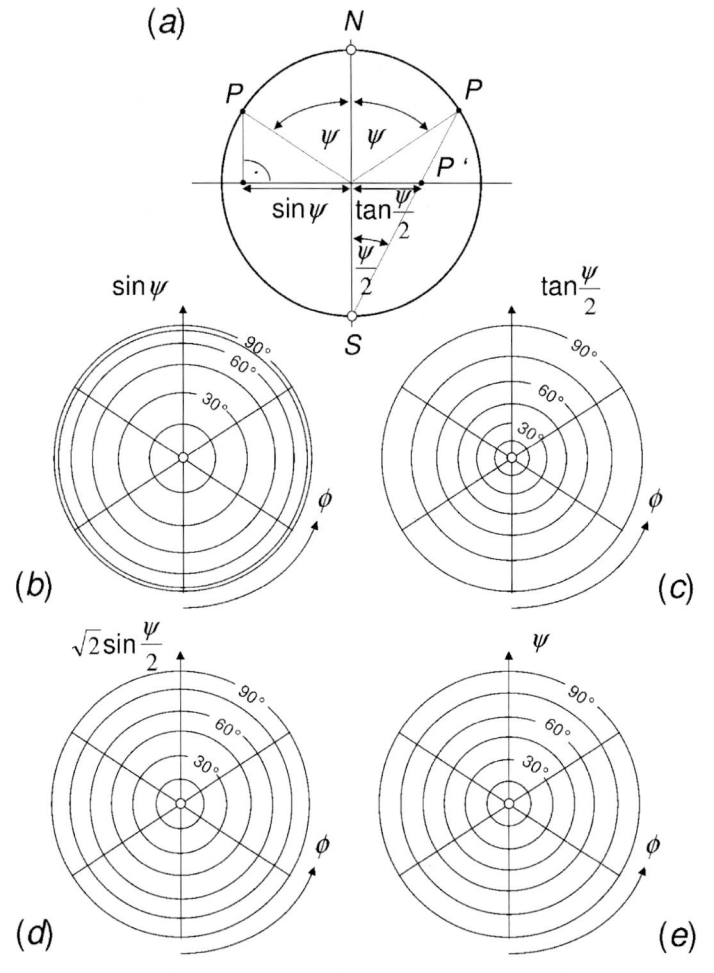

Figure 5.5 Several representations of the (ϕ, ψ) plane used in texture analysis. (a) Circular planes are derived from the projection of the unit sphere. (b) Perpendicular projection, (c) stereographic projection, (d) equal area projection, (e) constant $\Delta\psi$ projection.

Instrumental Box 6:

Euler Cradles

In a conventional $\theta/2\theta$ measurement the obtained structural information is not sufficient to derive the texture or the orientation distribution function of the grain ensemble. In a pole figure measurement two additional degrees of freedom are varied, which are the tilting angle ψ and the rotation angle ϕ (see Fig. i6.1). The variation of the azimuth ϕ accounts for the rotation of the sample around the substrate normal. Frequently, the tilt angle is denoted by χ instead of ψ, where both are related by $\chi = 90° - \psi$. A pole figure of a reflection (hkl) is monitored by measuring the integral intensity of the peak for all (ϕ, ψ) values.

A so-called Euler cradle realizes the rotation and tilt of the sample. This is a further goniometer stacked on the 2θ circle as visualized in Fig. i6.1. In most diffractometers carrying Euler cradles the usual θ circle is decoupled from the 2θ circle and may be varied independently. Instead of θ this further rotational degree of freedom is named ω and the complete system is called a four-circle diffractometer because of the four variable angles ω, 2θ, ϕ and ψ. For $\psi = \pm 90°$ the rotation axes of the three circles ω, 2θ and ϕ all coincide. Various types of Euler cradles are in use, which differ by the range of attainable ψ values. A Euler cradle with a closed ψ circle ($0 \leq \psi \leq 360°$) is shown in Fig. i6.1, but there are also ψ goniometers in use for which the maximum ψ is restricted to 180° or even 90° and which are called 1/2 and 1/4 circle cradles. For the high-symmetry crystal classes like the cubic ones measuring only a part of the full pole figures may solve the texture problem. In general, the crystal symmetry has the effect that the full texture information may already be given in subspaces of the full Euler space. In the cubic crystal system, for instance, the full ODF information is already contained in the $[0, \pi/2] \times [0, \pi/2] \times [0, \pi/2]$ cut of Euler space. Similarly, only fractions of the stereographic projection have to be displayed when fiber textures are to be presented by inverse pole figures. Figure i6.2 displays the symmetry-related areas of the stereographic projection for six crystal systems.

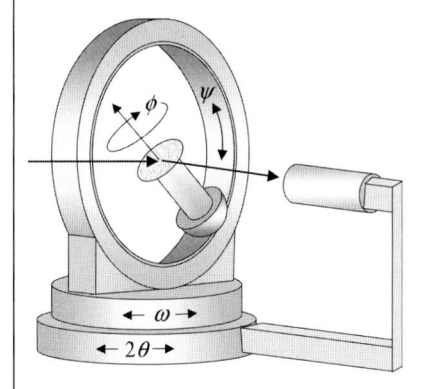

Figure i6.1 Four-circle Euler cradle with variable azimuth ϕ and tilt angle ψ.

The orientation distribution is sufficiently described by specifying $R(\Phi, \beta)$ in the shaded areas as it is often realized in the literature.

A broadening of the beam footprint on the sample surface occurs with the tilting of the sample and this defocusing increases with increasing $|\psi|$. If measurements have to be performed that make use of ψ tilts, the x-ray spot on the film surface has to be minimized to avoid too strong a defocusing. For investigations with an x-ray laboratory diffractometer the x-ray tube has to be adjusted to the point focus configuration. However, the confinement of the spatial extension of the probing x-ray beam to a small area causes a significant reduction in the number of crystallites scattering into the detector. Since a measured Bragg reflection at one (ϕ, ψ) point is only caused by a subset of all illuminated grains the effect may cause large statistical variations in the measured intensity [1]. Moreover, the effect may be especially severe in thin-film work where the number of measurable crystallites is further restricted by the film thickness t. In order to improve the counting statistics a translational oscillation of the sample should be carried out. This may be performed for most Euler cradles by virtue of a translation stage exhibiting two degrees of freedom, x and y, in the plane of the film surface. The sample then oscillates during the measurement of the pole figure $I_{hkl}(\phi, \psi)$ to evaluate a larger area of the sample surface. Before starting the measurement the sample height has to be adjusted for the purpose of which most Euler cradles enable a z variation of the sample table.

Pole figure measurements with point focus beams may be performed in the parafocusing geometry. It would be more advantageous, however, to make use of parallel-beam settings in order to reduce defocusing effects. For parallel-beam geometry in conjunction with a point focus configuration of the x-ray tube use should be made of the polycapillary optics as presented in Instrumental Box 5. Independent of what measures are taken to reduce defocusing the collected intensity data have to be corrected for it. This may be achieved by performing a ψ scan from a Bragg peak of a standard powder sam-

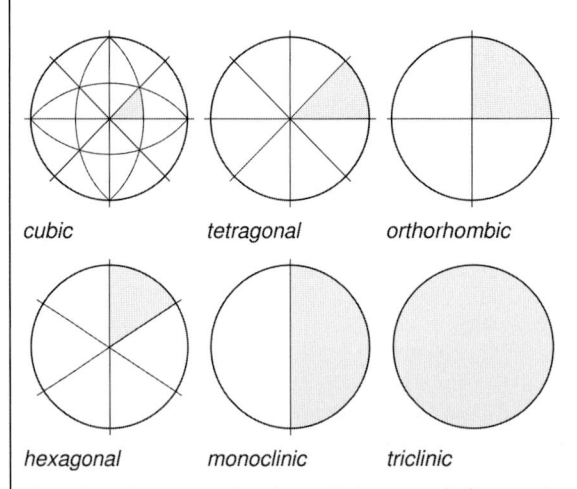

cubic *tetragonal* *orthorhombic*

hexagonal *monoclinic* *triclinic*

Figure i6.2 Symmetry-related areas in inverse pole figures of six crystal systems. The determination of the ODF is required only in the shaded subsets of orientation space.

ple with the same instrumental setting as applied to the thin-film measurement. The course of intensity as a function of ψ is given by the intensity for zero tilt $I(0)$ multiplied by the correction function $Corr(\psi)$. The latter should be used to correct the collected pole figure data.

The measurement of a pole figure may be performed by a successive sequence of (a) measurement of $I(2\theta)$ at fixed tilt angle ψ and azimuth ϕ, and (b) stepwise variation of the latter by increasing the actual value by $\Delta\psi$ and $\Delta\phi$, and repeating this sequence. Step widths of $\Delta\psi = 5°$ and $\Delta\phi = 5°$ are typically chosen. However, the setting of an equal step width in $\Delta\psi$ and $\Delta\phi$ is associated with a high density of measured data points close to the origin of the pole figure, while the information density would be diluted for areas of higher tilt angle ψ. For the measurement of equal data density pole figures so-called nearly equally distance (NED) grids were introduced enabling the determination of more evenly distributed intensity information from the scanned part of the (ϕ, ψ) plane [2].

References

Monographs relevant to this box

L. Wcislak, H.-J. Bunge, *Texture Analysis with a Position Sensitive Detector*, Cuvillier, Göttingen, 1996.

Special papers

[1] A. Mücklich, P. Klimanek, Experimental errors in quantitative texture analysis from diffraction pole figures, *Mater. Sci. Forum* **157–162** (1994) 275.
[2] K. Helming, A nearly equal distant grid of orientations for quantitative texture analysis, *Textures Microstruct.* **28** (1997) 219.

graphic projection for the representation of angles between macroscopic crystal planes for which the stereographic projection was originally developed. It became common practice that also $I_h(\phi, \psi)$ plots in the equal area and the constant tilt distance net are named pole figures and one should be aware of this fact. Intensity pole figures are displayed with respect to the sample coordinate system $\{s_i\}$. The angle of rotation ϕ has to be related to a defined direction, for the purpose of which the x axis is usually defined by the projection of the diffracted beam wave vector K on the sample surface in the coplanar configuration.

5.3
Measurement of Pole Figures

Pole figures of different Bragg reflections represent the basic experimental ingredients for the solution of the texture problem. Some important points have to be considered for their measurement.

The first point is concerned with the tilting of the sample by the angle ψ. The non-coplanar configuration deviates more from the parafocusing geometry as the tilting becomes larger. In consequence the x-ray footprint on the surface broadens and the x-ray beam becomes defocused. The question of how to correct for the effect of defocusing is dealt with in Instrumental Box 6. The tilted sample geometry in the pole figure measurement moreover necessitates the use of an absorption factor A_ψ different from those introduced for the $\theta/2\theta$ and the grazing incidence configuration. It is evident from the previous derivations that the variation of the azimuth angle ϕ will not cause a variation of absorption. Only the tilting of the sample by ψ will vary the path of the incident and diffracted beam in the sample. This variation may be accounted for by multiplying the configuration factor from the $\theta/2\theta$ configuration $k_{\theta 2\theta}$ by the factor $1/\cos\psi$

$$k_\psi = \frac{2}{\sin\theta \cos\psi} \quad (5.4)$$

The total path length 2ℓ of the x-ray beam within the sample becomes $2\ell = 2z/\sin\theta\cos\psi$, where z is the depth coordinate. The integral intensity is thus proportional to the integral over all depths z from the surface $z = 0$ to the interface $z = t$. Applying the same considerations as in the previous derivations it is obtained for the absorption factor in a pole figure measurement

$$A_\psi = \left[1 - \exp\left(-\frac{2\mu t}{\sin\theta \cos\psi}\right)\right] \quad (5.5)$$

In order to obtain integral intensities comparable with those of a powder pattern the measured I_h values have to be divided by this factor. Similar corrections have to be applied in the case where buried layers, multilayers or the pole figure of a covered substrate are to be elucidated (see Ref. [1] for a compilation of appropriate absorption factors).

The second point is related to the incompleteness of measured pole figures. It is evident from geometric restrictions that Bragg peak intensities cannot be evaluated for $\psi = 90°$. In fact, the defocusing increases steeply for values close to $\psi = 90°$ such that pole figures in reflecting geometry may reliably be measured only to a certain maximum tilt ψ_{max} the precise value depending on the actual beam optics. In the parafocusing geometry $\psi_{max} = 70°$ is often applied, while in parallel beam geometry ψ_{max} values of 85–89° may be achieved. The missing orientation information for $\psi > \psi_{max}$ in one pole figure $I_h(\phi, \psi)$ may be assessed via the measurement of further pole figures from other Bragg peaks. The additional information on the distribution of orientations stems from the crystallographic relation between the various lattice planes. For instance, a set of grains from pseudo-cubic BaTiO$_3$ that hypothetically might exhibit (100) planes occurring at $\psi = 90°$ may indirectly be monitored from the intensity of (111) reflections at $\psi = 90° - \angle(\mathbf{n}_{100}, \mathbf{n}_{111}) = 35.3°$. The many geometric relations that exist between different lattice planes allow for the compensation of incomplete orientation information from one pole figure $I_h(\phi, \psi)$ by the measurement of others. In Table 5.1 a listing of angles between the low-indexed planes in the cubic system is given. In lower symmetry crystal classes the angles will depend in a more complex fashion on unit cell parameters like the c/a ratio in the hexagonal system. The precise number of pole figures necessary for the determination of the ODF then depends on the crystal symmetry and the resolution desired [2]. For the cubic crystal system the evaluation of three pole figures is in most cases sufficient for determining the full ODF. Another possibility to supplement the missing information of an incomplete pole figure would be to also measure the reflection in transmission geometry. This will not be considered here, since in the majority of practical applications the necessary thinning of the substrate is undesirable, but one should be aware of this alternative.

Table 5.1 Angles between plane normals (**h**, **h'**) of low-indexed crystallographic lattice planes in the cubic crystal system.

h	h'	∠(h, h') (°)
100	100	0, 90
	110	45, 90
	111	54.7
	210	26.6, 63.4, 90
110	110	0, 60, 90
	111	35.3, 90
	210	18.4, 50.8, 71.6
111	111	0, 70.5
	210	39.2, 75
210	210	0, 36.9, 53.1, 66.4, 78.5, 90

In principle, texture measurements of thin films will proceed like those of volume material by evaluating several incomplete pole figures in the reflection mode. The course of the measurement of a pole figure as seen from the sample reference frame is shown schematically in Fig. 5.6. Usually, the measurement begins at the center $(\phi, \psi) = (0, 0)$ of the stereographic projection and subsequently changes to increasing tilt angles ψ_i, for each of which the azimuth is varied (ϕ_i, ψ_i) and the peak intensity is measured. Prior to the combined scanning in the (ϕ, ψ) plane the diffractometer is set to the Bragg peak position $\theta_B/2\theta_B$ under consideration. Pole figures from bulk samples are collected in the so-called open detector mode, i.e. a large receiving slit is inserted in front of the detector to safely monitor the full peak intensity when ϕ and ψ are varied. A linear background correction is employed by re-measuring the pole figure for scattering angles $2\theta_B - \Delta$ and $2\theta_B + \Delta$ well below and above the peak position and subtracting the mean of both intensities (see Fig. 5.7(a)). In total, the (ϕ, ψ) plane has finally been scanned three times, while the θ and 2θ positions were held constant.

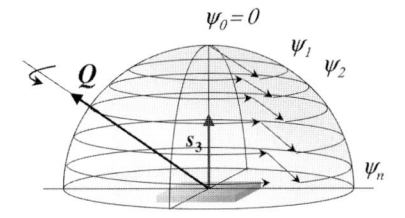

Figure 5.6 Course of the scattering vector during a pole figure measurement.

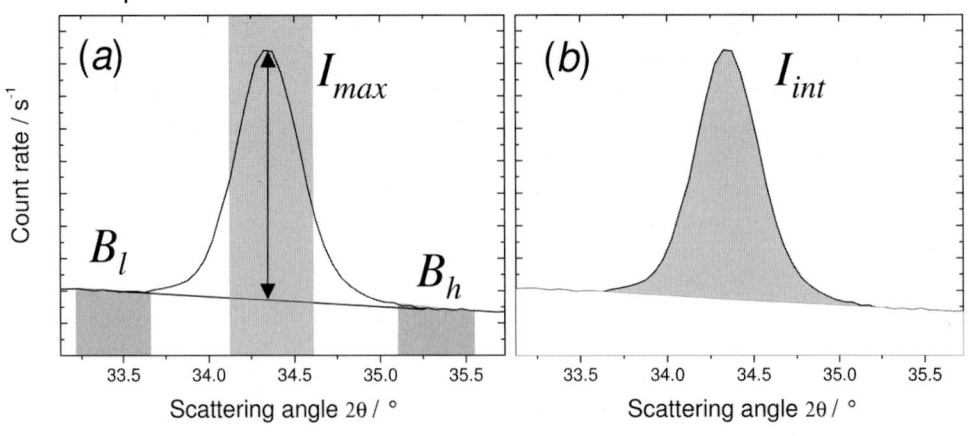

Figure 5.7 Peak intensity measurement in (a) conventional texture analysis and (b) as recommended for thin-film work.

It might be necessary to vary this procedure for the pole figure measurements of some thin-film samples. For one reason, the peak position may shift under varying ψ and ϕ due to residual stress in the film. The peak maximum $2\theta_0$ obtained for one (ϕ_i, ψ_i) couple may thus be found on the flank of the peak for another couple. For a second reason, the peak profile may vary in conjunction with orientation due to differences in film morphology parallel and perpendicular to the film normal. These problems are only reliably resolved by measuring the integral intensity of the Bragg peak for all points in the (ϕ, ψ) plane. The procedure is shown in Fig. 5.7(b). The reflection has to be integrated and the set of all integral intensities $I_h(\phi, \psi)$ represents the input for the pole figure plot. This procedure indeed requires more measurement time and expense on numerical analysis, but its use might be necessary in some cases of thin-film work to achieve reliable results.

As an example of pole figure measurements let us consider the investigation of thin-film samples of the Sr–Bi–Ta–Nb–O system from Saito et al. These authors made use of the pole figures for distinguishing between chemical phases in this complex material system [3]. The goal of this investigation was to identify low-temperature deposition conditions under which $SrBi_2(Ta_{1-x}Nb_x)O_q$ (SBT) can be prepared in the ferroelectric $A2_1am$ structure. Like other polar oxides such as $Pb_{1-x}La_x$-TiO_3, SBT has received attention for its possible use in nonvolatile ferroelectric random access memory (FeRAM) devices. The structure of ferroelectric SBT may be derived from perovskite $CaTiO_3$ by the vertical stacking of five unit cells, replacing the metal atoms in an ordered fashion with Sr, Bi and Ta and introducing oxygen vacancies. The unit cell thereby changes from tetragonal to orthorhombic symmetry having $a = 0.553$, $b = 0.554$ and $c = 2.50$ nm. The preparation of this chemically complex structure is complicated by the competition among several modifications. In the case of SBT, for instance, the excess of bismuth is associated with the formation of a fluorite phase (F) from the cubic space group $Fm3m$ having $a = 0.5446$ nm. In this structure metal atoms are distributed randomly over metal lattice sites and no ferroelectricity is observed which is understood from the centrosymmetry of $Fm3m$.

In the symmetric $\theta/2\theta$ configuration only a few Bragg peaks from the deposited Sr–Bi–Ta–Nb–O film can be observed, because the thin-film growth is optimized to yield highly preferred oriented grain ensembles (see Fig. 5.8(a)). The identification of chemical phases of the desired structure thus has to be performed from the few Bragg peaks dominating the diffractogram. In the case of the competing SBT and F phase system the 115 and 111 reflections occur at close-spaced 2θ values around 29° for Cu Kα radiation such that a phase identification becomes virtually impossible. The phase assignment becomes possible, however, by the measurement of the pole figures of this reflection that are shown in Fig. 5.8(b) and (c). The Bragg peaks observed in conventional $\theta/2\theta$ diffractograms correlate with the intensity at $(\phi, \psi) = (0, 0)$. The phase assignment is possible by consulting Table 5.1, where the angles between low-index lattice planes in cubic crystals are listed. It can be seen in Fig. 5.8(c) that in addition to the reflection intensity at the origin additional intensity occurs at tilt angles of about $\psi = 70°$. This corresponds with the angle between different (111) planes of 70.5° as given in Table 5.1. From this and additional considerations the authors could conclude that the sample represented by pole figure (c) consisted of the fluorite F phase, while the SBT phase was contained only in sample A, see Fig. 5.8(b).

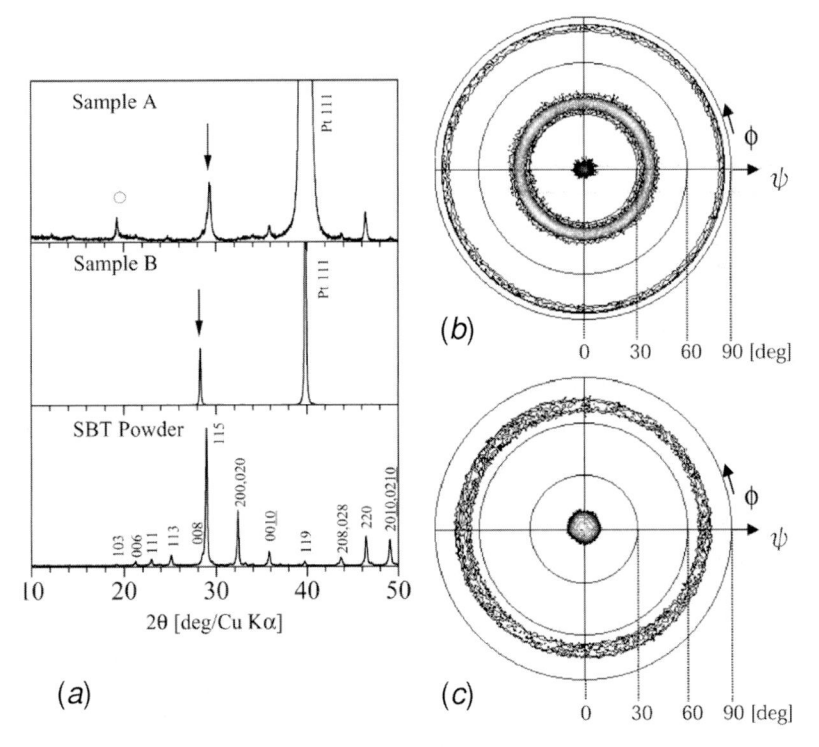

Figure 5.8: (a) $\theta/2\theta$ diffraction patterns and (b) and (c) pole figures from the Bragg reflection at 29° of polycrystalline Sr–Bi–Ta–Nb–O thin films (from Ref. [3]).

The pole figures in the plot are seen to be of rotational symmetry. This is a first example of a so-called fiber texture, the name of which is derived from the first observation of this texture symmetry in drawn-out and tempered metal wires or polymer fibers. The phenomenon is also named layer texture since it is often observed in thin films. In particular, if the deposition process is of rotational symmetry with respect to the substrate normal s_3 the preferred orientation of polycrystalline grains in most cases is of the same symmetry directing the fiber axis parallel to s_3.

5.4
Directions, Orientations and Inverse Pole Figures

The phrases orientation, orientation distribution function, direction, etc., have been frequently used in this chapter and their precise meaning was implicitly assumed to be understood. For the following considerations, however, it appears necessary to put them on a firmer basis. A unit sphere is shown in Fig. 5.9 for this purpose, which shall assist in clarifying the geometrical meaning.

Directions are represented by unit vectors starting at the origin and ending on the sphere surface. They are accounted for by two parameters, which might be given in the sample-related coordinate system $\{s_i\}$ by the angle of rotation ϕ and the tilt angle ψ having $0 < \phi < 360°$ and $-90 < \psi < 90°$. The more abstract object in Fig. 5.9, however, is that of an orientation. An orientation g refers to the location of one direction with respect to another one. Orientations can thus be represented by transformations between unit vectors that might be facilitated by virtue of Euler angles, see Fig. 5.10. As is well known from the mechanics of the rigid body, the three Euler angles φ_1, Φ and φ_2 transform a unit vector by a sequence of succeeding rotations from one coordinate system to another. The rotations are around the z axis, the new x axis and finally around the new z axis, which can be represented by the product matrix g of three matrices of rotat,ion A, B and C

$$g = ABC = \{\varphi_1, \Phi, \varphi_2\} \tag{5.6}$$

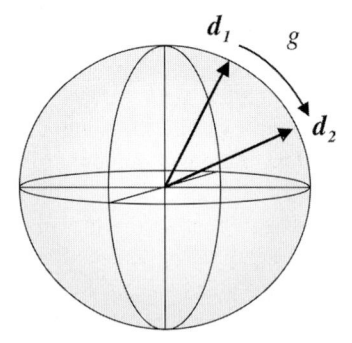

Figure 5.9 Visualization of direction vectors d_i and orientation g on the unit sphere.

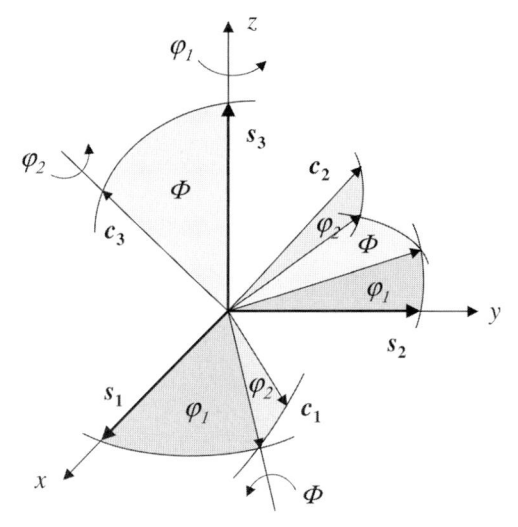

Figure 5.10 Schematic representation of the operation of Euler angles φ_1, Φ and φ_2 transforming the sample reference frame $\{s_i\}$ to one of the many crystallite reference frames $\{c_i\}_j$.

Orientations g thus have to be described by the three geometric parameters φ_1, Φ and φ_2 which is an important distinction from directions d_i. In fact, the rotation matrix g fully compares to the matrix a_{ij}^{SC} that transforms the sample reference frame $\{s_i\}$ into the crystallographic reference frame $\{c_i\}$ as it is introduced in Instrumental Box 1.

As has already been stated, in texture analysis one is mainly concerned with the orientation of individual crystallites from which a thin polycrystalline film is composed. We are now in a position to understand that the orientation distribution function $f(g)$ is a density function that relies on three parameters $f(g) = f(\varphi_1, \Phi, \varphi_2)$. It is evident from this fact that the ODF is not a simple geometrical object, but a function in a three-dimensional orientation space. The latter is also called Euler space, since it is built up from the three Euler angles. Euler space may be imagined as a rectangular box that is restricted by the definition ranges of φ_1, Φ and φ_2 being $0 \leq \varphi_1 < 360°$, $0 \leq \Phi < 180°$ and $0 \leq \varphi_2 < 360°$. The ODF has to be inscribed into this box and examples for general, three-dimensional textures are presented in later sections. The complexity reduces considerably if fiber textures are considered, since the orientation distribution becomes independent of φ_1 and thus only depends on Φ and φ_2. The latter angle is then transformed to the new angle coordinate β and both are related via $\beta = \varphi_2 + \pi/2$.

The ODF of fiber textures may therefore be described by a semipositive function above a two-dimensional area spanned by two orientation angles. We already faced the job of drawing such a function when looking for a graphical way of drawing pole figures $I_h(\phi, \psi)$ and it will thus use will be made of the same nets of orthogonal angle coordinates as introduced in Section 5.2. However, this time the graph is not in the sample reference frame $\{s_i\}$, but in the crystal reference frame $\{c_i\}$. The latter is an abstraction from all individual coordinate systems $\{c_i\}_j$ that may be assigned to each grain and that are aligned with the axes of crystallographic unit cells from which the crystallites are composed. On a first try the crystal coordinate system $\{c_i\}$ is difficult to imagine, since it gives an abstract space, whereas the coordinate systems $\{s_i\}$ and $\{c_i\}_j$ may be identified in real space as shown in Fig. 5.1.

The representation of orientation densities in the crystal coordinate system $\{c_i\}$ is named inverse pole figure, which can be understood from the following considerations. Normal pole figures are obtained by varying the sample's orientation with respect to the scattering vector Q by the variation of the goniometer angles ϕ and ψ. In the case of the inverse pole figure one takes a position in the crystal coordinate system $\{c_i\}$ and determines the directional angles Φ and β for which a grain diffracts the incoming beam into the detector. The volume density of grains over the angular (Φ, β) plane then yields the inverse pole figure $R(\Phi, \beta)$ for the crystal direction h when $h = (hkl)$ designates the Bragg reflection under consideration. It is shown in texture theory that the inverse pole figure $R(\Phi, \beta)$ represents the orientation distribution function of a fiber texture when the fiber axis is directed along the crystallographic direction h.

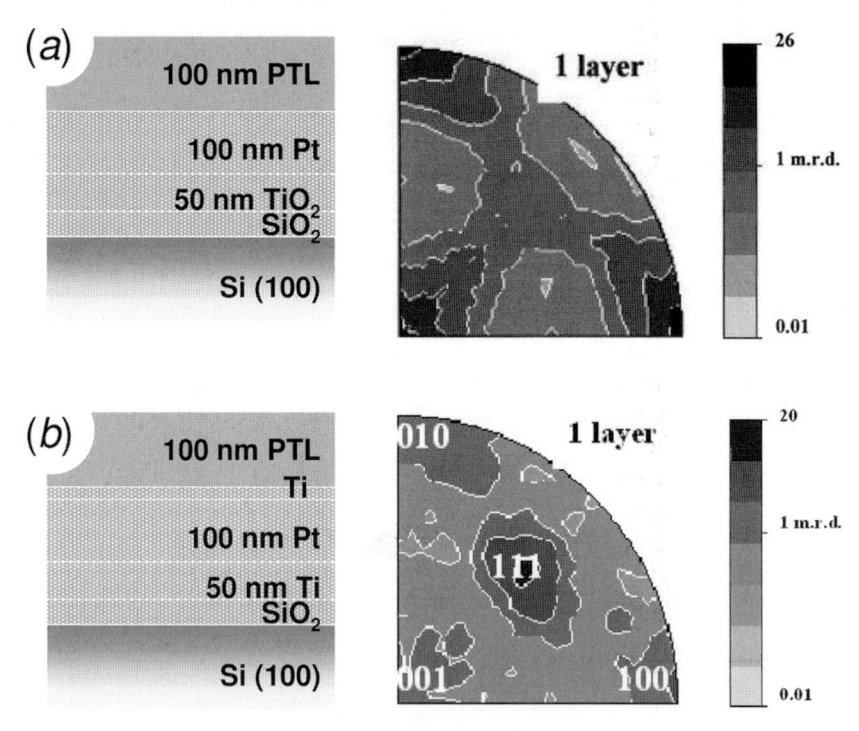

Figure 5.11 Inverse pole figures for the normal direction to the film surface of PLT films on (a) Pt/TiO$_2$ and (b) Ti/Pt/Ti (from Ref. [4]).

As an example for inverse pole figures the results obtained for La-modified PbTiO$_3$ films [4] are shown in Fig. 5.11. Because of the crystal symmetry or the sample symmetry it often suffices to restrict the (Φ, β) plane to a smaller area than the total plane. In the case shown here limiting to a quarter circle was chosen for drawing of the inverse pole figure with respect to the sample normal s_3 (equal area projection).

The thin Pb$_{0.88}$La$_{0.08}$TiO$_3$ (PLT) films were prepared by chemical solution deposition on different substrates. In the case shown in Fig. 5.11(a) the deposition was applied on Pt/TiO$_2$/SiO$_2$/Si, while a Ti layer of only 10 nm thickness was additionally sputtered onto the Pt layer in the case of Fig. 5.11(b). Platinum buffer layers are often used in the preparation of thin ferroelectric films because of the high oxidations resistance of Pt at the high temperatures of subsequent processing. As-deposited and amorphous PLT was crystallized in a temper step at 650 °C after the deposition. The crystallization was assumed to start from the PLT interface with the metallic buffer layer.

Ferroelectric thin films in general and PLT films in particular are intended for different applications depending on their preferred orientation. On the one hand, for FeRAMs applications the polar axis preferably lies in the film plane because of the absence of high remanent polarization and abrupt switching behavior. On the other hand, for pyroelectric sensor applications a ⟨001⟩ fiber texture is desired exhibiting the polar axis in the direction of the substrate normal. PLT crystallizes in the tetragonal perovskite structure and the polar axis is directed along ⟨001⟩. It can be seen from Fig. 5.11 that PLT deposition directly on Pt and subsequent crystallization causes a mixed ⟨100⟩⟨001⟩ fiber texture having both the unit cell a and c axes aligned s_3. The introduction of a Ti interlayer, however, leads to a ⟨111⟩ preferred orientation. The ODF values are plotted as iso-contour lines in units of multiples of the random distribution (m.r.d.). This dimensionality is widely used in the representation of textures, where an untextured sample would correspond to a constant value of 1 m.r.d. in the whole orientation space. Preferred orientation in part of the (Φ, β) plane must be balanced by missing ODF amplitudes in other parts which is expressed in the normalization condition

$$\int R(\Phi,\beta)\sin\Phi d\Phi d\beta = 1 \tag{5.7}$$

The example demonstrates how distinct preferred orientations may be designed for ferroelectric thin-film applications. The authors argue that in the case of the sample of Fig. 5.11(a) the mixed ⟨100⟩⟨001⟩ fiber evolves because of the lowest Pt–PLT interface energy for PLT (100) planes. Since PLT is in the cubic paraelectric state at the process temperature of 650 °C populations of a-, b- and c-oriented crystallites evolve in equal amounts and separate in crystallographically distinct fractions by crossing the critical temperature T_c during cooling. The significance of the chemical nature of the underlayer for the preparation of textured thin films becomes very obvious from this study.

5.5
Fiber Textures or Layer Textures

A rotational symmetry is obeyed in many depositions of thin films on flat substrates with the symmetry axis oriented along the substrate normal. Examples are given by the magnetron sputtering from circular targets or by plasma-enhanced chemical vapor deposition (CVD) processes, where the precursor gas is dissociated between two circular electrodes with one of them serving as substrate holder. Also in other deposition geometries, where no explicit cylindrical symmetry is obeyed, the substrate normal s_3 is often the one singled-out direction of the process while the directions in the film plane s_1 and s_2 are equivalent. In all of these cases, the observed texture often turns out to be axisymmetric with respect to the substrate normal, which is named fiber texture. Due to the large number of thin-film studies in which the phenomenon was observed it has also been named layer texture.

The occurrence of an axisymmetric texture considerably simplifies the formalism necessary to determine the orientation distribution function (ODF) from measured diffraction data. Also the requirements for data collection are significantly reduced if the preferred orientation is safely identified to be of axial symmetry. It then suffices to collect the Bragg peak intensities $I_h(\psi)$ only depending on tilt angle ψ instead of an evaluation of the full pole figure $I_h(\phi, \psi)$. The set of ψ-scanned peak intensities $I_h(\psi)$ for several Bragg peaks h serves as the experimental input for the evaluation of the fiber texture ODF. By elucidating $R(\Phi, \beta)$ one may also derive the texture factors T_h presented in Section 5.1. Three different approaches for solving this task are outlined in the following.

5.5.1
Harmonic Method

The first technique presented is known as the harmonic method. Two different notations are prevalent in the literature that go back to Bunge and Roe who independently developed and advocated the technique. Here, use is made of the Bunge notation, but it should be noted that Roe's approach also is widely used. The transition between both nomenclatures is given in Refs. [5, 6].

In the harmonic analysis the ODF is expanded into a sum of orthogonal functions such that the sum extends over two or three indices depending on whether the ODF accounts for a fiber texture or a general three-dimensional texture. The orthogonal basis functions depend on the Euler angles. In the case of a fiber texture the ODF reads

$$R(\Phi, \beta) = \sum_{l=0}^{\infty} \sum_{\mu=1}^{M(l)} C_l^{\mu} k_l^{\mu}(\Phi, \beta) \qquad (5.8)$$

where the $k_l^{\mu}(\Phi, \beta)$ are symmetrized spherical harmonic functions that depend on crystal symmetry and can be derived from the normalized spherical harmonics k_l^m

$$k_l^m(\Phi, \beta) = (2\pi)^{-1/2} e^{im\beta} \overline{P}_l^m(\cos \Phi) \qquad (5.9)$$

Here, $\bar{P}_l^m$ is a normalized associated Legendre polynomial defined by

$$\bar{P}_l^m(x) = \sqrt{\frac{(l+m)!}{(l-m)!} \frac{2l+1}{2}} \frac{(-1)^l}{2^l l!} (1-x^2)^{-m/2} \frac{d^{l-m}}{dx^{l-m}} (1-x^2)^l \tag{5.10}$$

The coefficients C_l^μ have the meaning of weighting factors accounting for the significance each individual $k_l^\mu(\Phi, \beta)$ function has for the ODF. In the framework of the harmonic analysis the complete solution of the texture problem consists in determining the set of C_l^μ coefficients.

The set of C_l^μ coefficients is constrained by the normalization condition of Eq. (5.7) that the fiber texture ODF $R(\Phi, \beta)$ has to obey. The normalization ensures that $R(\Phi, \beta)$ is directly given in units of m.r.d., i.e. in multiples of the random distribution. For a random OD all C_l^μ coefficients vanish except for the first, which reads

$$\text{Random orientation: } C_l^\mu = \begin{cases} (2\pi)^{-1/2} & l=0 \\ 0 & l \geq 1 \end{cases} \tag{5.11}$$

It is evident that in practical applications the ODF expansion must be truncated beyond a maximum l value that is denoted as the rank of the harmonic expansion l_{max}. The number $M(l)$ of nonvanishing basis function $k_l^\mu(\Phi, \beta)$ within each order l depends on the crystal symmetry. In the most general case, i.e. for triclinic crystal lattice symmetry, $M(l) = 2l + 1$ holds for each l, but the number decreases significantly for the high-symmetry crystal classes like the cubic or the hexagonal ones. This point is taken up again in Section 5.6.2.

For the application of the concept use is made of the decomposition of measured pole figures in a set of basis functions. In the case of fiber-textured samples the one-dimensional pole figures $I_h(\psi)$ can be approximated by a weighted sum of normalized Legendre functions $P_l(\psi)$

$$I_h(\psi) = \sum_{l=0}^{l_{max}} F_l(h) P_l(\psi) \tag{5.12}$$

with weighting coefficients $F_l(h)$. The set of $F_l(h)$ coefficients may be derived by fitting the measured pole figures with the sum of Eq. (5.12) in the known manner. These coefficients may be regarded as the condensate of measured data introduced into the analysis.

It is shown in the theory of harmonic texture analysis that pole figure coefficients $F_l(h)$ and ODF coefficients C_l^μ are related via

$$F_l(h) = \sqrt{\frac{2}{2l+1}} \sum_{\mu=1}^{M(l)} C_l^\mu k_l^\mu(h) \tag{5.13}$$

This system of equations represents the basic relation between the experimental input and the theoretical description of a fiber texture ODF.

The number of equations depends on the number of measured pole figures $I_h(\psi)$, on the order of the ODF expansion l_{max} and via $M(l)$ on the crystal symmetry. Of course, the number of equations also depends on the measured range of tilt angle ψ, which is in many cases restricted to values significantly smaller than 90°. For some cases the measurement of incomplete pole figures with rather small maximum tilt angles like 30–50° may cause an increase in the number of required pole figures. The system of equations may either be solved by inverting it or by performing a numerical regression in terms of the usual minimization of quadratic deviations.

Various computer programs are available for texture analysis in the framework of harmonic analysis. By specifying the crystal symmetry the symmetry-adapted basis functions $k_l^\mu (\Phi, \beta)$ are automatically generated and also the minimum number of pole figures that have to be measured is quoted by these programs. User input is demanded, however, for specifying the maximum order l_{max} up to which the ODF and the other equations given above are expanded. This point is not uncritical and deserves attention for a reasonable planning of the experiment. On the one hand, a fundamental criticism of harmonic analysis is concerned with the truncation error introduced by possibly too small a choice of l_{max}. On the other hand, procedures to estimate the truncation error are available. The question has not fully been resolved in the theoretical discussion, but in practice the concern is met by using large l_{max} values in the range of about 20 and even above.

A parameter accounting for the degree of texture may easily be derived from the ODF of harmonic analysis. This is the texture index J that is calculated from the sum of squares of C_l^μ coefficients

$$J = \frac{1}{4\pi} \sum_{l,\mu} \left[C_l^\mu \right]^2 \tag{5.14}$$

For a polycrystalline sample of random grain orientation $J_r = 1$, but for an increasing texture J takes increasing values. In case of a single crystal the texture index diverges versus infinity.

By virtue of C_l^μ coefficients one is also in the position to specify a theoretical expression for the course of the tilt intensity function that is based on the ODF. By combination of Eqs. (5.12) and (5.13) one obtains for the reflection of order h

$$I_h(\psi) = \sum_{l,\mu} \sqrt{\frac{2}{2l+1}} C_l^\mu k_l^\mu (h) P_l(\psi) \tag{5.15}$$

The expression can also be applied to calculate tilt intensities of unmeasured reflections once all ODF coefficients are known. Additionally, the texture factors T_h introduced in Section 5.1, Eq. (5.3), may now be derived from Eq. (5.15) that display the influence of preferred orientation on the integral intensities in a $\theta/2\theta$ diffraction pattern. The texture factors T_h are deduced from the fact that the tilt is zero $\psi = 0$ during the measurement of a $\theta/2\theta$ pattern and we thus have

$$T_h = I_h(0) = \sum_{l,\mu} C_l^\mu k_l^\mu (h) \tag{5.16}$$

This factor enters the product describing the integral intensity of a Bragg reflection to account for the effect of preferred orientation. In the case of a random orientation all T_h are equal to unity.

The availability of C_l^μ coefficients also allows specification of the volume share within the sample that belongs to a certain orientation. In the majority of applications one is interested in the volume share that is aligned along the fiber axis h_f. Since the ODF is a continuous distribution function the determination of volume fractions can only be derived by integrating the ODF in that special part of orientation space for which the appropriate sample volume is required. It therefore holds for the volume share that is oriented between the fiber axis and tilt angle ψ

$$V_f(\psi) = \int_0^\psi I_{h_f}(\tilde{\psi})\sin\tilde{\psi}\,d\tilde{\psi} \tag{5.17}$$

By inserting Eq. (5.15) into Eq. (5.17) the course of $V_f(\psi)$ may be determined which describes how strongly crystallites are aligned to the fiber. $V_f(\psi)$ in any case increases from zero at $\psi = 0$ to unity at $\psi = \pi/2$. The result $V_f(\pi/2) = 1$ is obtained for any arbitrary $I_h(\psi)$, which follows from the normalization and describes the inclusion of the total sample volume. An important parameter to be derived from $V_f(\psi)$ is the half population tilt $\psi_{1/2}$ indicating that tilt angle at which half the crystallite population is included. The smaller the $\psi_{1/2}$ value the more the crystallites are orientated to the fiber axis. Figure 5.12 displays the course of $V_f(\psi)$ and shows $\psi_{1/2}$ values for a hypothetical sharp and flat fiber texture and for a random orientation of crystallites.

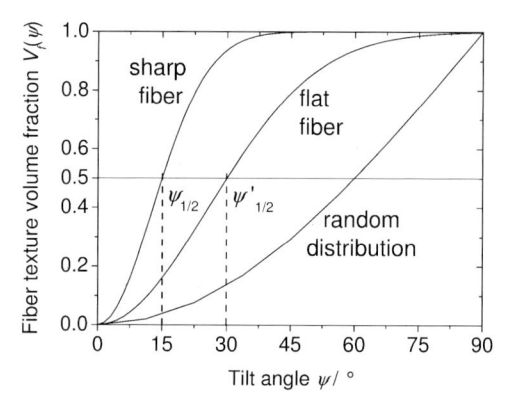

Figure 5.12 Fiber texture volume fraction $V_f(\psi)$ and half population angles $\psi_{1/2}$ for a sharp and a flat fiber texture and a random orientation distribution of crystallites.

5.5.2
Whole Pattern Techniques

A further strategy to evaluate the degree of preferred orientation is offered by whole pattern fitting, which has already been introduced with respect to microstructure in Section 3.5. In particular, if multilayer systems are under investigation the measurement of complete pole figures $I_h(\phi, \psi)$ or simple ψ scans $I_h(\psi)$ might be prohibited by strong peak overlap from layers of different sample depths. The advantage of the whole pattern fitting approach is that it principally allows elucidation of a combination of interesting parameters on the multilayer architecture as well as on the microstructure and texture of individual layers. A set of $\theta/2\theta$ scans under varying tilt angle ψ is performed in this method and the collected intensity data $I(2\theta, \psi)$ are subjected to a numerical regression as given in Eq. (3.71). For this technique the most significant quantities are again the texture factors T_h that represent the effect of preferred orientation on integral intensity. For each crystal symmetry there already exists a large set of angular relations (h, h') between the crystallographic lattice planes. The total fit of a symmetric $\theta/2\theta$ pattern would therefore rely on a variety of boundary conditions that have to be obeyed by the texture-including fit function. In an extension of Eq. (3.71) texture factors $T_h(\psi)$ are included in the model function

$$y(2\theta,\psi) = B(2\theta,\psi) + \sum_{i,h} T_h(\psi) I_h(2\theta) A_i(\theta,\psi) \tag{5.18}$$

where $B(2\theta,\psi)$ is the background function and the sum now extends over all observed Bragg peaks h of the set of all layers (index i). The absorption factor reads

$$A_i(\theta,\psi) = \left[1 - \exp\left(-\frac{2\mu_i t_i}{\sin\theta\cos\psi}\right)\right] \prod_{j=1}^{i-1} \exp\left(-\frac{2\mu_j t_j}{\sin\theta\cos\psi}\right) \tag{5.19}$$

It has to be emphasized that the texture factors in Eq. (5.18) now appear in their ψ-dependent form. This is different from their initial meaning introduced in Section 5.1, but may be intuitively understood to give the density of ψ-oriented crystallites in units of m.r.d. There are different strategies to solve for the texture factors. One possibility would be to analyze them in the framework of harmonic theory and then Eq. (5.18) would apply. However, other approaches have also been developed.

If $I(2\theta, \psi)$ data are available the degree of preferred orientation may also be derived by fitting with a model function that describes the textured volume fraction. A variety of model functions have been tested to be included in Rietveld fitting schemes, from which the model function introduced by March was found to be the most suited [7]. For the investigation of plastically deformed quartz rocks March developed a model, which assumes the crystallites in the sample to take either the form of flat disks or of elongated cylinders [8]. These forms derive from the nature of deformation processes and are characterized by the reduction parameter $r = t_i/t_f$ that depends on the thickness of crystallites before and after the deformation (see Fig. 5.24). In the case of rolling, r ranges between 0 and 1, and r values close to zero account for a high degree of texture. In the case of wire drawing, the deformation in fact results in an elongation of grains and $r > 1$. Both cases have in common that a fiber axis evolves that is oriented (a) in-plane in the first case (rolling texture) and (b) out-of-plane in the second one (fiber texture).

March moreover assumed the textured crystallites to be embedded in an untextured matrix and abbreviated their respective volume fraction by G and $(1 - G)$. The untextured fraction causes a constant background to be observed as a part of the fiber Bragg reflection at all tilt angles ψ. The March distribution function $T(G, r, \psi)$

now describes the density of crystallites in a certain direction which forms the angle ψ with the fiber axis having

$$T(G,r,\psi) = \frac{G}{(r^2 \cos^2 \psi + r^{-1} \sin^2 \psi)^{3/2}} + (1-G) \qquad (5.20)$$

The March function depends on three parameters: textured fraction G, reduction factor r and tilt angle ψ. It obeys several conditions, which are essential for every texture distribution function, like the normalization in the tilt angle range $0 \leq \psi \leq \pi/2$. The March function has found applications for a large set of material systems and proved to be of value in the quantitative description of textures. It has been applied in geology for texture formation in deformed rocks, in metallography for plastically deformed sheets and wires and recently for ceramic volume materials [9].

Distinctively deformed work pieces may be accounted for by the March function when appropriate values for the reduction parameter r are chosen. The degree of texture varies in an increasing degree as r tends towards the extreme values 0 and ∞. For thin films with pronounced fiber textures r values in the range $r < 1$ will mostly be relevant. In this sense the limits for $\psi = 0$ and $\psi = \pi/2$ may also be understood

$$T(G,r,0) = \frac{G}{r^3} + (1-G)$$
$$T\left(G,r,\frac{\pi}{2}\right) = Gr^{3/2} + (1-G) \qquad (5.21)$$

the precise value of which are governed by the texture degree G. The distinct exponents in r for the cosine and the sine term follow from the fact that there are two orthogonal directions present in the film plane and only one out-of-plane.

A valuable example of the whole pattern technique is represented by the investigation of a Ca-doped PbTiO$_3$ film deposited on a layer stack substrate [10] comparable to the stacking displayed in Fig. 5.11(a). The multilayer architecture to be refined finally becomes Pb$_{0.76}$Ca$_{0.24}$TiO$_3$/Pt/TiO$_2$/SiO$_2$/Si(100). The most severe overlap in the $\theta/2\theta$ pattern of this layer system appeared for the 111 reflection of the (Pb,Ca)TiO$_3$ and the Pt layer such that a pole figure measurement of the (Pb,Ca)TiO$_3$ 111 peak was prohibited. The combined approach applied in this study allowed for the quantitative determination of textures in both the (Pb,Ca)TiO$_3$ and the Pt layer of nominal thicknesses of 400 and 50 nm, respectively. Moreover, the crystallographic structure, layer thickness, average crystallite size and microstrain were refined simultaneously. Figure 5.13 displays the set of diffraction patterns obtained for various tilt angles and pole figures as recalculated from the refined parameters [10].

The model functions are moreover of interest, since they may establish relations between the observed texture and the physical cause for its formation. One may thus leave the purely descriptive investigation of texture to its physical understanding and technical manipulation. At the moment there have been only few investigations of thin films that made use of the March function, although it can be expected to (a) represent a simple and appropriate model for efficiently evaluating

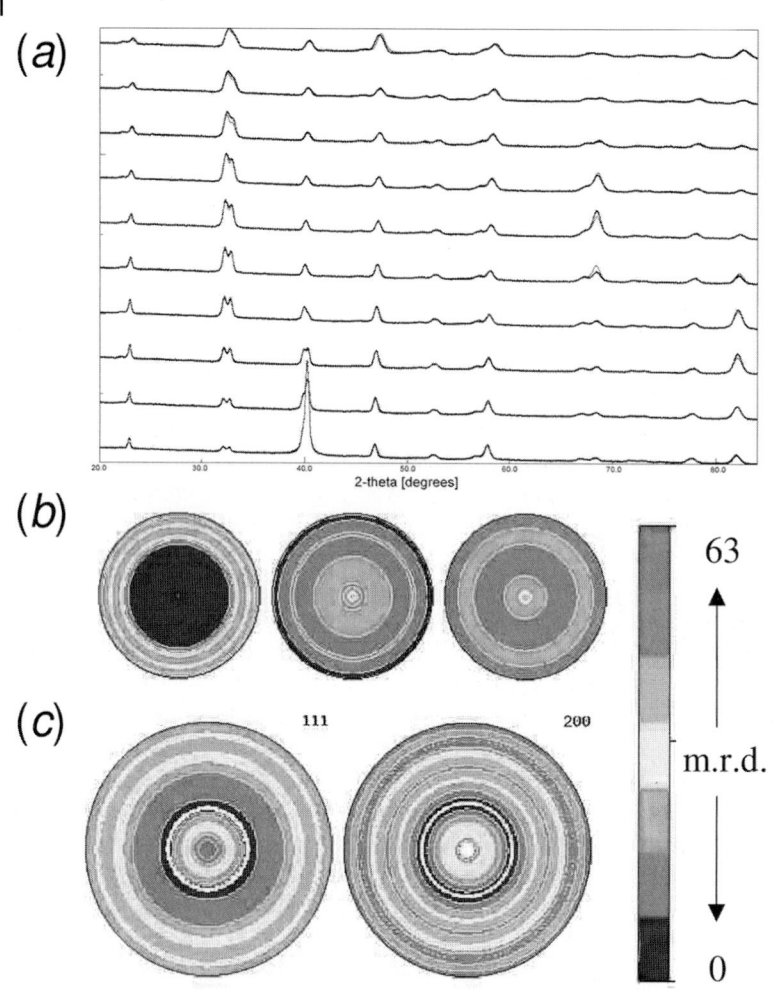

Figure 5.13 (a) Experimental (dotted lines) and refined patterns (solid lines) from a (Pb,Ca)TiO$_3$/Pt/TiO$_2$/SiO$_2$/Si(100) layer system for various tilt angles. Pole figures as recalculated from the obtained parameters for (b) 001, 100 and 111 reflections from (Pb,Ca)TiO$_3$ and (c) 111 and 200 from Pt layer, logarithmic intensity scale (from Ref. [10]).

thin-film textures and (b) to yield interesting results with respect to an improved understanding of texture evolution in thin films. The introduction of a distribution function like Eq. (5.20) into Rietveld refining schemes has the advantage that only few additional parameters are needed. The distribution of crystallite orientations in combination with the crystallographic relations represents an enormously strong boundary condition. Rietveld programs like MAUD (Ref. [56] in chapter 3) and GSAS [11] are available that include the regression of texture coefficients T_h into the total model function. However, the application of these techniques to thin films measured by x-ray diffraction has just begun.

5.5.3
Rocking Curves (ω Scans)

The whole pattern fitting approach presented above may valuably be applied in cases when only modest textures and/or strong peak overlap occur. For strong textures and diffraction patterns with only a few freestanding reflections another approach may be applied, which is the ω scan or rocking curve technique. Advantageously, this method does not require a Euler cradle and may thus be performed by virtue of any powder diffractometer that can operate with decoupled θ and 2θ circles.

The measurement of a rocking curve is performed such that the detector on the 2θ circle is fixed to the centroid position $2\theta_0$ of the Bragg peak under investigation, while the sample is tilted on the θ circle ("rocked") in the vicinity of the Bragg angle θ_0. The θ circle and the 2θ circle are thus decoupled and the θ angle is called ω in the case of its independent variation. There are two nomenclatures for ω in use. In the first notion, ω is measured with respect to the sample surface, i.e. $\cos\omega = (-K_0)(-s_1)/K_0$. This compares to the definition of the incidence angle α in grazing incidence configurations. The second convention relates ω to the surface normal s_3 and defines it as the deviation from half the scattering angle $\omega = \theta - 2\theta_0/2$ of the symmetric position. Both conventions are so widely in use that it apparently makes no sense to favor one over the other. In this section, use will be made of the second convention. It should be noted that the ω scan is frequently denoted as a θ scan, which is fully justified by the fact that θ is the varied angle during a rocking curve measurement.

A scheme of the measurement setup is shown in Fig. 5.14. It is seen that for fixed $2\theta_0$ the scan angle ω is restricted to the range between $-\theta_0$ and $+\theta_0$. For ω angles exceeding these limits the incoming/exiting beam would decline beneath the sample surface and no scattering information from the thin film would be collected. The result of the rocking curve measurement thus is a set of intensities $I_{rc}(2\theta_0, \omega)$ or in short $I_{rc}(\omega)$ measured at constant scattering angle $2\theta_0$ and varying ω. Similar to the symmetric $\theta/2\theta$ scan or a GIXRD measurement the rocking curve operates in the coplanar geometry, since s_3 remains in the same plane as K_0 and K. In con-

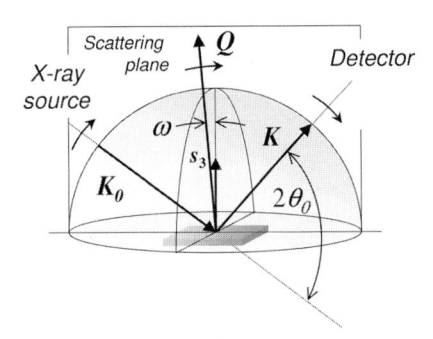

Figure 5.14 Rocking curve measurement (ω scan) from the point of view of the sample reference frame. The scattering angle is fixed to the centroid of the reflection, $2\theta_0$, while the rocking angle ω is tilted around θ_0.

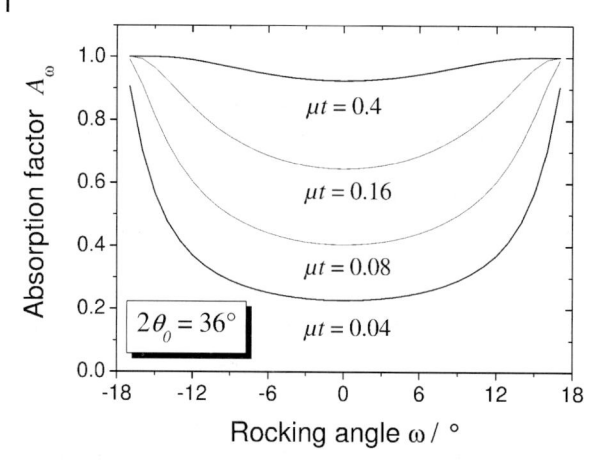

Figure 5.15 Absorption coefficient A_ω in the rocking curve configuration for samples of various thickness t, represented by varying μt products; calculated for scattering angle $2\theta_0 = 36°$.

sequence, the geometry factor k_ω and the absorption factor A_ω for the ω scan fully compare with the k_α and A_α factors of the GIXRD configuration except for the fact that the angles θ and α have to be replaced by θ_0 and ω in Eqs. (4.1) and (4.4). Figure 5.15 shows the course of some absorption factors A_ω as a function of rocking angle ω for a set of thin films exhibiting different μt products. It is seen from the plot that the attenuation may significantly influence the shape of the rocking curve and may cause an artificial broadening of $I_{rc}(\omega)$.

What a shape might be expected for a rocking curve $I_{rc}(\omega)$ of a fiber-textured sample when the Bragg peak of preferred oriented lattice planes is scanned? On the one hand, $I_{rc}(\omega)$ can be expected to be symmetric around $\omega = 0$. On the other hand, it is evident that the decrease of lattice plane density with increasing distance from θ_0 will cause a decrease of intensity in either direction. The shape of $I_{rc}(\omega)$ will thus resemble the intensity scan of a Bragg peak $I_h(\psi)$ for varying tilt angle ψ and both scans should comprise of the same information on preferred orientation. Similarly, one might expect to derive the fiber texture orientation function $T_h(\omega)$ comparable to $T_h(\psi)$. However, the loss of scattered intensity with increasing $|\omega|$ will also be due to defocusing. The intensity will thus decline as $|\omega|$ increases independent of the degree of texture. The FWHM obtained from a rocking curve may thus not be simply assumed to give an FWHM similar to the tilt angle $\psi_{1/2}$ within which half of the total grain population of the specimen is concentrated. This restriction is valid least when the diffractometer system is operated in the parafocusing mode. The effect of defocusing might be deconvoluted from the rocking curve by normalizing $I_{rc}(\omega)$ with respect to the rocking curve of an untextured sample $I_{rand}(\omega)$. The texture profile would then be derived from

$$T_h(\omega) = \frac{I_{rc}(\omega)}{I_{rand}(\omega)} \tag{5.22}$$

It can be concluded that in order to deal with defocusing a sample of random orientation has to be measured in addition to the specimen with preferred orientation. This appears a simple recipe, albeit in practice it might pose a severe restriction for the application of the rocking curve technique to thin films, because samples of fully random orientation are seldom available. It should also be noted that both the textured and the randomly oriented films would have to exhibit the same thickness in order for Eq. (5.22) to apply, otherwise a correction for their distinct absorption coefficient would have to be included.

In a series of papers Vaudin et al. recently introduced a method that makes use of Eq. (5.22) by normalizing with $I_{rand}(\omega)$ that derives from a measurement of the same sample in the symmetric $\theta/2\theta$ mode [12–14]. The approach is based on a thorough consideration of the x-ray beam divergence, which causes different absorption effects to occur in the symmetric $\theta/2\theta$ and the rocking curve mode. The method thus requires in addition to $I_{rc}(\omega)$ a conventional $\theta/2\theta$ scan $I_{pk}(2\theta)$ of the reflection under investigation. Both scans have to be collected with the same instrumental settings.

The basic idea of the approach is that the instrumental defocusing in the ω scan can be derived from an analysis of the Bragg peak measured in a conventional $\theta/2\theta$ scan. Following Vaudin et al., the rocking curve of a sample with random crystallite orientation can be derived by correcting the $I_{pk}(2\theta)$ scan by dividing by the absorption coefficient $A_{\theta 2\theta}$ of the $\theta/2\theta$ configuration and multiplying by the absorption coefficient A_ω of the ω scan. Due to the parafocusing beam geometry, however, one may not simply correct by multiplying with the $A_\omega/A_{\theta 2\theta}$ ratio, but both absorption factors have to account for the divergence of the x-ray beam. Because both scans are performed in the coplanar configuration it only has to be considered the in-plane divergence δ_p, while the axial divergence δ_a can be neglected. The geometry of the rocking curve scan with a divergent x-ray beam is visualized in Fig. 5.16. If the deviation of one beam path from the central beam is denoted by δ, we may write for the geometry factor $k(\omega, \delta)$ of this beam

$$k(\omega,\delta) = \frac{1}{\sin(\omega+\delta)} + \frac{1}{\sin[2\theta_0 - (\omega+\delta)]} \tag{5.23}$$

According to the usual integration along the depth coordinate it follows for the absorption factor valid for all beams that deviate by δ from the central beam

$$A(\omega,\delta) = 1 - \exp[-\mu t k(\omega,\delta)] \tag{5.24}$$

The same line of arguments is applied to derive the divergence-dependent geometry factor $k(2\theta, \delta)$ and the absorption factor $A(2\theta, \delta)$ in the symmetric $\theta/2\theta$ configuration for any beam that deviates by δ from the ideal $\theta/2\theta$ line

$$k(2\theta,\delta) = \frac{1}{\sin(\theta+\delta)} + \frac{1}{\sin(\theta-\delta)}$$
$$A(2\theta,\delta) = 1 - \exp[-\mu t k(2\theta,\delta)] \tag{5.25}$$

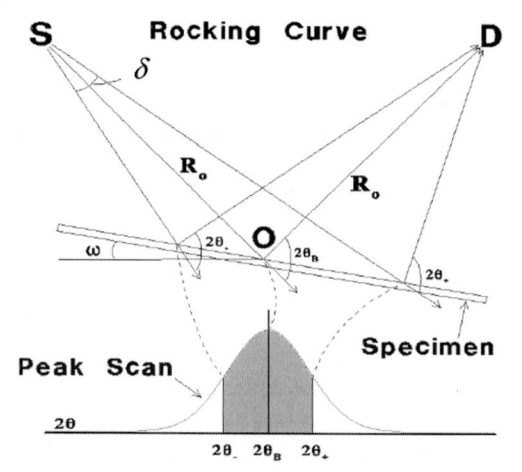

Figure 5.16 Schematic of the derivation of the divergence-dependent absorption coefficient $A(\omega, \delta)$ in the ω scan (from Ref. [13]).

In this form the absorption coefficients have to be used in order to correct adequately for defocusing. The corrected rocking curve is subsequently obtained by multiplying the measured $I_{pk}(2\theta)$ profile by the $A(\omega, \delta)/A(2\theta, \delta)$ ratio and integrating over all divergence angles δ defined by the slit setting, i.e. in the range between $-\delta_p/2$ and $+\delta_p/2$

$$I_{rand}(\omega) = \frac{1}{\delta_p} \int_{-\delta_p/2}^{+\delta_p/2} I_{pk}(2\theta) \frac{A(\omega,\delta)}{A(2\theta,\delta)} d\delta \qquad (5.26)$$

Before inserting $I_{pk}(2\theta)$ the intensity profile has to be relieved from any background intensity. By application of this procedure the rocking curve of a sample with a hypothetical random orientation of crystallites is achieved that can be inserted into the normalization Eq. (5.22) to derive $T_h(\omega)$. The normalized texture profile $T_h(\omega)$ follows from the condition

$$\int_0^{\pi/2} T_h(\omega) \sin\omega \, d\omega = 1 \qquad (5.27)$$

Finally, the texture profile has completely been determined from a rocking curve measured with a finite beam divergence. The formulas may be evaluated by numerical procedures available from Vaudin [14].

As an example Fig. 5.17 displays a symmetric $\theta/2\theta$ scan and three rocking curves from a thin $Ba_{0.7}Sr_{0.3}TiO_3$ (BST) film deposited on Pt/Si [13]. The film was seen from the $\theta/2\theta$ scan to exhibit a preferred orientation of (100) and (110) oriented grains and the pronounced reflections were selected for a closer examination by the ω scan technique. The results are given in the plot, where all rocking curves are normalized to the same maximum intensity. It is seen that (100) and (200) rocking curves are very similar, which is expected since they stem from the same lattice planes. Moreover, the FWHM of (h00) curves are distinct from that of (110) while their absolute values of about 2.8° and 3.6° signifies the (h00) fiber to be sharper than the preferential (110) orientation. Due to the very sharp texture fibers the FWHM of the texture profiles compared with the FWHM of the rocking curves as was revealed by applying the analysis technique outlined above.

It is concluded that the rocking curve technique may be applied to derive the texture profiles of sharp axisymmetric ODFs from rocking curves and a single peak scan. A Euler cradle is not required. Since the absorption factors depend on the μt product of the investigated sample, the precision by which μt is known sets the lower limit to the precision by which the texture profile may be derived by this method. The texture profiles derived may be tested with respect to physical model functions like the March distribution function presented in the previous section.

At this point the introduction of axisymmetric textures and methods for elucidating them come to a preliminary end. In the following section we are concerned with more complex, three-dimensional textures. It turns out that fiber textures also play a role in this field as will soon be realized and that we have made a valuable step into the field by developing the concepts to elucidate this phenomenon.

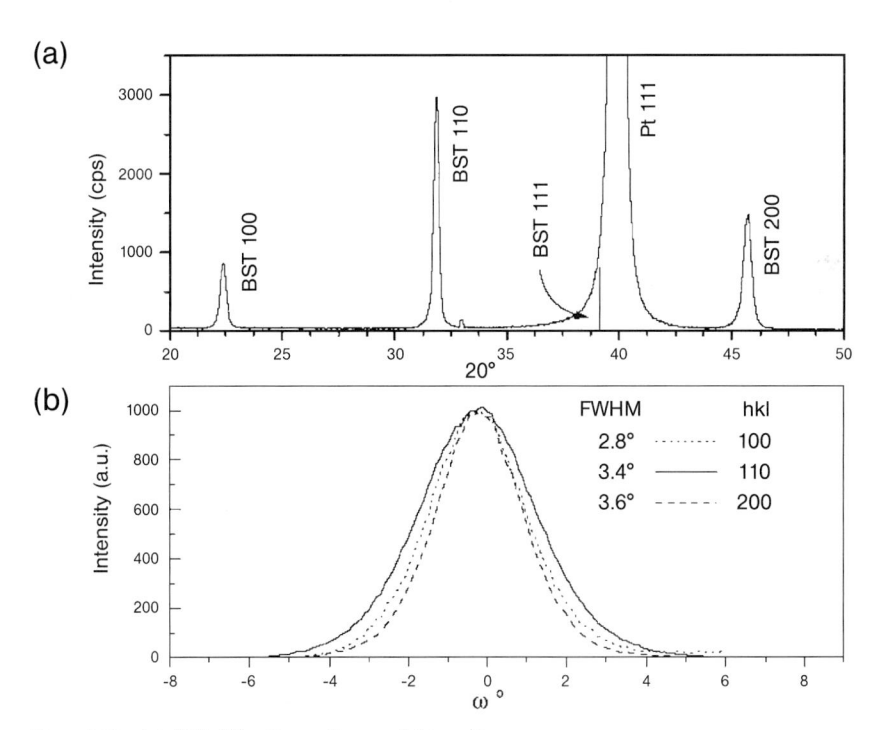

Figure 5.17 (a) $\theta/2\theta$ diffraction pattern and (b) rocking curves from thin $Ba_{0.7}Sr_{0.3}TiO_3$ (BST) film (from Ref. [13]).

5.6
Biaxial and Fully General Textures

It has already been mentioned that texture may be considered as a measure for the transition of a specimen from the polycrystalline to the single-crystalline state. For the case of a fiber texture the crystallites are ideally oriented along a common axis in the sample reference frame accounted for by a range of tilts $\Delta\psi$ to which the majority of the grain population is confined (see Fig. 5.18(a)). There remains an orientational degree of freedom for each crystallite, which is the azimuth ϕ or the angle of rotation around the fiber axis. An even higher state of orientational order is achieved if the azimuth is also restricted to a certain value ϕ_0 or to a small range $\Delta\phi$ of allowed azimuths as shown in Fig. 5.18(b). There are some important examples in thin-film technology where such a high degree of ordering in a – still polycrystalline – grain ensemble is important for the suitable functioning of the device. It is possible to attempt the preparation of such biaxially textured samples. Luckily, in thin-film deposition the first axis may often automatically be given by the substrate normal. Then, measures only have to be taken to define the second axis of preferred orientation. This may be achieved by several techniques. For instance, an inclined beam of positive ions may be directed to a biased substrate or the technique of oblique deposition may be applied. Ideally, the preparations are carried out upon substrates or layers that already exhibit a certain degree of biaxial texture. According to the details of the deposition process and the material system in use one may end up with a polycrystalline film where the orientation distribution of crystallites is confined to small ranges of twist and tilt, $\Delta\phi$ and $\Delta\psi$, as these structural parameters are denoted.

A recent example of such a thin-film system is given by high-T_c superconducting YBa$_2$Cu$_3$O$_7$ (YBCO). The material is developed as a thin-film deposit on metallic tape used as superconducting cable, cryoelectronic data transmission wire and in

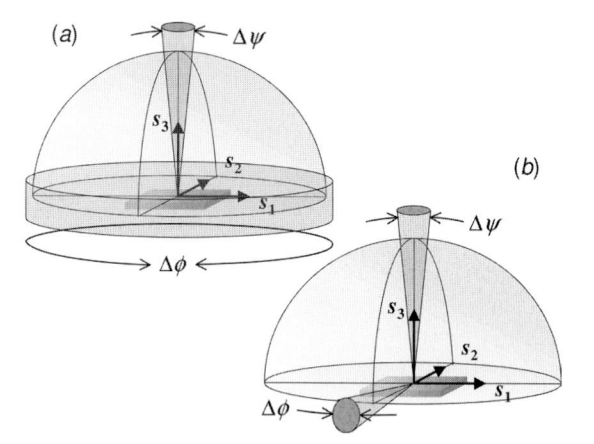

Figure 5.18 Schematic representation of the two most prominent textures in thin films given in the sample reference frame:
(a) fiber or layer texture and (b) biaxial texture.

5.6 Biaxial and Fully General Textures

other applications. A structural presupposition that has to be fulfilled to make use of the superconducting properties of YBCO films is that the set of crystallites has to be highly aligned. The highly parallel alignment is necessary for the superconducting current to surmount the grain boundaries in the polycrystalline material. In the case of too strong a misalignment the grain boundaries act as a weak link such that the current density is strongly damped for large distances. Figure 5.19(a) shows the dependence of the critical current density J_c in a thin YBCO film as a function of the average misorientation angle.

It is evident from Fig. 5.19 that the ideal YBCO cable would be a single-crystalline film. Obviously, such a material of a length of some kilometers cannot be produced and alternative preparation procedures have to be established. Thin-film preparation techniques for high-temperature superconductors thus aim at creating a highly ordered polycrystalline grain ensemble with minimized twist and tilt, $\Delta\phi$ and $\Delta\psi$. Two main approaches are currently under investigation that both make use of depositing the superconducting layer on a textured template assisting the intended biaxial growth. While the first approach starts with the deposition of increasingly textured oxide layers, the second approach makes use of an already textured metallic tape as substrate. Figure 5.18(b) displays schematically the material stacking in the first approach realized by the ion beam-assisted deposition (IBAD) technique [16]. A set of oxide buffer layers is deposited upon the substrate to accommodate

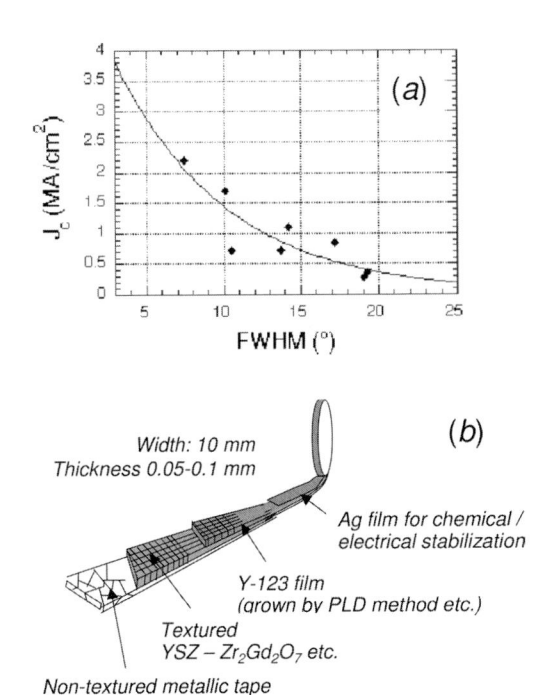

Figure 5.19 (a) Critical current density J_c in YBCO thin film as a function of average misorientation (from Ref. [15]). (b) Schematic view of YBCO-coated conductor formed by ion beam-assisted deposition (IBAD) (compare Ref. [16]).

the YBCO lattice parameters with the degree of crystallographic orientation thereby increasing within each layer of the composite. In the following, examples from the investigation of these material systems will be presented to introduce the x-ray diffraction techniques that may be applied to specify biaxial and fully general textures.

5.6.1
Azimuthal Scans (ϕ Scans)

Figure 5.20(a) displays a symmetrical $\theta/2\theta$ scan of a YBCO film exhibiting a high degree of c-axis alignment [17]. Practically all crystallites in the sample have their c axis oriented perpendicular to the substrate surface as is seen from solely the 00l reflections occurring in the diffraction pattern. Figure 5.20(b) shows the corresponding ICDD histogram 00-038-1433 of $YBa_2Cu_3O_7$ given in the $a < b < c$ setup of the orthorhombic unit cell with a logarithmic intensity scale. It can be seen by comparison of the two patterns that a highly selective growth process is operative excluding all hkl planes to be oriented parallel to the substrate plane if the condition $hkl = 00l$ is not obeyed.

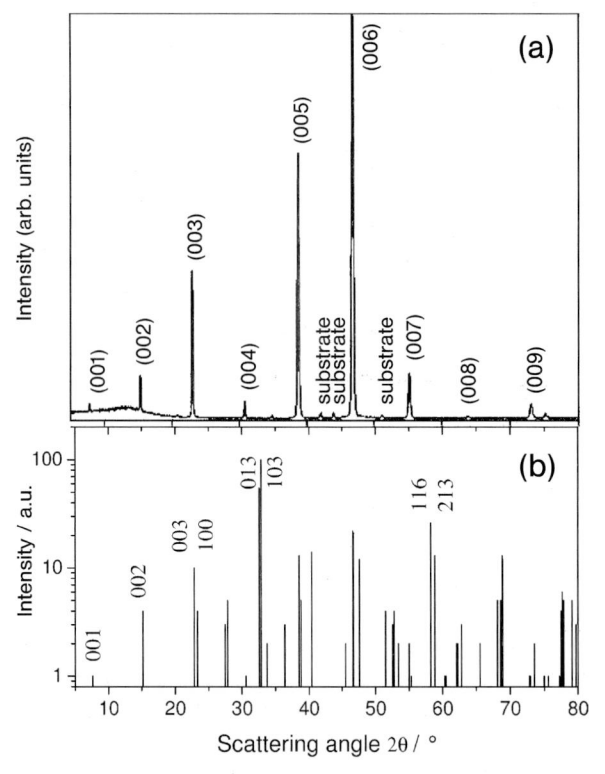

Figure 5.20 (a) $\theta/2\theta$ diffraction pattern of highly c-axis oriented YBCO film (from Ref. [17]). (b) ICDD stick pattern 00-038-1433 of YBCO with some of the Bragg peaks accounted for by their Miller indices.

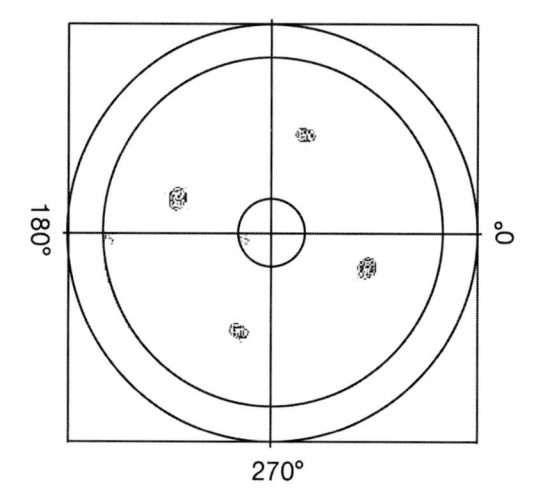

Figure 5.21 Pole figure of YBCO 103 Bragg reflection observed for $2\theta = 32.8°$. Intensity projected onto an equal-area net (from Ref. [15]).

The measurement of symmetric diffraction patterns is the first step in the structural characterization of YBCO films with an intentional high degree of texture. If only $00l$ reflections are present in the pattern one has successfully overcome the first obstacle. However, such a pattern may also signify a perfect layer texture with a random azimuthal orientation of single crystallites. Since also the azimuthal misorientation of crystallites is a critical feature techniques have to be applied that enable the determination of the azimuthal degree of ordering.

A method that might be applied for this purpose is the pole figure measurement of an $h0l$ reflection. For a perfectly $00l$-textured YBCO film this reflection might only be assessed in a tilted configuration, i.e. $\psi \neq 0$. This is evident from the $\theta/2\theta$ pattern displayed in Fig. 5.20, where the 103 peak is missing, but which would occur in a Cu Kα powder pattern for $2\theta = 32.8°$. The angle between [001] and [103] may be calculated from the YBCO lattice constants in a way comparable to Table 5.1 and is found to amount to $\arctan(c/3a) = 45.1°$. If a pole figure is collected for $2\theta_{103}$ the reflections will accordingly be observed for this tilt angle ψ. This is seen from Fig. 5.21, where such a pole figure is given for a YBCO film deposited by pulsed laser deposition on a MgO buffered metallic substrate [16]. In contrast to the fiber texture pole figure shown in Fig. 5.8, the YBCO 103 reflection intensity is not evenly distributed on a ring around the origin, but is concentrated to four spots accounting for the different set of lattice planes (103), (013), ($\bar{1}$03) and (0$\bar{1}$3). It can be concluded from the intensity distribution that the polycrystal is biaxially textured. In addition, the width of the intensity peaks $\Delta\phi$ and $\Delta\psi$ represent a measure of the degree of parallel alignment in the crystalline ensemble.

Instead of measuring complete pole figures one often proceeds by measuring only that part of the complete (ϕ, ψ) plane from which the information is desired. This might be performed by setting the tilt angle ψ of the diffractometer to a constant value appropriate for the $h0l$ reflections under study and solely varying the azimuth angle ϕ. Such a measurement of intensity as a function of ϕ for constant ω, 2θ and ψ is called a ϕ, scan. The diffractometer movement for a ϕ scan in the sample reference system fully compares with the ϕ averaging displayed in Fig. 2.10, except for the distinction that in a ϕ scan the intensity is collected as a function of azimuth angle $I(\phi)$. One always has to consider carefully the setting of divergence and detector slits and the question of defocusing when experiments in nonparallel beam optics are performed. The same recommendations hold for a ϕ scan as were given above and in Instrumental Box 6 for the measurement of pole figures.

Figure 5.22 shows the ϕ scan of a 103 YBCO reflection exhibiting only a small value of $\Delta\phi \approx 7.4°$ for the full width at half maximum [15]. This parameter serves as a quality parameter in the preparation of YBCO thin films. As a rule of thumb it has been established that critical current densities in excess of 10^6 A cm^{-2} can be achieved when the FWHM of a 103 ϕ scan with Cu$K\alpha$ is less than 10°.

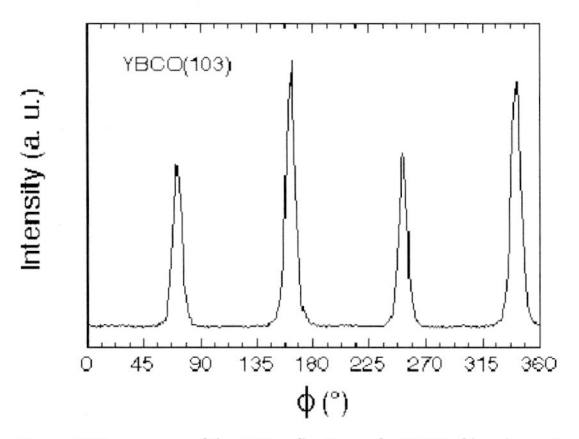

Figure 5.22 ϕ scan of the 103 reflection of a YBCO film deposited on an oxide buffer layer (from Ref. [15]).

5.6.2
General Orientation Distribution

In the two texture types presented so far rather restricted preferred orientations occur. In the layer texture a certain lattice plane vector n_h ideally is perfectly aligned along the substrate normal s_3 and in the biaxial texture a further lattice plane vector $n_{h'}$ is preferentially oriented within the layer plane (s_1, s_2). Much more general types of texture may arise in a polycrystal that require the full theoretical texture apparatus. In the case of a general texture the solution of the texture problem is by specifying the orientation distribution function in full Euler space. Various possibilities exist for the choice of suitable Euler angles. Use will be made here of the convention given in Fig. 5.10, i.e. the three Euler angles φ_1, Φ, φ_2 account for the transition from $\{s_i\}$ to $\{c_i\}$. The ODF will be symbolized by $f(\varphi_1, \Phi, \varphi_2)$. The Euler space is a rectangular box with the three Euler angles φ_1, Φ and φ_2 that give the finite orthogonal coordinates (see Fig. 5.10). In the general case the coordinate axes extend to 360°, 180° and 360°. According to the actual symmetry constraints, subsections of Euler space may suffice to represent the complete orientation information. For the highly symmetric cubic crystal classes, for instance, 1/32 of the full Euler space is sufficient to account for the full orientation distribution, since it is related via symmetry operations to the subset in $0 \leq \varphi_1$, Φ, $\varphi_2 < 90°$.

The ODF to be inscribed into Euler space may assume rather complex and complicated patterns. Figure 5.23 gives some special ODFs like the random distribution, a biaxial texture and a fiber texture. For the random case all orientations are of equal probability and have be represented by a homogeneous density. A biaxial texture is described by a three-dimensional Gauss function with standard deviations accounting for twist $\Delta\phi$ and tilt $\Delta\psi$. A fiber texture appears as a cylinder in Euler space. For more complicated distributions the ODF is often given in the form of iso-contour plots that are inscribed into plane sections cut from the Euler space by setting one of the angles constant. In many studies the Euler space sections are given for constant φ_2 and the iso-contour lines are displayed for varying φ_1 and Φ.

It might appear a triviality, but it has to be emphasized that the sample reference frame has to be defined by the analyst when an ODF analysis is to be employed. In the metallographic branch of texture analysis a convention of naming the axes RD, TD and ND has become widespread. The notation derives from the investigation of cold-rolling of metal sheets, which were found to exhibit pronounced rolling textures with respect to a sample reference frame defined by the rolling direction (RD), the transverse direction (TD) and the normal direction (ND) as is visualized in Fig. 5.24. The sequence of Euler angle rotations in this frame starts by a rotation φ_1 around ND and proceeds further as shown in Fig. 5.10, when s_1, s_2 and s_3 are identified with RD, TD and ND. In Figure 5.25(a) a 111 pole figure from a rolled Ag ribbon is shown that had been plastically deformed by $t_i/t_f = 15/0.34$ or 98% [18]. The intensity distribution is realized to be symmetric with respect to RD. Rolling textures are denoted by $\{hkl\}\langle uvw\rangle$ symbols specifying the plane $\{hkl\}$ that is preferentially oriented parallel to the surface and the direction $\langle uvw\rangle$ in the rolling direction. In the stereographic projection in Fig. 5.25(b) the intensity maxima are indicated that would theoretically be observed in a 111 pole figure for several types of rolling textures $\{hkl\}\langle uvw\rangle$. By comparing the two parts of Fig. 5.25 it can be seen that the deformed Ag ribbon mainly exhibits a $\{110\}\langle 011\rangle$ rolling texture, but reflected intensity also appears for $\{110\}\langle 001\rangle$ and $\{110\}\langle 112\rangle$ components.

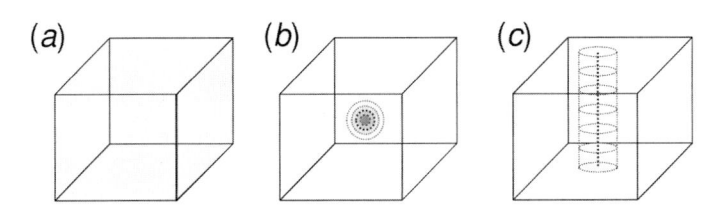

Figure 5.23 Special types of texture displayed in Euler space: (a) random orientation, (b) biaxial texture and (c) fiber texture.

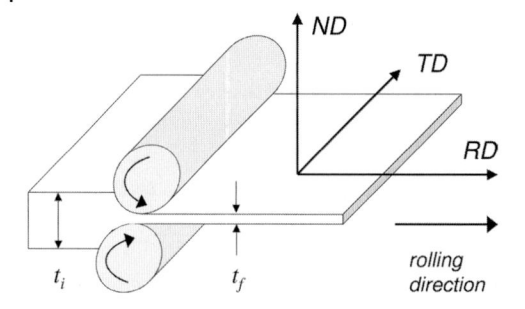

Figure 5.24 Definition of directions RD, TD and ND in metallographic texture analysis.

A useful description of complex textures can be obtained in the framework of harmonic analysis that has already been introduced in Section 5.5.1 for fiber textures. However, in the case of a general texture the set of orthogonal basis functions has to be adopted to apply to full Euler space instead of the spherical projection into which the inverse pole figure was inscribed. Again, use is made of the formalism in the notation of Bunge, but the reader is also referred to an equivalent representation introduced by Roe. The ODF is decomposed into generalized spherical harmonics $T_l^{mn}(\varphi_1, \Phi, \varphi_2)$ with

$$T_l^{mn}(\varphi_1, \Phi, \varphi_2) = e^{im\varphi_2} P_l^{mn}(\cos\Phi) e^{in\varphi_1} \qquad (5.28)$$

where $P_l^{mn}(\cos\Phi)$ is a generalized associated Legendre polynomial defined by

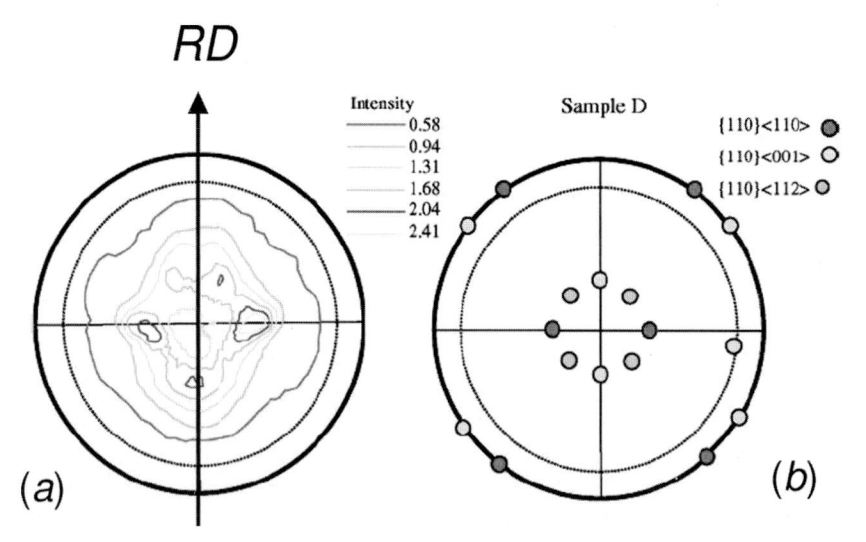

Figure 5.25 (a) 111 pole figure of 98% deformed Ag ribbon and (b) positions of theoretical intensity maxima for several types of rolling texture $\{hkl\}\langle uvw\rangle$ for 111 pole figure of fcc metal (from Ref. [18]).

5.6 Biaxial and Fully General Textures

$$P_l^{mn}(x) = \sqrt{\frac{(l-m)!(l+n)!}{(l+m)!(l-n)!}} \frac{i^{n-m}(-1)^{l-m}}{2^l(l-m)!} (1-x)^{-\frac{n-m}{2}} (1+x)^{-\frac{n+m}{2}}$$
$$\frac{d^{l-n}}{dx^{l-n}} \left[(1-x)^{l-m}(1+x)^{l+m} \right] \tag{5.29}$$

The ODF is given as a linear combination of the generalized spherical harmonics weighted by coefficients C_l^{mn}

$$f(\varphi_1, \Phi, \varphi_2) = \sum_{l=0}^{\infty} \sum_{m,n=-l}^{l} C_l^{mn} T_l^{mn}(\varphi_1, \Phi, \varphi_2) \tag{5.30}$$

The coefficients do not have a special physical or geometrical meaning. Again, the ODF expansion has to be truncated beyond the rank of harmonic expansion l_{max}. The solution of the texture problem is then equivalent to determining the number of $(2l_{max} + 1)^2$ ODF coefficients. The choice of l_{max} may have important consequences for the ODF resolution in Euler space and the associated errors.

The fact that the sum extends over three indices corresponds to the dependence of the ODF on three Euler angles. While the index n is related to φ_1, the index m is correlated with φ_2. Both angles describe rotations around the z axis, but φ_1 is around the sample z axis of the sample reference frame $\{s_i\}$ and φ_2 around the z axis of the crystal reference frame $\{c_i\}$. One therefore identifies the n- and m-dependent parts of T_l^{mn} to belong to sample properties and crystal properties, respectively. Both the shape of the sample and the crystal lattice introduce symmetry constrains to the ODF that lead to a reduction in the number of basis functions for each separate order l. Symmetry-adapted basis functions are introduced that are symbolized by $T_l^{\mu\nu}$ and derive from the T_l^{mn} functions defined in Eq. (5.28) by the linear transformation

$$T_l^{\mu\nu} = \sum_{m,n=-l}^{l} A_l^{m\mu} A_l^{n\nu} T_l^{mn} \tag{5.31}$$

The $A_l^{m\mu}$ coefficients that enable the transition to the crystal symmetry-adapted basis functions are listed for some crystal classes in Table 5.2. It is seen for the listed symmetries that the crystal symmetry acts as a "selection rule" on the set of T_l^{mn} functions. Of course, also in the case of the cubic crystal class new symmetry-adapted functions have to be formed, but in this case more complex transformation relations are valid. In all these cases, the symmetry elements of the crystal lattice cause a reduction in the number of required basis function for each separate order l, such that less than $(2l + 1)$ coefficients have to be determined for higher than triclinic crystal symmetry. The course of $M(l)$ as a function of expansion order l is given in Fig. 5.26 for various crystal symmetries. These results also apply on the expansion of the fiber texture ODF, where this point has already been mentioned.

The transition to the symmetry-adapted polynomials also causes a significant reduction in the number of ODF coefficients. They are symbolized by $C_l^{\mu\nu}$ instead of C_l^{mn} in order to signify that they account for the symmetry-adapted decomposition of the ODF. Frequently, a number of points is given above the $T_l^{\mu\nu}$ symbol in ac-

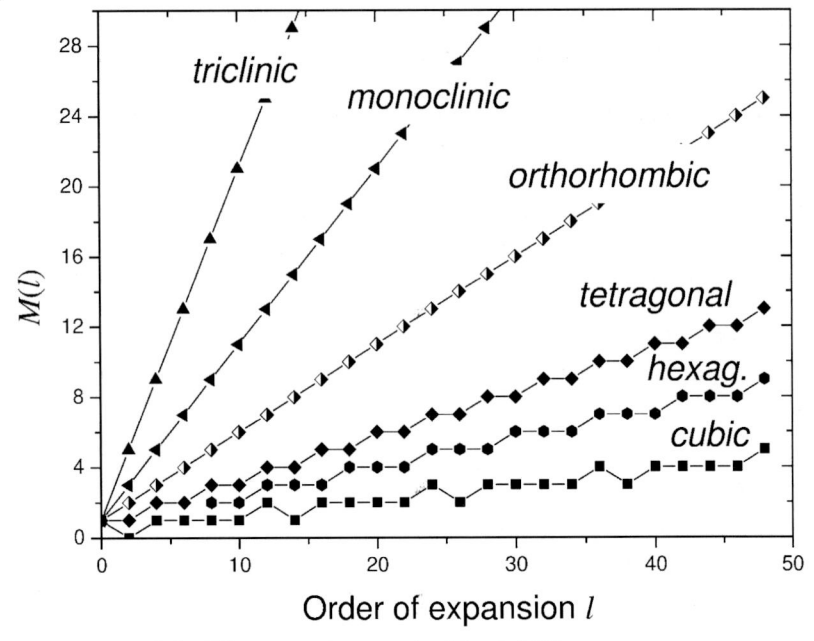

Figure 5.26 M(l) for different crystal symmetries (after Ref. [2]).

cordance with the number of symmetry elements, but here this notation is not used. This symmetry-adapted form of the ODF finally reads

$$f(\varphi_1, \Phi, \varphi_2) = \sum_{l=0}^{l_{max}} \sum_{\mu=1}^{M(l)} \sum_{\nu=1}^{N(l)} C_l^{\mu\nu} T_l^{\mu\nu}(\varphi_1, \Phi, \varphi_2) \tag{5.32}$$

The complete solution of the texture problem within the harmonic expansion is achieved by specifying all $C_l^{\mu\nu}$ coefficients up to the order l_{max}.

As has already been introduced for layer textures an index J may be defined that describes the sharpness of texture. This is calculated from the sum of squares of $C_l^{\mu\nu}$ coefficients according to

$$J = \sum_{l,\mu,\nu} \frac{1}{2l+1}\left(C_l^{\mu\nu}\right)^2 \tag{5.33}$$

Table 5.2 $A_l^{m\mu}$ for different crystal symmetries.

$A_l^{m\mu}$	Orthorhombic (D_{2h})		Tetragonal (D_{4h})		Hexagonal (D_{6h})	
	m condition	μ	m condition	μ	m condition	μ
1	$m = 0$	1	$m = 0$	1	$m = 0$	1
$1/\sqrt{2}$	$\|m\|$ mod $2 = 0$	$1 + m/2$	$\|m\|$ mod $4 = 0$	$1 + m/4$	$\|m\|$ mod $6 = 0$	$1 + m/6$
0	$\|m\|$ mod $2 \neq 0$	./.	$\|m\|$ mod $4 \neq 0$	./.	$\|m\|$ mod $6 \neq 0$	./.

Again, for a random orientation $J = 1$ holds. In this case all $C_l^{\mu\nu}$ vanish, except for C_0^{11}, which is set equal to 1 in order to guarantee normalization. For increasingly sharp textures the texture index J diverges to ever-larger values.

The determination of textures by x-ray diffraction always yields a volume-weighted distribution of orientations. As has already been outlined in the previous chapters on chemical phase identification and peak broadening this is caused by the fact that the integral intensity of a Bragg peak scales with the scattering volume (see Eq. (1.43)). The scattered intensity from a small volume element dV of orientation $(\varphi_1, \Phi, \varphi_2)$ thus scales with the value of its orientation distribution function $f(\varphi_1, \Phi, \varphi_2)$ multiplied by the volume element of Euler space $\sin\Phi d\varphi_1 d\Phi d\varphi_2$. The ODF normalization affords the scaling to be chosen such that the integration over the sample volume equals the integral of f in full Euler space. This condition can be expressed by

$$\frac{dV}{V} = \frac{1}{8\pi^2} f(\varphi_1, \Phi, \varphi_2) \sin\Phi d\varphi_1 d\Phi d\varphi_2 \tag{5.34}$$

which restricts the set of all $C_l^{\mu\nu}$ coefficients and ensures that the ODF is directly given in units of m.r.d.

5.6.3
Determination of Fully General Texture

The solution of the general texture problem starts from an adequate representation of the measured pole figures that are indexed by their distinct Miller indices $\mathbf{h}_i$. Since pole figures are continuous and continuously differentiable functions of azimuth ϕ and tilt angle ψ they may be expanded into a sum of spherical harmonics $k_l^\nu(\phi, \psi)$

$$I_{\mathbf{h}_i}(\phi, \psi) = \sum_{l=0}^{\infty} \sum_{\nu=1}^{N(l)} F_l^\nu(\mathbf{h}_i) k_l^\nu(\phi, \psi) \tag{5.35}$$

Here, the decomposition has again been symmetry-adapted to the sample, which is indicated by the use of coefficient ν instead of n. The coefficients $F_l^\nu(\mathbf{h}_i)$ describe the relative weight by which the spherical harmonic of order (l, ν) enters into the pole figure expansion. It is shown by Bunge that the pole figure coefficients are related via

$$F_l^\nu(\mathbf{h}_i) = \frac{4\pi}{2l+1} \sum_{\mu=1}^{M(l)} C_l^{\mu\nu} k_l^\mu(\mathbf{h}_i) \tag{5.36}$$

with the $C_l^{\mu\nu}$ coefficients of the ODF [2]. Also in this equation the spherical harmonics k_l^μ appear in the symmetry-adapted variant, but this time they are adapted to the crystal symmetry. The two Eqs. (5.35) and (5.36) relate the measured pole figure intensities and the ODF coefficients $C_l^{\mu\nu}$. They represent a system of equation that may be solved for the unknown $C_l^{\mu\nu}$ by the usual procedures of minimization of the square of differences. Obviously, for this purpose as many pole figures have to be measured as necessary for the solution of the ODF. In the usual case that only incomplete pole figures can be determined this could necessitate the measurement of additional pole figures than it might appear from crystal symmetry considerations alone.

Evidently, the setting of the rank of harmonic expansion l_{max} in Eq. (5.35) is decisive for the precision and accuracy of the determined ODF. If too small a value is chosen the coefficients determined may be associated with large errors, since for an adequate description terms beyond l_{max} would be afforded. Too large an l_{max} value could inadequately extend computation time compared to improvement in precision. Typically, values between 10 and 25 are inserted for l_{max}, where the choice sensitively depends on the crystal symmetry of the investigated sample. For the thoroughly investigated cubic materials the size of $C_{l_{max}}^{\mu\nu}$ coefficients turned out as a valuable criterion for choosing l_{max}: a reasonably accurate description of the ODF is generally arrived at if values of $C_{l_{max}}^{\mu\nu}$ are smaller than 10% of $C_2^{\mu\nu}$ coefficients. A further test may be performed by analyzing the texture index J as a function of l_{max}. If J has attained a stable value that is not subject to change by increasing l_{max}, the rank of harmonic expansion can be assumed to be sufficiently large [2]. Considerations of error propagation are especially important in thin-film work, since only a few investigations of the full solution of texture have been performed so far and the small intensity in pole figures from thin films allows for larger statistical fluctuations than in bulk materials for which this formalism has initially been developed.

Again, as in the case of layer texture, the ODF coefficients can be used to specify the texture factors T_h for the correction of integral intensities in a $\theta/2\theta$ diffractogram as presented in Section 5.1. They are simply obtained by inserting $(\phi, \psi) = (0, 0)$ in Eq. (5.35) and one arrives at

$$T_h = I_h(0,0) = \frac{4\pi}{2l+1} \sum_{l=0}^{l_{max}} \sum_{\mu=1}^{M(l)} \sum_{\nu=1}^{N(l)} C_l^{\mu\nu}(h) k_l^{\mu}(h) k_l^{\nu}(0,0) \tag{5.37}$$

These coefficients represent the scaling factors for the comparison of diffraction patterns of a textured sample and a sample of random crystallite orientation like an ICDD data set. In contrast to the purely phenomenological formula of Eq. (5.3) this T_h representation is now put on a firm basis, since it includes the full expansion of the ODF, which enters by the $C_l^{\mu\nu}$ coefficients. In many cases, the T_h coefficients may carry rather complex information on the orientation of the polycrystal in a very condensed manner. It then might appear hopeless to derive any valuable information on preferred orientation by elucidating them from a $\theta/2\theta$ pattern. These arguments were the reason why the texture coefficients T_h were regarded with some skepticism in Section 5.1. It may be stated, however, that the criticism of texture factors appears only justified for fully general textures as they are observed in bulk materials. In thin-film samples, the determination of T_h coefficients may indeed yield valuable information on the kind and degree of texture that might be useful for technological aspects like an optimization of the deposition process. In the case that a thorough and well-founded determination of the texture in the specimen is deserved, however, the measurement of several pole figures and the application of the full formalism have to be applied.

In fact, the availability of analysis tools for the full elucidation of texture has significantly progressed. Even for thin-film samples various numerical techniques and computer programs have been developed for the solution of the general texture problem by calculating ODF coefficients from measured pole figure data like popLA [19], MulTex [20], BEARTEX [21], LaboTex [22], TexEval [23] and POFINT [1] to mention only a few (see Ref. [1] for a compilation). For some programs detailed input data on the number and Miller indices of pole figures to be elucidated are already quoted during the programming of the measurement. These specifications are checked with respect to consistency and the user may be requested to adopt a higher number of pole figures such that the gathered information is sufficient for solving the ODF. These programs are either distributed by the diffractometer manufacturers or have been written as public domain software.

So far, the main areas of application for full texture analysis have been in the materials science of mechanically deformed metals like rolled sheets or drawn wires and in geological investigations of plastically deformed rocks [24]. The investigation of textures of rolled metal sheets elucidated various types of textures that typically occur in fcc and bcc metals [24]. At the time of writing there have only been a few thin-film investigations of fully general texture carried out and, in fact, the analysis of preferred orientation in thin films in the framework of layer texture is sufficient in most cases. Interestingly, the results on rolling textures recently found application in the preparation of thin-film superconductors on metallic ribbons. As mentioned above, one pathway in obtaining biaxial textured high-temperature superconducting films makes use of metallic templates with highly aligned crystallites. Metals that are currently under investigation for these applications include nickel, nickel-based alloys, silver and others [18, 25, 26]. Similarly as in IBAD, a set of oxide buffer layers is used to increase the texture with increasing thickness such that the required low twist $\Delta\phi$ and tilt $\Delta\psi$ values are achieved for the finally deposited high-temperature superconducting layer.

Also metallic ribbons are under development on which the thin superconducting film shall be directly coated upon with a high degree of biaxial orientation. This might be achieved by introducing a rolling texture and subsequently recrystallizing the ribbons at higher temperatures. Suo et al. investigated mechanically reinforced Ag composite [25]. As an example of their work Fig. 5.27(a) displays the 111 pole figure of the Ag ribbon top layer and Fig. 5.27(b) the three ODF sections from Euler space for constant φ_2 of 0°, 25° and 45°, while Φ and φ_1 extend from 0 to 90°. In contrast to Fig. 5.25, the pole figure in Fig. 5.27 displays a sharp $\{110\}\langle011\rangle$ rolling texture, while further components are almost absent. The iso-intensity lines are chosen as 2, 4, 8, 16, 32 and 64 in arbitrary units to emphasize the small secondary $\{110\}\langle112\rangle$ component. This enhancement of the $\{110\}\langle011\rangle$ orientation compared to Fig. 5.25 is mainly caused by the recrystallization step. In the ODF plot of Euler space sections the iso-ODF lines are given for increasing strength of 4, 8, 12, 16, 24 and 40 m.r.d. Also in the ODF sections the $\{110\}\langle011\rangle$ orientation is recognized as the major component attaining a peak value as high as 45 m.r.d. The pattern in the center of the $\phi_2 = 0°$ projection can be seen to reproduce at the bottom of the $\phi_2 = 45°$ plot, which is due to the symmetry constraints of the cubic crystalline structure.

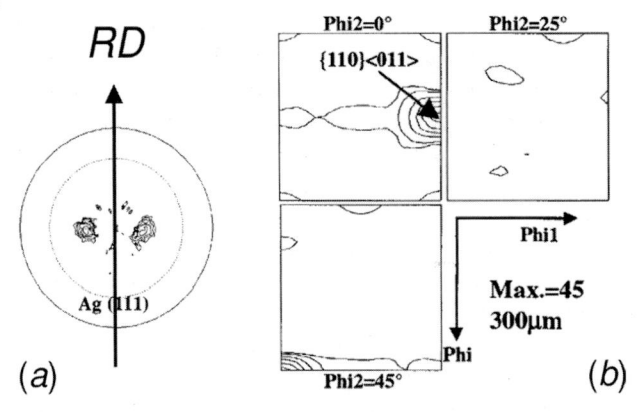

Figure 5.27 (a) Ag 111 pole figure and (b) derived ODF sections from Ag top layer of 300 μm Ag/Ni composite ribbon (from Ref. [25]).

The only minor occurrence of orientations other than {110}⟨011⟩ achieved in this study is advantageous for the subsequent deposition of high-temperature superconducting thin films. The in-plane misorientation, i.e. the twist $\Delta\phi$, in the top Ag layer was calculated to amount to only 15°. The example demonstrates the interesting results that may be obtained from a full ODF analysis for the optimized preparation of substrate templates. One important feature of the full ODF analysis is that it allows for the determination of a specifically oriented volume fraction of the sample by integrating the ODF in that part of Euler space [27].

It should finally be pointed out that in addition to the harmonic analysis basically different approaches have been developed to deal with fully general textures. One of these is the method of ideal components [28, 29], which appears well suited for thin films exhibiting pronounced layer and biaxial textures. In the case of the layer texture one ideal component would be directed along s_3, while in biaxial textures the second component would be within the film plane. It will be interesting to elucidate the potential of this technique for thin-film analysis more intensively in the future, which might be simplified by the fact that numerical tools for the analysis of pole figure data in the framework of this method are already available [20].

5.7
Depth Dependence of Thin-Film Textures

An interesting feature of thin-film textures is that they are often identified as exhibiting a pronounced depth dependence. It appears no exaggeration to state that the degree of preferred orientation in thin polycrystalline films was almost always found to increase with increasing thickness, whenever the investigation allowed for a depth resolution. Several techniques have been introduced for an experimental determination of texture gradients. The simplest one consists in preparing a set of samples under identical deposition conditions, but of various thicknesses and measuring each sample separately. The use of synchrotron radiation sources allows for *in situ* monitoring of selected Bragg peaks or even whole diffraction patterns during the deposition process and valuable investigations have been made with this tool in the last few years. However, the technique requires the deposition system to be connected to the synchrotron beam line and thus is only of limited use for most thin-film technologists. A very appealing approach to elucidate a texture gradient within a sample makes use of the procedures of depth monitoring by grazing incidence techniques. By varying the incidence angle of the incoming beam, various depths of the sample are exposed to different x-ray intensities and so is the diffracted beam. The information depth τ and the related Laplace property transform are introduced in Chapter 4 for this purpose.

Having arrived at this point, it is realized that during the measurement of a simple pole figure or a ψ scan the information depth is varied. This point has not been considered so far, but it might severely affect the interpretation of the results derived from ψ-dependent measurements. The effect can be quantified by calculating the information depth τ according to Eqs. (4.7) and (4.8) that are adopted to the actual configuration encountered by inserting the appropriate configuration factor k_ψ (see Eq. (5.4)). Fully compatible with Eq. (4.8) it is derived for the average information depth for ψ-tilted configurations

$$\bar{\tau}_\psi = \frac{1}{\mu k_\psi} + \frac{t}{1 - \exp(\mu t k_\psi)} \qquad (5.38)$$

The influence of the tilting angle on the average information depth is demonstrated in Fig. 5.28 for a set of thin BaTiO$_3$ films of various thickness or μt products. Exposing the samples to Cu Kα radiation, the attenuation coefficient becomes $\mu = 0.10631$ μm^{-1}. Layer thicknesses were chosen for the simulation of 0.5, 1 and 2 μm, while a reflection at $2\theta = 32°$ was assumed for the analysis. The general tendency is observed in the plot that the average information depth $\bar{\tau}_\psi$ in-

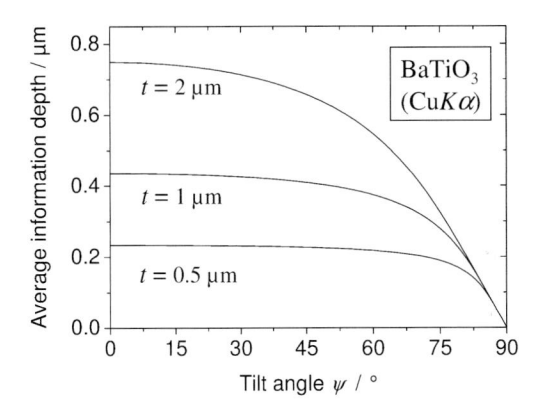

Figure 5.28 Average information depth $\bar{\tau}_\psi$ as a function of tilt angle ψ for a $2\theta = 32°$ reflection from thin BTO films. Layer thicknesses were assumed as indicated.

creases with increasing film thickness. It should be remembered that the limiting values is half the thickness $t/2$ which would be attained for $\theta = 90°$. Close to zero tilt the course of $\bar{\tau}_\psi$ stays nearly constant but diminishes as the tilt angle approaches 90°. This is easily understood, since in the vicinity of $\psi = 90°$ the diffraction is mainly due to layers from the topmost part of the sample.

It can be concluded from these considerations that a possible texture gradient would mostly affect the measurement when large tilt angles ψ were chosen. It is thus recommended to always estimate the variation of the information depth prior to the planned measurements. According to this result it might be preferred to measure more pole figures with small ψ_{max} instead of few pole figures with high ψ_{max}. A method for measuring pole figures under constant information depth has been devised by Bonarski et al. [30]. These authors pointed to the fact that for samples with texture gradients the set of pole figures would contain inconsistent information on preferred orientation from different depths and that even pole figures would become inconsistent in themselves (compare Fig. 5.28). The method facilitates a constant information depth by varying ψ and ω simultaneously (Exercise 12).

5.8
Applications

The occurrence of texture is such a characteristic phenomenon in thin-film growth that the preparation of untextured layers is the exception rather than the rule. It may easily be argued in favor of performing a texture analysis prior to any other structural investigation of a polycrystalline thin film by x-ray diffraction. For one reason, the state of texture ultimately has to be considered for any residual stress analysis as will be seen in Chapter 6. Also, the microstructural analysis as outlined in Chapter 3 may be severely affected by the orientation distribution of crystallites.

In the field of metallic thin films a large number of studies has been performed on texture evolution in combination with basic investigations regarding the active mechanism of thin-film growth. Many of the recent studies were concerned with applied research focusing on metallic interconnects in Si semiconductor technology. Al played a dominant role in the field for a long time, but the focus shifted in accordance with the recent introduction of Cu for these applications. For both metals, the texture and microstructure were identified to have an effect on lifetime-degrading electromigration and thus turn out as features of high economic relevance [31–34]. An important aspect that has been emphasized by these studies is the post-deposition structural reordering processes that often occur in thin films [32]. The usefulness of determining the full ODF has been demonstrated for Cu interconnects electroplated into submicrometer trenches of a low-k dielectric. The ODF approach allowed for the representation of inverse 111 pole figures with respect to the bottom and the sidewall of the trench, revealing a substantial component of (111) sidewall texture and giving insight into the evolution of grain growth of damascene Cu [31].

Several examples from the group of oxides have already been given in this chapter for dielectric and superconducting compounds. Also the buffer layers for high-temperature superconducting composite ribbons are made from oxides like MgO, CeO_2, Y-stabilized Zr_2O_3 (YSZ) and $Gd_2Zr_2O_7$ to mention only the most important ones. The evolution of a biaxial texture in these buffer layers is supported by oblique deposition or ion-beam irradiation techniques [16, 35]. The degree of texture may remarkably improve with increasing thickness t as is described by the course of the misorientation angle $\Delta\phi$ versus t (see Fig. 5.29 for an example). In this case, the enhancement of texture with increasing thickness is technologically applied to mediate between the too large twists and tilts of the metallic ribbons in order to arrive at the high degree of texture as required for the high-temperature superconducting layer. Further techniques for texture engineering were employed for oxide compounds like the deposition of a thin PrO_2 film of only monolayer thickness on $LaAlO_3$ (100) that resulted in an almost exclusively a-axis oriented growth of a subsequently deposited YBCO layer [36]. The effect of defocusing and ψ-dependent absorption on the measurement of pole figures was investigated for thin YBCO films on single-crystalline substrates [37, 38]. Only a small fraction of in-plane misaligned crystallites was revealed to cause a strong decrease of J_C in epitaxially grown YBCO films on *MgO* substrates [39]. Also the introduction of a $\{211\}\langle111\rangle$ copper-type rolling texture into an amorphous precursor film has been demonstrated for yttria-stabilized zirconia in a combined process of oblique ion bombardment, oxidation and thermally induced crystallization [40]. The possibility of influencing the state of texture by ion beam irradiation has long been recognized [41, 42], although the explanation of the effect is still controversial.

ZnO and its derivatives may be applied for various optoelectronic thin-film applications, among them the use of Al-doped ZnO for transparent electrodes in solar cells and liquid crystal displays. Wurtzite-structured ZnO was identified as ex-

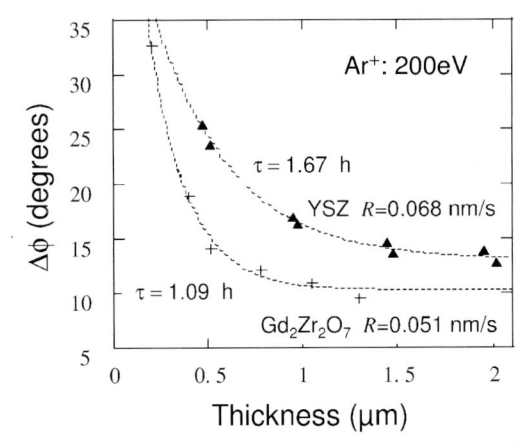

Figure 5.29 Thickness dependence of twist $\Delta\phi$ for YSZ and $Gd_2Zr_2O_7$ films. Growth rates are indicated by R (compare Ref. [16]).

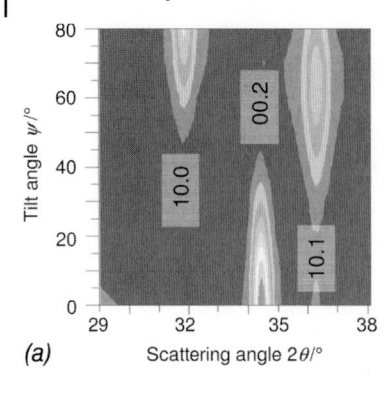

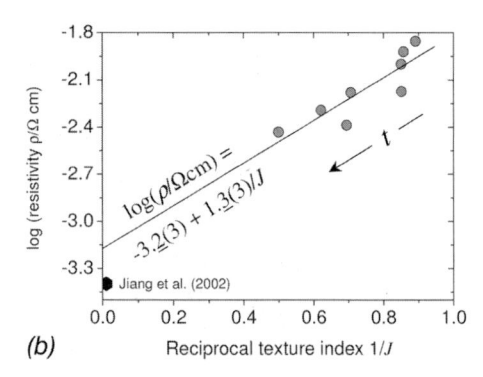

Figure 5.30 (a) $I(2\theta, \psi)$ scan of a nominally 500 nm thin ZnO:Al film; hk.l triples indicate Miller indices of Bragg reflections. Intensity increases towards individual centers of intensity islands [43]. (b) Resistivity of thin ZnO:Al films versus reciprocal texture index $1/J$ [44].

hibiting an almost universal tendency to develop a c-axis layer texture in many different deposition processes [44–46]. Figure 5.30(a) shows a scan of the $I(2\theta, \psi)$ plane from the same sample from which the symmetrical $\theta/2\theta$ pattern is given in Fig. 5.2 (also shown on the cover of this book). A pronounced (00.l) layer texture having $c \parallel s_3$ is clearly recognized from this mapping [43]. In one of the few texture studies where the C_l^μ coefficients of the harmonic analysis were elucidated for thin-film samples, the texture index was found to be directly related to the resistivity of ZnO:Al layers [44]. Figure 5.30(b) displays the resistivity ρ of a set of eight ZnO:Al samples from a thickness series versus the inverse texture index $1/J$ derived from C_l^μ coefficients. A structure–function relationship could be established for both parameters, relating the resistivity via $\rho = A\exp(B/J)$ to the inverse texture index. According to a grain boundary scattering model a limiting value of $\rho = 6.9 \times 10^{-4}$ Ω cm would be predicted for perfectly aligned ZnO:Al crystallites having $1/J = 0$. This value corresponds within the margins of errors with one of the lowest resistivity values measured in a highly textured thin ZnO:Al film [46] shown as a black hexagon in Fig. 5.30(b).

In the field of semiconducting layers preferred orientation has also extensively been investigated. Thin polycrystalline silicon films (poly-Si) as used for contacting the transistor gate oxide were early described to exhibit a pronounced layer texture changing from a (hh0) preferred orientation observed at medium temperatures to (h00) at higher temperatures [47–50]. The effect was also observed for microcrystalline thin Si films intended for solar cell applications, for which the depth dependence was accounted for by a power law revealing the layer texture to increase with a power law close to $t^{1/6}$ [51]. Also the combined analysis approach was applied to nanocrystalline Si films revealing an anisotropic crystalline grain growth with elongated crystallites along $\langle 111 \rangle$ directions independent of the crystallographic fiber and the degree of preferred orientation [52, 53].

Diamond may be an ideal semiconductor due to its large band gap associated with a small electronic noise at room temperature and the nearly ideal thermal con-

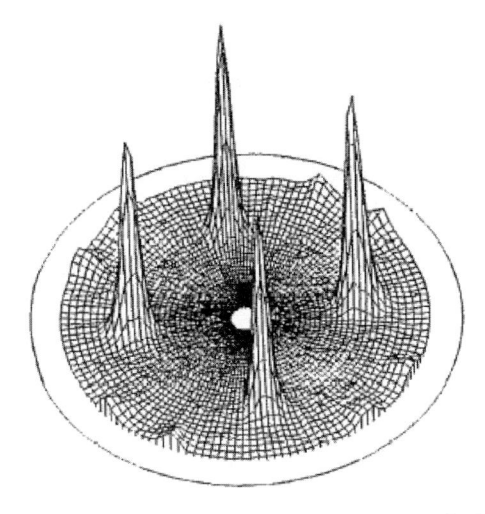

Figure 5.31 Three-dimensional representation of pole figure of asymmetric 111 reflection from 2 μm thin diamond film on Si (100) (from Ref. [55]).

ductivity. Its application in semiconductor applications is still restricted by the unavailability of (affordable) large single crystals. As an alternative the fabrication of thin diamond films on appropriate substrates has been a constant issue. The task is even more challenging than for the biaxial growth of thin high-temperature superconducting films, since even smaller values for twist and tilt have to be achieved. Progress in this direction could be demonstrated by using Ir buffer layers epitaxially grown on SrTiO$_3$ (100) substrates that were subsequently coated by thin diamond films prepared by CVD. Minimum values of 0.17° and 0.38° were arrived at. The deposition was performed by a CVD process at about 700 °C and the detachment of some 10 μm thick layers was observed during cooling. This phenomenon may offer a promising route for the preparation of single-crystalline diamond wafers [54, 55]. The texture of thin diamond films on Si 100 and 111 wafers was investigated by the component method [56], which assumes the ODF to decompose in a finite set of Gauss-peak components. Figure 5.31 shows the pole figure from the 111 reflection of a thin diamond film in a three-dimensional representation with measured intensities plotted above the (ϕ, ψ) plane from this investigation. The component technique may advantageously be applied to multiphase material with overlapping pole figures and is thus well suited for the analysis of thin films with a high degree of texture approaching an epitaxial layer. Information on the randomly oriented phase content and twinning provided a valuable quality control for the nucleation step of the diamond film [56].

Last, but not least, the growth of hard nitride coatings like TiN, TaN, AlN, (Ti,Al)N, etc., by plasma-assisted processes was found to be associated with strong layer textures. These textures exhibit a strong interplay with residual stress and a discussion of this group of materials is therefore postponed to Chapter 6.

Exercises

5.1 Give a sketch for the derivation of the configuration factor k_ψ when a Euler cradle is employed.

5.2 Derive the absorption correction $A^T_{\theta,\psi}$ in transmission geometry by analogous considerations to those given in Chapters 1 and 4, where the absorption correction is evaluated for symmetric $\theta/2\theta$ and grazing incidence x-ray diffraction.

5.3 Explain why the tetragonal splitting of $BaTiO_3$ in the perovskite structure causes the occurrence of three reflections 210, 021 and 012 instead of one 012 for the cubic paralelectric phase. Derive a general condition from this special case that the Miller indices hkl have to obey for two- and threefold tetragonal splitting.

5.4 Consider a thin film of $BaTiO_3$ in its tetragonal modification exhibiting a $\langle ００l \rangle$ layer texture along s_3. Give the first ten $k_l^\mu(\mathbf{h})$ for $\mathbf{h} = (111)$ and (200).

5.5 How do $V_f(\psi)$ and $\psi_{1/2}$ (Fig. 5.12) derive from $I_h(\psi)$ for a random crystallite distribution?

5.6 Calculate the crystal symmetry-adapted ODF basis functions $T_l^{\mu n}$ from T_l^{mn} in the case of (a) a twofold axis, (b) a threefold axis of rotational symmetry and (c) a mirror plane.

5.7 Describe the features of orthorhombic splitting in the powder pattern of YBCO. To which reflection from the one-elemental bcc structure is the 013/103 couple related?

5.8 Assume that a ϕ scan is to be statistically be verified with respect to the occurrence of a fiber texture. The scan consists of N intensity values I_i on background intensities B_i measured with an integration time t'. The zero hypothesis H_0 is that the scan accounts for a random sampling of constant intensity. The alternative hypothesis H_1 states that $I(\phi_i)$ is modulated by an n-fold rotational symmetry of amplitude A around the substrate normal. A and its estimated standard deviation σ_A are obtained by fitting $I(\phi_i)$ with a model function. Which integration time t' has to be adjusted for each ϕ step to reject H_1 on a 95% confidence level?

5.9 Describe the effect of the orthorhombic splitting, $a \neq b$, in YBCO on the peaks in a ϕ scan from an $h0l$ reflection. How is the interpretation of $\Delta\phi$ affected by the splitting and which instrumental measures should be taken to minimize the effect?

5.10 The product of the three Euler matrices given in Eq. (5.6) is an oversimplification. In a mathematically correct derivation the multiplication of the second (third) rotation B (C) has to be performed in the transformed coordinate system [57], such that $T' = ((AB)C(AB)^{-1})(ABA^{-1})A$. Prove that the multiplication in T' (fortunately) ends up with the same matrix elements as in $T = ABC$.

5.11 Give the ODF of a Si(111) wafer in Euler space. Comment on the minimum symmetry unit $0 \leq \phi_1, \Phi, \phi_2 \leq 90°$, in which all the orientational information should be contained for cubic crystals.

5.12 Derive the geometry factor $k_{\psi\omega}$ for the measurement of a thin-film Bragg peak under various tilt angles, but constant information depth $\bar{\tau}$ (compare Ref. [30]).

5.13 Derive the structure factor of the paraelectric $BaTiO_3$ phase.

References

Monographs and conference proceedings relevant to this chapter

U. F. Kocks, C. N. Tomé, H.-R. Wenk, *Texture and Anisotropy: Preferred Orientations in Polycrystals and Their Effect on Materials Properties*, Cambridge University Press, 1998.

K. Helming, *Texturapproximation durch Modellkomponenten*, Cuivillier Verlag, Göttingen, 1995.

H.-J. Bunge, *Texture Analysis in Materials Science*, Butterworth, London, 1982. English translation and extended version of H.-J. Bunge, *Mathematische Methoden der Texturanalyse*, Akademie-Verlag, Berlin, 1969.

International Conference on Texture of Materials (ICOTOM-13), 2002, Seoul, ed. Dong Nyung Lee, Materials Science Forum **408-412**, Transtech Publication, Zürich.

ICOTOM-12, 1999, Montreal, ed. J. A. Szpunar, 2 Volumes, NRC Research Press, Ottawa.

ICOTOM-11, 1996, ed. Xian Z. Liang, L. Zuo, Y. Chu, International Academic Publishers, Beijing.

Special papers

[1] D. Chateigner, Combined Analysis: structure-texture-microstructure-phase-stresses-reflectivity determination by x-ray and neutron scattering, available at: www.ecole.ensicaen.fr/~chateign/texture/combined.pdf (2005).

[2] H.-J. Bunge, *Texture Analysis in Materials Science*, Butterworth, London, 1982. English translation and extended version of H.-J. Bunge, *Mathematische Methoden der Texturanalyse*, Akademie-Verlag, Berlin, 1969.

[3] K. Saito, M. Mitsuya, N. Nukaga, I. Yamaji, T. Akai, H. Funakubo, Method of distinguishing $SrBi_2Ta_2O_9$ phase from fluorite phase using x-ray diffraction reciprocal space mapping, *Jpn. J. Appl. Phys.* **39** (2000) 5489.

[4] J. Ricote, R. Poyato, M. Alguero, L. Pardo, M. L. Calzada, D. Chateigner, Texture development in modified lead titanate thin films obtained by chemical solution deposition on silicon-based substrates, *J. Am. Ceram. Soc.* **86** (2003) 1571.

[5] J. S. Kallend, Determination of the orientation distribution from pole figure data, in *Texture and Anisotropy*, p. 102.

[6] U. F. Kocks, The representation of orientations and texture, in *Texture and Anisotropy*, p. 44.

[7] W. A. Dollase, Correction of intensities for preferred orientation in powder diffractometry: application of the March model, *J. Appl. Cryst.* **19** (1986) 267.

[8] A. March, Mathematische Theorie der Regelung nach der Korngestalt bei affiner Deformation, *Z. Kristallogr.* **81** (1932) 285.

[9] M. M. Seabaugh, M. D. Vaudin, J. P. Cline, G. L. Messing, Comparison of texture analysis for highly oriented a-Al_2O_3, *J. Am. Ceram. Soc.* **83** (2000) 2049.

[10] M. Morales, D. Chateigner, L. Lutterotti, J. Ricote, X-ray Combined QTA using a CPS applied to ferroelectric ultrastructure, in *ICOTOM 13*, p. 113.

[11] A. C. Larson, R. B. von Dreele, General structure analysis system (GSAS), available at: www.ncnr.nist.gov/programs/crystallography/software/gsas.html.

[12] M. D. Vaudin, M. W. Rupich, M. Jowett, G. N. Riley, J. F. Bingert, A method for crystallographic texture investigations using standard x-ray equipment, *J. Mater. Res.* **13** (1998) 2910.

[13] M. D. Vaudin, Accurate texture measurements on thin films using a powder X-ray diffactometer, in *ICOTOM 12*, p. 186.

[14] M. D. Vaudin, Crystallographic texture in ceramics and metals, *J. Res. NIST* **106** (2001) 1063, (eMail: mark.vaudin@nist.gov).

[15] M. Li, B. Ma, R. E. Koritala, B. L. Fisher, K. Venkataraman, U. Balachandran, Pulsed laser deposition of YBCO thin films on IBAD-YSZ substrates, *Supercond. Sci. Technol.* **16** (2003) 105.

[16] Y. Iijima, K. Kakimoto, Y. Yamada, T. Izumi, T. Saitoh, Y. Shiohara, Research and development of biaxially textured IBAD-GZO templates for coated superconductors, *MRS Bull.* **X** (2004) 564.

[17] R. P. Reade, P. Berdahl, R. E. Russo, Ion-beam nanotexturing of buffer layers for near-single-crystal thin-film deposition: application to $YBa_2Cu_3O_{7-\delta}$ superconducting films, *Appl. Phys. Lett.* **80** (2002) 1352.

[18] J. J. Wells, J. L. MacManus-Driscoll, J.-Y. Genoud, H. L. Suo, E. Walker, R. Flükiger, {110}<110> textured Ag ribbons for biaxially aligned $YBa_2Cu_3O_{7-x}$ coated conductor tapes, *Supercond. Sci. Technol.* **13** (2000) 1390.

[19] T. Mason, popLA, available at: www.lanl.gov/orgs/mst/cms/poplalapp.html.

[20] K. Helming, MulTex, available at: www.textur.de.

[21] H.-R. Wenk, S. Matthies, J. Donovan, D. Chateigner, BEARTEX: a Windows-based program system for quantitative texture analysis, *J. Appl. Cryst.* **31** (1998) 262.

[22] K. Pawlik, P. Ozga, LaboTex, available at: www.labosoft.com.pl.

[23] Bruker AXS, TexEval V2.3 Manual, Karlsruhe.

[24] U. F. Kocks, C. N. Tomé, H.-R. Wenk, *Texture and Anisotropy: Preferred Orientations in Polycrystals and Their Effect on Materials Properties*, Cambridge University Press, 1998.

[25] H. Suo, J. Y. Genoud, M. Schindl, E. Walker, R. Flukiger, Reinforced {110} textured Ag/Ni composite ribbons for biaxially textured YBCO coated tapes, *Supercond. Sci. Technol.* **14** (2001) 854.

[26] A. Goyal, M. Parans Paranthaman, U. Schoop, The RABiTS approach: using rolling-assisted biaxially textured substrates for high-performance YBCO superconductors, *MRS Bull.* (2004) 552.

[27] L. Zuo, J. Muller, C. Esling, Volume fractions of texture components in polycrystalline materials, *J. Appl. Cryst.* **26** (1993) 422.

[28] K. Helming, *Texturapproximation durch Modellkomponenten*, Cuivillier Verlag, Göttingen, 1995.

[29] K. Helming, Some applications of the texture component model, in *ICOTOM 13*, p. 363.

[30] J. T. Bonarski, H. J. Bunge, L. Wcislak, K. Pawlik, Investigation of inhomogeneous surface textures with constant information depth: 1: Fundamentals, *Textures Microstruc.* **31** (1998) 21.

[31] C. Lingk, M. E. Gross, W. L. Brown, X-ray diffraction pole figure evidence for (111) sidewall texture of electroplated Cu in submicron damascene trenches, *Appl. Phys. Lett.* **74** (1999) 682.

[32] J. M. E. Harper, C. Cabral, P. C. Andricacos, L. Gignac, I. C. Noyan, K. P. Rodbell, C. K. Hu, Mechanism for microstructure evolution in electroplated copper thin films near room temperature, *J. Appl. Phys.* **86** (1999) 2516.

[33] L. D. Madsen, E. B. Svedberg, D. B. Bergstrom, I. Petrov, J. E. Greene, Texture of Al thin films deposited by magnetron sputtering onto epitaxial W(001), *J. Appl. Phys.* **87** (2000) 168.

[34] Y. H. Cheng, B. K. Tay, S. P. Lau, X. Shi, H. S. Tan, Influence of the substrate bias on the microstructure and internal stress in Cu films deposited by filtered cathodic vacuum arc, *J. Vac. Sci. Technol. A* **19** (2001) 2102.

[35] P. N. Arendt, S. R. Foltyn, Biaxially textured IBAD-MgO templates for YBCO-coated conductors, *MRS Bull.* (2004) 543.

[36] L. Tröger, D. K. Fork, J. B. Boyce, *a*-axis growth of $YBa_2Cu_3O_7$ thin films on $LaAlO_3$ induced by monolayer buffers of Pr_2O_3, *J. Appl. Phys.* **72** (1992) 4816.

[37] D. Chateigner, P. Germi, M. Pernet, Texture analysis by the Schulz reflection method: defocalization corrections for thin films, *J. Appl. Cryst.* **25** (1992) 766.

[38] D. Chateigner, P. Germi, M. Pernet, X-ray texture analysis of thin films by the reflection method: intermediate regime in defocusing corrections, *J. Appl. Cryst.* **27** (1994) 278.

[39] M. Pernet, D. Chateigner, P. Germi, C. Dubourdieu, O. Thomas, J. P. Sénateur, D. Chambonnet, C. Belouet, Texture influence on critical current density of YBCO films deposited on (100)-MgO substrates, *Physica C* **235–240** (1994) 627.

[40] P. Berdahl, R. P. Reade, J. Liu, R. E. Russo, L. Fritzemeier, D. Buczek, U. Schopp, Oblique ion texturing of yttria-stabilized zirconia: the {211}<111> structure, *Appl. Phys. Lett.* **82** (2003) 343.

[41] B. Rauschenbach, K. Helming, Implantation-induced texture, *Nucl. Instr. Meth.* **B42** (1989) 216.

[42] L. Dong, D. J. Srolovitz, Mechanism of texture development in ion-beam-assisted deposition, *Appl. Phys. Lett.* **75** (1999) 584.

[43] M. Birkholz, F. Fenske, unpublished results (2003).

[44] M. Birkholz, B. Selle, F. Fenske, W. Fuhs, Structure–function relationship between preferred orientation of crystallites and electrical resistivity in thin polycrystalline ZnO:Al films, *Phys. Rev. B* **68** (2003) 205414.

[45] N. Fujimura, T. Nishihara, S. Goto, J. Xu, T. Ito, Control of preferred orientation for ZnO_x films: control of self-texture, *J. Cryst. Growth* **130** (1993) 269.

[46] X. Jiang, C. L. Jia, B. Szyska, Manufacture of specific structure of aluminium-doped zinc oxide films by patterning the substrate surface, *Appl. Phys. Lett.* **80** (2002) 3090.

[47] T. I. Kamins, T. R. Cass, Structure of chemically deposited polycrystalline-silicon films, *Thin Solid Films* **16** (1973) 147.

[48] G. Harbeke, L. Krausbauer, E. F. Steigmeier, A. E. Widmer, H. F. Kappert, G. Neugebauer, Growth and physical properties of LPCVD polycrystalline silicon films, *J. Electrochem. Soc.* **131** (1984) 675.

[49] R. Bisaro, J. Maraino, N. Proust, K. Zellama, Structure and crystal growth of atmospheric and low-pressure chemical vapor-deposited silicon films, *J. Appl. Phys.* **59** (1986) 1167.

[50] H.-R. Wenk, M. Sintubin, J. Huang, G. C. Johnson, R. T. Howe, Texture analysis of polycrystalline silicon films, *J. Appl. Phys.* **67** (1990) 572.

[51] M. Birkholz, B. Selle, E. Conrad, K. Lips, W. Fuhs, Evolution of structure in thin microcrystalline silicon films grown by electron-cyclotron resonance CVD, *J. Appl. Phys.* **88** (2000) 4376.

[52] M. Morales, Y. Leconte, R. Rizk, D. Chateigner, Anisotropic crystallite size analysis of textured nanocrystalline silicon thin films probed by X-ray diffraction, *Thin Solid Films* **450** (2004) 216.

[53] M. Morales, Y. Leconte, R. Rizk, D. Chateigner, Structural and microstructural characterization of nanocrystalline silicon thin films obtained by radio-frequency magnetron sputtering, *J. Appl. Phys.* **97** (2004) 034307.

[54] M. Schreck, H. Roll, B. Stritzker, Diamond/Ir/$SrTiO_3$: a material combination for improved heteroepitaxial diamond films, *Appl. Phys. Lett.* **74** (1999) 650.

[55] M. Schreck, F. Hörmann, H. Roll, J. K. N. Lindner, B. Stritzker, Diamond nucleation on iridium buffer layers and subsequent textured growth: a route for the realization of single-crystal diamond films, *Appl. Phys. Lett.* **78** (2001) 192.

[56] K. Helming, S. Geier, M. Schreck, R. Hessmer, B. Stritzker, B. Rauschenbach, Texture analysis of chemical vapor deposited diamond films on silicon by the component method, *J. Appl. Phys.* **77** (1995) 4765.

[57] W. I. Smirnow, *Lehrgang der höheren Mathematik, Teil III/1*, VEB Deutscher Verlag der Wissenschaften, Berlin, 1976.

6
Residual Stress Analysis

Mario Birkholz and Christoph Genzel

Thin films may contain high mechanical stresses of up to a few hundred MPa or even several GPa. These are large values when compared to the ambient pressure of 100 kPa to which we are exposed at the ground level of our planet's atmosphere. Such large pressures may be measured at depths of several kilometers in the oceans. There exists, however, an important distinction between the pressure in the deep sea and residual stresses contained in thin films. Hydrostatic pressure exerts a constant isotropic force while mechanical stresses in thin films are strongly anisotropic.

Mechanical stresses in thin films are mostly measured after they are introduced by a specific process or a post-treatment. They may be due to a mismatch of thermal expansion coefficients between the substrate and the deposit, caused by an ion bombardment during the deposition process or other factors. Since the stresses are in most cases determined after being incorporated they are named residual stresses. In the case that the stresses are measured during loading they are named load stresses, but these will not be considered here.

The incorporation of stress can be associated with a large amount of mechanical energy, which might be released by crack generation, crack propagation and peeling of the layer from the substrate. Stored mechanical stress may thus have enormous deleterious effects on the film–substrate system. A delamination, which the thin-film grower painfully has to observe sometimes, often serves as the starting point for investigations of residual stresses contained in the film. Various analysis techniques may be used for this purpose and this chapter focuses on x-ray stress analysis (XSA). The presentation will proceed by increasing complexity as is visualized by the "ladder of complexity" in Fig. 6.1. Important results may already be obtained by assuming the thin film to be composed of isotropic material. Since the mechanical response of a polycrystal is isotropic only to a good approximation, diffraction elastic constants (DECs) have to be introduced to account for the true anisotropy of elastic behavior of crystalline matter. Complexity increases when the occurrence of texture has to be considered, which may hardly be avoided in thin-film analysis. Also other thin-film phenomena like the depth dependence of many growth-related properties including stress and the existence of porosity necessitate the development of a rather sophisticated methodology. It should not be forgotten finally that mechanical stresses stored in the film are in most cases accompanied by mechanical stresses stored in the substrate. It cannot always be avoided to develop an understanding of the full system instead of investigating only the layer component.

The measurement of residual stresses by x-ray diffraction proceeds on a length scale given by the inverse linear attenuation coefficient μ of the material under investigation. As has frequently been outlined this translates to lengths of several to a few hundred micrometers. The in-plane dimension of the irradiated volume, however, which is defined by the size of the beam spot and the angle of incidence, usually reaches several square millimeters. The stresses on these length scales are called macrostresses or stresses of the first kind. There also exist stresses of the second and third kind which are named microstresses and which are covered in Chapter 3. Microstresses typically level out over length scales that are much smaller than the x-ray penetration depth, but some aspects of the distinction between microstresses and macrostresses are addressed in this chapter. In order to cope with the following presentation the reader is assumed to be familiar with basic elasticity theory of crystalline matter, albeit some introductory remarks will be made.

Historically, residual stress analysis (RSA) by diffraction techniques was preceded by the requirements of the steel industry. Most of the methods we use today were developed to control the strength and mechanical properties of steel or other metallic or alloyed work pieces. Detailed information on the history of XSA and a comprehensive outline of the field can be found in Hauk et al. [1].

Compared to the residual stress analysis in metallography the same technique is still in a phase of methodological development for thin-film materials. Although intrinsic stresses were early recognized as an important structural feature in thin solid films and an enormous amount of work has already been done, the development of coherent models to describe the generation, distribution and evolution of stresses in thin films is still in progress. One reason for the complexity of the issue is due to the dynamic nature of the phenomenon. It has been revealed by various *in situ* investigations that thin-film stresses vary significantly during the process of

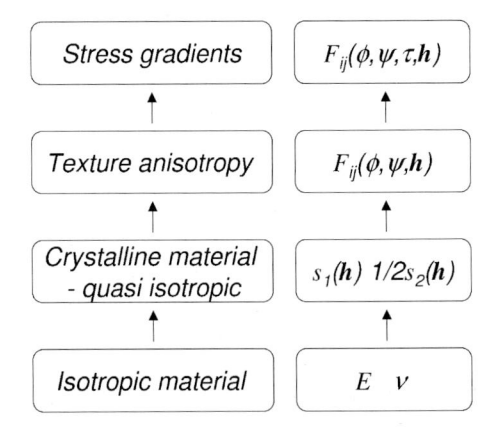

```
┌──────────────────┐      ┌──────────────────┐
│ Stress gradients │      │ F_ij(φ,ψ,τ,h)    │
└──────────────────┘      └──────────────────┘
         ↑                         ↑
┌──────────────────┐      ┌──────────────────┐
│Texture anisotropy│      │ F_ij(φ,ψ,h)      │
└──────────────────┘      └──────────────────┘
         ↑                         ↑
┌──────────────────┐      ┌──────────────────┐
│Crystalline material     │ s_1(h)  1/2 s_2(h)│
│ - quasi isotropic│      │                  │
└──────────────────┘      └──────────────────┘
         ↑                         ↑
┌──────────────────┐      ┌──────────────────┐
│Isotropic material│      │    E    ν        │
└──────────────────┘      └──────────────────┘
```

Figure 6.1 The "ladder of complexity" in x-ray residual stress analysis.

deposition – and even after the deposition is finished. This phenomenon evidently indicates a microstructural reorganization of the bulk of the film and is related to grain growth and grain boundary migration that occur on an extended timescale compared to the primary, more surface-related growth phenomena. Also other microstructural features like porosity and cavity networks, via which a postoxidation may occur, can significantly influence the stress state and its dynamics.

The illustrative examples presented in this chapter are selected from the group of nitrides like TiN, CrN, BN, etc., that have found application in the field of hard and protective coatings. They are used in widespread applications like diffusion barriers in semiconductor technology or for abrasive wear protection on mechanical cutting tools. The most popular example is probably represented by gold-colored TiN films, with which drilling tools and other mechanical working tools are coated to protect the steel via the mechanically hard nitride layer. Although even harder and more wear-resistant thin-film systems have been developed, a TiN topcoat is still widely used, since customers got used to the golden TiN color. The structure and some information on the application of these materials are given in Structure Box 7.

A discussion of the detectors used in x-ray diffractions instruments is given in Instrumental Box 7.

6.1
Ceiiinnosssttuv

The starting point of all XSA is Hooke's law. Although well known, it should be restated that it connects the induced elastic strain ε with the applied mechanical stress σ

$$\varepsilon = \frac{1}{E}\sigma \tag{6.1}$$

where the proportionality constant E is the elastic modulus. The archives of science history report that in 1677 Hook condensed his discovery in the anagram "ceiiinnosssttuv" by alphabetically ordering the Latin expression *ut tensio sic vis* [2]. It turns out that the formula given above is not sufficient to account for the full deformation process that is not only associated with an elastic strain in the direction of the applied stress. The effect is accounted for by the Poisson ratio ν describing the spatial variation of the body perpendicular to the load direction. In addition, the tensor properties of stress and strain were realized as shown in Fig. 6.2, where the nomenclature of exerting stresses with respect to the exposed surfaces is given for a coating on a rectangular-shaped substrate. The off-diagonal stress components σ_{ij}, $i \neq j$, are denoted as shear components. Evidently, both tensors are symmetric, $\sigma_{ij} = \sigma_{ji}$ and $\varepsilon_{ij} = \varepsilon_{ji}$, leaving each with only six independent components. The tensor character of stress and strain allows a reformulation of Hooke's law for each tensor component

Structure Box 7:

Rock Salt Structure

The hardness of materials is found to scale with the length of interatomic distances and to increase for smaller bond lengths. The phenomenon might be motivated by the steeper potential well to which the atoms are confined, i.e. by a relative increase of interatomic repulsion at short bond lengths. Therefore, hard coatings are mainly formed by compounds of light elements like boron, carbon, nitrogen and oxygen. The group of nitride compounds is by far the most significant one in coating technology. Widely used are transition metal nitrides like TiN, CrN, TaN, etc., and mixtures of them, (Ti,Cr)N, (Ti,Ta)N, etc. Some of them are listed in Table s7.1.

Table s7.1 Notations and symmetries in the rock salt structure.

Name	Space group (number)	Atomic positions		Point group of lattice points	Examples (unit cell edge a (nm))
Rock salt or halite structure (B1)	$Fm\bar{3}m$ (225)	000 ½0½ ½00 00½	0½½ ½½0 0½0 ½½½	O_h O_h	TiN (0.4242) CrN (0.4148) TaN (0.4340) HfN (0.4525)

The majority of these compounds crystallize in the rock salt structure, which first was revealed for the prototype compound NaCl (rock salt or halite). The space group is $Fm\bar{3}m$ (no. 225) and the Strukturbericht designation is B1, since it was the first structure of a binary compound solved by x-ray diffraction. The structure factor of a binary compound AB in rock salt structure reads

$$F_{rocksalt}(hkl) = \begin{cases} 4(f_A + f_B) & \text{all even} \\ 4(f_A - f_B) & \text{all odd} \\ 0 & \text{otherwise} \end{cases} \quad (s3.1)$$

It appears similar to that of the of the zinc blende structure, mostly because all $F(hkl)$ vanish for mixed hkl triple with both even and odd integers. The close relation between both structures stems from the fact that the rock salt and zinc blende structure are both composed from two interpenetrating fcc lattices. But while the shift between A and B type atoms is ¼¼¼ in ZnS it amounts to ½½½ in NaCl. In consequence as a characteristic feature to distinguish between the two pattern in a $\theta/2\theta$ scan the (hkl)-all-odd reflections are of low intensity – $16(f_A - f_B)^2$ – in rock salt structure compounds and of strong intensity – $16(f_A^2 + f_B^2)$ – in zinc blende-type structures. As a further consequence of the shift of ½½½ between A and B fcc lattices both atoms are octahedrally coordinated to their nearest neighbors in the rock salt structure obeying a point symmetry of O_h, which is visualized in Fig. s7.1.

Other relevant structures in hard nitride coatings are the wurtzite structure, observed for AlN and hexagonal BN (h-BN). There also exists a cubic polymorph of boron nitride, c-BN, which crystallizes in the zinc blende structure. In accordance with the bond length

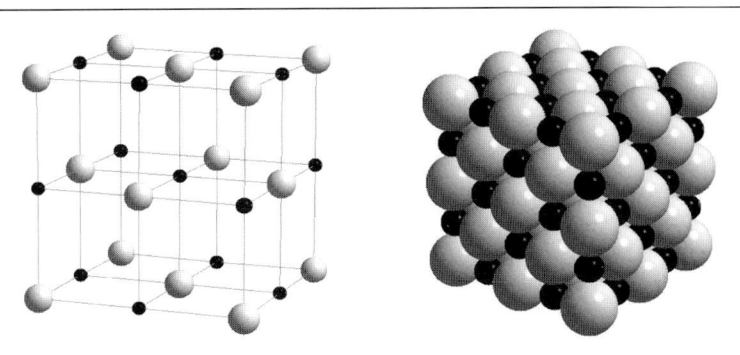

Figure s7.1 Two representations of the cubic rock salt structure.

rule given above, the hardness of c-BN is even higher than that of h-BN, because the B–N bond length is reduced in c-BN. Due to its enormous hardness c-BN receives great attention and is under development for hard coating applications. However, all nitride coatings suffer from severe adhesion problems and these problems have so far hampered the preparation of c-BN layers with thickness in the micrometer range. A further characteristic feature of the hard nitride compounds is the broad compositional range in which the structure is adopted. It is reported in the literature that titanium nitride films in the rock salt structure can be obtained for N/Ti ratios in a range between 0.7 and 1.3.

When applying these data in thin-film work, the values should be used with caution. Consult the references and further sources to learn about stoichiometry range of phases, polymorphism, metastable phases and variation of unit cell edges with applied stress. In principle, this remark holds for all compounds presented in the structure boxes. The problems, however, might become especially severe in the case of nitrides and can affect the interpretation of measured data.

$$\varepsilon_{ij} = \frac{1+v}{E}\sigma_{ij} - \delta_{ij}\frac{v}{E}\sum_{k}\sigma_{kk} \tag{6.2}$$

This is the generalized form of Hooke's law, from which the empirically well-known fact can be deduced that the compression (or tension) of an object causes its extension (or shrinking) normal to the load direction. It cannot be overemphasized that Eq. (6.2) is the basis for all the following derivations and it will be a valuable exercise trying to recognize Hooke's law from the following equations.

A second fundamentally important point is that x-ray *stress* analysis is founded on the measurement of *strains*. By carefully measuring the position of a Bragg reflection the interplanar spacing d can be obtained with high precision. The strain is derived by comparing with the interplanar spacing d_0 of the unstressed case. Figure 6.3 shows the 422 reflection from one of the first XSA investigations of thin TiN films [3]. The reflection was measured under various tilt angles ψ of the sample with respect to the scattering plane and the Bragg peak is seen to exhibit pro-

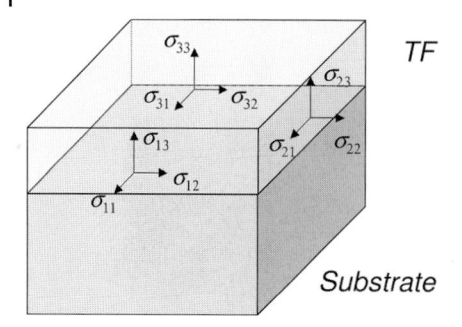

Figure 6.2 Definition of sample-related stress tensor σ_{ij} (TF = thin film, load direction along e_j and surface normal direction along e_i). For reasons of clarity only the stresses on visible surfaces are indicated.

nounced shifts with varying orientation. The induced strain ε is accordingly the measurement quantity, while the stress σ can only be derived by inserting a reliable proportionality constant. A great deal of work is related to an appropriate modeling of this proportionality as will be seen. In those cases where allowed, the analysis may be restricted solely to strain, since this is the more clearly defined quantity. This might be a useful strategy whenever an optimization of the deposition process for properties like film–substrate adhesion is required. In other cases a full stress analysis is unavoidable, which holds in particular when a thorough understanding of the stress and strain fields within the layer and the relation of both to the underlying growth mechanisms are required.

Strains and strain fields in thin films are measured by rotating and tilting the sample with respect to the scattering vector in a manner comparable to texture analysis. For various azimuth angles ϕ and tilt angles ψ the strain $\varepsilon_{\phi\psi}$ is determined from the normalized lattice spacing difference $\Delta d/d_0$ according to

$$\varepsilon_{\phi\psi} = \frac{d_{\phi\psi} - d_0}{d_0} \tag{6.3}$$

The strain may be larger or less than zero causing a peak shift to either lower or higher scattering angles (higher or lower d values). In the majority of deposition experiments the process geometry is axially symmetric as are the structural properties of the film. An important example is the occurrence of fiber textures exhibiting a preferred crystallite orientation of rotational symmetry around the substrate normal as introduced in Chapter 5. This kind of sample symmetry is also often observed in XSA, where either the so-called biaxial stress model or the biaxial strain model accounts for the phenomenon. In these cases it suffices to monitor the strain solely as a function of tilt angle ψ

$$\varepsilon_\psi = \frac{d_\psi - d_0}{d_0} \tag{6.4}$$

For stress fields in thin films that follow the biaxial stress model a single in-plane stress component $\sigma_\parallel$ is sufficient to account for the full stress tensor. This in-plane

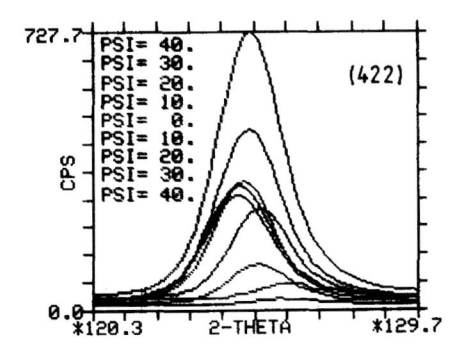

Figure 6.3 Basic working principle of XSA demonstrated for a 422 reflection of a TiN film on stainless steel. For various sample tilts ψ the peak shifts in accordance with the different strains to which the lattice planes are subjected (from Ref. [3].)

stress may either be compressive ($\sigma < 0$) or tensile ($\sigma > 0$) which is associated with a contraction or an elongation of lattice spacings along in-plane directions. The notation is visualized in Fig. 6.4, where it can be seen that an in-plane contraction occurs concomitantly with an out-of-plane elongation and vice versa and the coupling, again, is determined by Poisson's ratio ν. The schematic in the figure visualizes this effect by a bowing of the complete substrate-film system. This bending may also serve in deriving the residual stress by measuring the curvature and applying Stoney's formula (predicting the bending curvature R to scale with the in-plane stress σ as $\sigma R = Et^2_{Sub}/(1-\nu)/t/6$). However, it should be noted that the occurrence of compressive or tensile stresses is not necessarily accompanied by a convex or a concave bowing of the sample [4]. This fact is due to the different kinds of mechanical stresses as will be shown. In many cases, however, a bowing as depicted in Fig. 6.4 is indeed observed in accordance with compressive and tensile stresses and the picture may be memorized as a visualization of the nomenclature and the lattice spacing variations.

Finally, it is pointed out that XSA makes fundamental use of Hooke's law by evaluating residual stresses, while Hooke's law gives the stress–strain relation under

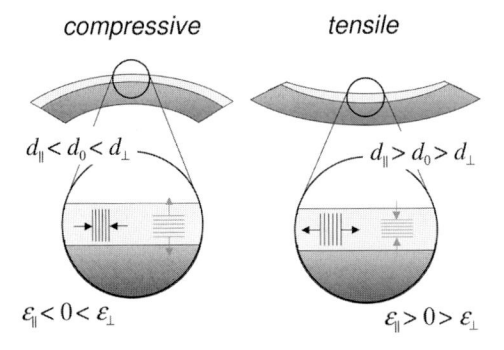

Figure 6.4 Typically bowed film–substrate composite under compressive and tensile stress in the film plane (exaggerated).

load. This is only a seeming contradiction, since one considers the internal stresses as a load to which the material system is subjected.

6.2
Fundamental Equation of XSA

The fundamental equation of XSA is obtained by performing a tensor transformation followed by the insertion of the generalized form of Hooke's law. By measuring the strain for a selected set of azimuth ϕ and tilt angle ψ a strain value is obtained

$$\varepsilon_{33}^L = \varepsilon_{\phi\psi} \qquad (6.5)$$

that specifies the $ij = 33$ component of the strain tensor in the laboratory reference frame $\{l_i\}$. We are interested, however, in the ε_{ij}^S components of the strain tensor in the sample reference frame $\{s_i\}$. The orthogonal matrix a_{ij}^{LS} that facilitates the transformation from the laboratory to this sample system has already been introduced and is of the form

$$a_{ij}^{LS} = \begin{pmatrix} \cos\phi\cos\psi & \sin\phi\cos\psi & -\sin\psi \\ -\sin\phi & \cos\phi & 0 \\ \cos\phi\sin\psi & \sin\phi\sin\psi & \cos\psi \end{pmatrix} \qquad (6.6)$$

According to the rules of tensor algebra the 33 component of a second-rank tensor transforms by twofold multiplying with the third column of a_{ij}^{LS} and adding over all products

$$\varepsilon_{33}^L = \sum_{i,j} a_{3i}^{LS} a_{3j}^{LS} \varepsilon_{ij}^S \qquad (6.7)$$

Insertion of the transformation matrix and ordering them by ε^S components yields

$$\varepsilon_{33}^L = \begin{matrix} \varepsilon_{11}^S \cos^2\phi \sin^2\psi & +\varepsilon_{12}^S \sin 2\phi \sin^2\psi & +\varepsilon_{13}^S \cos\phi \sin 2\psi \\ +\varepsilon_{22}^S \sin^2\phi \sin^2\psi & +\varepsilon_{23}^S \sin\phi \sin 2\psi \\ +\varepsilon_{33}^S \cos^2\psi \end{matrix} \qquad (6.8)$$

This equation is of general validity for any transformation of strains measured in the laboratory system $\{l_i\}$ to those in the sample reference frame $\{s_i\}$. By use of this equation the relation between the sample-related stresses σ_{ij}^S and the strain measured in the laboratory frame ε_{33}^L may be established by inserting the stress–strain relation of each tensor component, Eq. (6.2), which results in

$$\varepsilon_{33}^L = \left(\frac{1+\nu}{E}\right) \begin{bmatrix} \left(\sigma_{11}^S \cos^2\phi + \sigma_{22}^S \sin^2\phi + \sigma_{12}^S \sin 2\phi\right)\sin^2\psi \\ +\left(\sigma_{13}^S \cos\phi + \sigma_{23}^S \sin\phi\right)\sin 2\psi + \sigma_{33}^S \cos^2\psi \end{bmatrix} \\ + \left(-\frac{\nu}{E}\right)(\sigma_{11}^S + \sigma_{22}^S + \sigma_{33}^S) \qquad (6.9)$$

This is the so-called fundamental equation of x-ray residual stress analysis. In the case of the biaxial stress model the stress tensor in $\{s_i\}$ is of the form

$$\sigma_{ij}^S = \begin{pmatrix} \sigma_\| & 0 & 0 \\ 0 & \sigma_\| & 0 \\ 0 & 0 & 0 \end{pmatrix} \tag{6.10}$$

which indicates that all residual stresses vanish except the in-plane component. For this case the fundamental equation of XSA reduces to

$$\varepsilon_\psi = \frac{1+v}{E}\sigma_\| \sin^2\psi - \frac{2v}{E}\sigma_\| \tag{6.11}$$

The strain ε in the biaxial stress model is seen to depend only on tilt angle ψ and to be independent of azimuth ϕ. If the strain or the normalized difference of lattice spacing is plotted versus $\sin^2\psi$, the in-plane stress $\sigma_\|$ can be extracted from the slope of the curve when the elastic modulus and the Poisson ratio are inserted. This is the well-known $\sin^2\psi$ technique, which is usefully performed at the beginning of any XSA. The technique gives a first estimate of the in-plane stress, but its accuracy should not be overestimated in thin-film work.

Such a d–$\sin^2\psi$ relation is also named a d distribution, and Fig. 6.5 displays two of them for hypothetical TiN coatings that were calculated for a compressive stress (left) and a tensile stress (right) of $\sigma_\| = \pm 1$ GPa. In these simulations the d values were assumed to account for 420 reflections measured with Cu Kα radiation. For the compressively stressed film the d values are seen to decrease with increasing $\sin^2\psi$, which can be understood from an inspection of Fig. 6.4. In the left-hand plot the interplanar spacing d is seen to extend in the out-of-plane direction, $d_\perp > d_0$, and to shrink along the in-plane directions, $d_\| < d_0$. With an increase of the tilt angle ψ from 0 to 90° the diffractometer switches from the measurement of one set of spacings to the other. The course of d distribution in case of the tensile stressed film can be understood along the same line of argument.

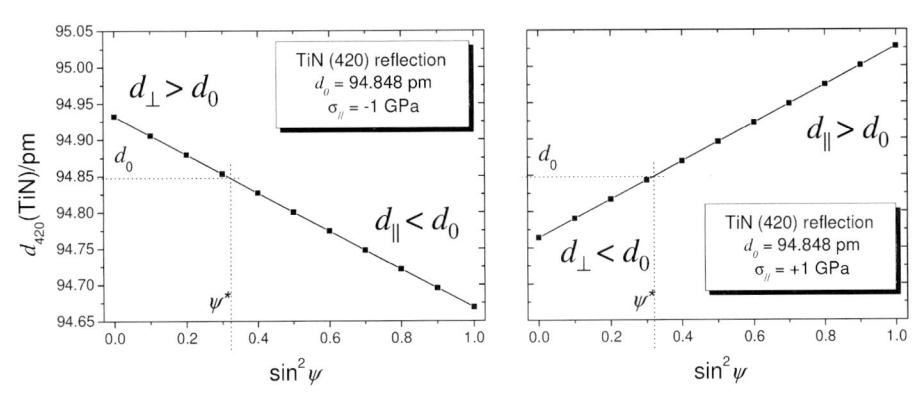

Figure 6.5 Course of simulated $d(\psi)$ distribution of thin TiN film under compressive (left) and tensile (right) stress. The in-plane stress was assumed to be 1 GPa.

A special pair of values, d_0 and ψ^*, have been emphasized in the plots. They both account for the strain-free direction, i.e. for this tilt angle ψ the film is oriented with respect to the scattering plane such that the scattering vector probes the same interplanar spacing as in an unstrained sample. An asterisk marks the tilt angle at this special position. Since $d_\psi = d_0$ holds the strain vanishes for reflecting lattice planes tilted by ψ^*. It is derived from the fundamental Eq. (6.11) that for the biaxial stress model being valid the strain-free tilt angle is related to the Poisson ratio of the measured layer via

$$\sin\psi^* = \sqrt{\frac{2\nu}{1+\nu}} \tag{6.12}$$

Again, this relation only holds for isotropic material and has to be improved if crystalline anisotropy is included. The equation may also be used for the determination of the Poisson ratio when the unstrained lattice spacing d_0 is exactly known and the measured d values are highly precise. It has to be emphasized, however, that the precise determination of d_0 is one of the difficult tasks in XSA.

In the previous section the biaxial stress model is introduced *ad hoc*, but the adequacy of the approach can be justified by the equilibrium conditions the stress tensor has to obey. Since we are dealing with residual stresses there are no sources of the stress field, which translates into the equation

$$\sum_{j=1}^{3} \frac{\partial \sigma_{ij}}{\partial x_j} = 0 \tag{6.13}$$

and reads for each component

$$\frac{\partial \sigma_{11}}{\partial x} + \frac{\partial \sigma_{12}}{\partial y} + \frac{\partial \sigma_{13}}{\partial z} = 0$$

$$\frac{\partial \sigma_{12}}{\partial x} + \frac{\partial \sigma_{22}}{\partial y} + \frac{\partial \sigma_{23}}{\partial z} = 0$$

$$\frac{\partial \sigma_{13}}{\partial x} + \frac{\partial \sigma_{23}}{\partial y} + \frac{\partial \sigma_{33}}{\partial z} = 0 \tag{6.14}$$

Regarding the third equation and taking additionally into account the boundary condition for free external surface (see Section 6.8.1), $\sigma_{33} = 0$ must evidently hold for the topmost layer, since the film–substrate system is in equilibrium and no tractions are acting on its surfaces. There is no reason to assume why σ_{33} should deviate from zero over the total thickness of the film and thus also the derivative of σ_{33} with respect to z will vanish. From these boundary conditions one may also deduce the other σ_{i3} components to vanish, $\sigma_{13} = \sigma_{23} = 0$. We have now argued five out of nine components to equal zero, or – if the symmetry of the stress tensor is considered – three out of six components to vanish. We are thus left only with the three components σ_{11}, σ_{22} and σ_{12}. It is clear from the cylindrical symmetry around the substrate normal that σ_{11} and σ_{22} have to be equal. The shear component σ_{12} will only deviate from zero in these cases when surface finishing procedures with traction components parallel to the film's surface are applied, but in most cases $\sigma_{12} = 0$ is valid. We are finally left with a single in plane stress component $\sigma_{11} = \sigma_{22} = \sigma_\parallel$.

The biaxial stress model will account for most of the examples presented in this chapter. Extensions of this approach will be given with respect to an accompanying texture or a stress gradient within the film. Samples with nonzero shear stress components σ_{ij}, $i \neq j$, will only briefly be considered, because they are rarely of relevance in thin-film work. It is just mentioned that nonvanishing shear components cause a splitting of the $\sin^2 \psi$ curve into an upper and a lower branch both of which form an ellipsoid in the $\sin^2 \psi$ graph. For further information the reader is referred to the particular literature mentioned at the end of the chapter. Our restriction to the biaxial stress model is associated with a further simplification: in the most general form of stress analysis a further coordinate system of principal stresses has to be considered next to the three reference frames $\{s_i\}$, $\{l_i\}$ and $\{c_i\}$ already in use. Here, the coordinate system of principal stresses will be assumed to coincide with the sample reference frame.

6.3
Measurement of d_ψ Distributions

Before the x-ray residual stress analysis is accomplished for the sample of interest, the operator should be aware of the precision and accuracy in d value determination that may be achieved with the instrument. A pre-analysis can be performed by measuring a standard sample in the symmetric $\theta/2\theta$ configuration and comparing measured and expected d values as a function of scattering angle 2θ for each of the standard's Bragg reflections hkl. Also the determination of d values of the standard from $\theta/2\theta$ diffraction patterns under varied tilt angles ψ is strongly recommended. An imprecise diffractometer may introduce large systematic errors into the strain values to be evaluated. In the subsequent stress analysis of a thin-film sample some constraints should be considered regarding the choice of an appropriate reflection. In the classic XSA of voluminous steel samples reflections from the high scattering angle range 2θ of about 160° are selected. For this purpose higher wavelength tubes like Cr Kα (0.229 nm) or Co Kα (0.179 nm) are applied which additionally reduce the Fe x-ray fluorescence compared to Cu Kα. By choosing larger wavelength tubes the whole reflection pattern is expanded to higher scattering angles such that many of the classic XSA investigations deal with the ferrite 211 or 310 reflections that appear at 156° for Cr Kα and 161° for Co Kα. The shift of reflections to higher scattering angles is a reliable method for infinitely thick polycrystals, since a high accuracy in the measured d value is desired and some systematic errors of the peak position decrease with increasing scattering angle (see Chapter 2). Also the Lorentz polarization factor rises around 160° above its minimum level at 90° and enables high-count-rate measurements in this backscattering configuration. In measuring thin films, however, reflections at high scattering angles are critical due to the strong damping caused by the absorption factor and finite film thickness and due to peak broadening caused by imperfect microstructure ($\sim 1/\cos\theta$). These two effects might cause thin-film Bragg reflections at high scattering angles to become hardly measurable.

One should therefore begin with the measurement of a complete $\theta/2\theta$ pattern in the symmetric configuration with average resolution. If possible, the wavelength should be adopted to the film–substrate system by choosing an appropriate x-ray tube. Even in thin-film work high scattering angles might be a reliable choice sometimes and the user should check this question with other x-ray tubes when available. Evidently, this point depends on the material under investigation and it should also be considered that the λ^3 dependence of the integral intensity might help reflections to become measurable at higher 2θ values. From the thin-film powder pattern the strongest reflection should be selected that can safely be measured in an adequate amount of time with a reliable signal-to-noise ratio and for a ψ angle range as large as possible. The chosen peak should be scanned with high resolution in the coupled $\theta/2\theta$ mode for various tilts, starting with $\psi = 0$. The set of tilt angles ψ should yield an equidistant net in the $\sin^2\psi$ plot (see Table 6.1). Peak fitting is accomplished with the usual model functions in order to determine precise peak centroids from which the d values are calculated. In XSA of bulk materials other centroid-determining procedures like the parabola method, etc., are in use, but the use of these techniques might introduce systematic errors and should be avoided in thin-film work. Instead, the peak fitting functions and procedures as outlined in Chapter 3 should be applied. In particular, the division of measured intensity data by the Lorentz polarization factor Lp and the absorption factor A may be required to correct for asymmetries in measured reflection profiles. Finally, the set of d values for different tilt angles ψ then has to be plotted versus $\sin^2\psi$ as shown in Fig. 6.5. It is sometimes common in XSA of cubic materials to convert measured d values into unit cell edges a, causing the ordinate to expand by the factor $\sqrt{h^2 + k^2 + l^2}$. Of course, this has no effect on the stress to be evaluated, but it points to the fact that any d versus $\sin^2\psi$ or a versus $\sin^2\psi$ plot should be documented by indicating the Miller indices (hkl) for which the measurement was performed.

Table 6.1 Tilt angle values ψ for equidistant $\sin^2\psi$ plot.

$\sin^2\psi$	0.1	0.2	0.3	0.4	0.5	0.6	0.7	0.8	0.9
ψ (°)	18.43	26.57	33.21	39.23	45	50.77	56.79	63.43	71.57

An important point in the stress analysis is related to the determination of the strain-free direction ψ^*. A common procedure is based on the presupposition that a straight line can safely approximate the d–$\sin^2\psi$ distribution. In this case d_0 is determined by the value of $\sin^2\psi^*$ according to the formula for the strain-free direction, Eq. (6.12), which allows its calculation by inserting the Poisson ratio v of the material. With the knowledge of d_0 the d distribution might be converted into a strain distribution ε_ψ according to Eq. (6.4). Measurement results are sometimes presented as a strain versus $\sin^2\psi$ function, but it should be borne in mind that the assumption of knowing the material parameter v has already entered this kind of presentation. The plot of d–$\sin^2\psi$ distribution as a real measurement quantity is thus preferred, since it is free of any underlying assumptions. In addition, it is good practice in XSA to evaluate in addition to the $d(\psi)$ value also the integral intensity $I_h(\psi)$ of a reflection and to display a combination of both quantities, i.e. interplanar spacings d and integral intensities I_h as a function of $\sin^2\psi$. An example from a 422 Bragg reflection of a 4.9 μm thick TiN layer [5] is given in Fig. 6.6. This sample was identified in prior pole figure measurements to exhibit a 111 layer texture leading to a pronounced modulation of 422 peak intensities as shown in the lower plot.

It is recommend to measure at least 10 points of a $d(\psi)$ curve, since the d distribution is only rarely approximated by a straight line in thin-film analysis. Rather often the strain versus tilt plot exhibits a pronounced bending or oscillatory behavior as may also be realized from the upper plot in Fig. 6.6. Such a bending would not be expected for an elastically isotropic material, but it is a phenomenon frequently encountered with thin-film samples as outlined below. In these cases the density of measured points should be enlarged in the ranges of nonlinear behavior.

Asymmetric deposition processes or postdeposition surface treatments may introduce off-diagonal components, $\sigma_{ij} \neq 0$ for $i \neq j$, into the residual stress tensor. It is evident from the inspection of Eq. (6.9) that this will be associated with a dependency on the azimuth ϕ in the measured strain component. Off-diagonal components in the intrinsic stress tensor may thus be revealed by shifts in the peak position with varying rotation angle ϕ. In order to elucidate such effects the procedure has evolved of measuring the peak position for both positive tilt ψ_+ and negative tilt ψ_- (also indicated in Fig. 6.6). The two sets of d values obtained correspond to measurements for $\phi = 0$ and $\phi = 180°$. For nonvanishing off-diagonal stress components σ_{ij} the two $d-\sin^2\psi_\pm$ distributions will separate into an upper half and a lower half

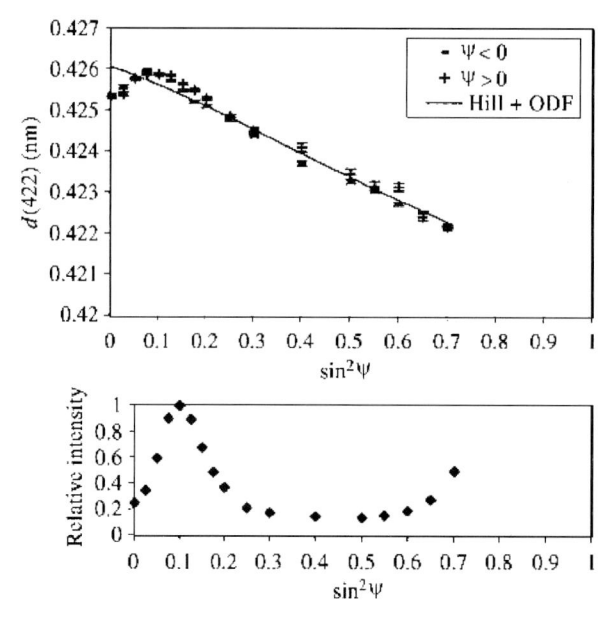

Figure 6.6 Lattice spacing d and relative intensities of 422 reflection from a 4.9 μm thin TiN film (from Ref. [5]).

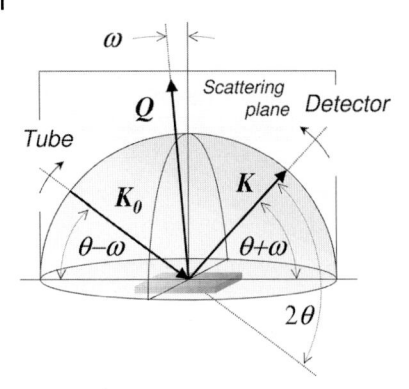

Figure 6.7 Bragg reflection measurement in the Ω mode with θ axis offset by ω, seen from the sample reference frame.

exhibiting the appearance of a bent ellipse in the $(\sin^2\psi, d)$ plane. In routine XSA work strain distributions should thus be monitored for positive and negative tilt angles ψ_+ and ψ_-. For the biaxial stress model being valid the two strain distributions will coincide on one single line and the occurrence of shear stress can safely be excluded.

As in texture analysis the determination of strain may be performed in either the Ψ mode or the Ω mode of the diffractometer. And again, the measurement may be severely influenced by the effect of defocusing with increasing deviation of the diffractometer configuration from parafocusing geometry. Also in x-ray residual stress analysis the use of parallel beam geometry is therefore highly recommended. A polycapillary lens is most advantageously applied in the Ψ mode, whereas a graded multilayer mirror or a polycapillary could be used in the Ω mode.

Strain measurements in the Ω mode do not proceed via the measurement of a rocking curve, for which the scattering angle is fixed to twice the Bragg angle $2\theta_B$ of the reflection under investigation, while the diffraction angle θ is rotated from smaller values $\theta_B - \omega$ than θ_B to larger ones $\theta_B + \omega$. Regarding the Ω mode in strain analysis a 2θ scan is employed, but this time the angle of sample rotation is offset from half the scattering angle $2\theta/2$ by an amount ω. This offset is varied and various $(\theta + \omega)/2\theta$ scans are made for each of which the position of the Bragg reflection θ_B is determined. Instead of a set of d–$\sin^2\psi$ data the XSA measurement in the Ω mode yields a d–$\sin^2\omega$ distribution. The Ω mode configuration is visualized in Fig. 6.7. It can be seen that the scattering angle is tilted by the magnitude of the offset angle ω from the substrate normal, albeit the configuration is still a coplanar one. For vanishing offset, $\omega = 0$, a conventional 2θ scan of the reflection in the symmetrical $\theta/2\theta$ configuration is performed. The variation of ω in the Ω mode fully compares to the variation of ψ in the Ψ mode, although the attainable tilt angle range is larger in the Ψ mode. In the latter any tilt angle between 0° and 90° may be chosen, while the Ω mode allows for maximum offset angles of about the Bragg angle, $\omega_{max} < \theta_B$. This sometimes restricts the usable tilt angles to unacceptable small ranges, which may hold especially for reflections at low scattering angles $2\theta_B$.

One should be aware of the variation of penetration depth during the tilting of the sample in both measurement modes. The effect is of particular concern when the intrinsic stress is suspected to exhibit a pronounced depth gradient. Penetration depths $\tau_{1/e}$ and information depth $\bar{\tau}$ can be calculated by virtue of formulas derived in Chapter 4. For this purpose the appropriate configuration factors k_ψ and k_ω have to be inserted that have already been in given in Eqs. (5.4) and (5.22). Figure 6.8 displays the penetration depth $\tau_{1/e}$ for the 200 reflection of a cubic TaN layer measured in both diffractometer configurations using Cr Kα radiation, where the Bragg angle of the peak is 63.72°. In the Ω mode the measurement range is restricted to tilt angles $\omega < 2\theta_B/2$ and $\tau_{1/e}$ is seen to decrease steeply to very low values when approaching ω_{max}. For tilt angles close to half the scattering angle the Ω mode becomes rather surface sensitive as is made use of in the grazing incidence x-ray diffraction (GIXRD) technique. Larger tilt angle ranges are accessible in the Ψ mode and the penetration depth remains nearly constant for a wide range of tilt angles. The variation of penetration depth realized from the plot points to a stress gradient as one possible cause for a nonlinear d–$\sin^2\psi$ distribution: because the x-rays reach different sample depths for various tilt angles ψ various stress states are probed causing the linear term in Eq. (6.11) to vary with ψ. The stress variation would cause a convex (concave) bending of the d–$\sin^2\psi$ distribution for an in-plane stress $\sigma_\parallel$ monotonically increasing (decreasing) with depth. If the stress gradient is not too pronounced a linear fit might reveal a reliable approximation to the correct stress–strain relation. The problem is considered in more detail in Section 6.9.

It is concluded that the determination and analysis of a $\sin^2\psi$ plot gives interesting initial results regarding the stress state in a thin film. Although the measured points of the strain distribution do not always yield a straight line, a linear fit should be performed anyway. The in-plane stress is approximately derived from the slope of this fit and the known isotropic elastic constants E and ν. If a straight line

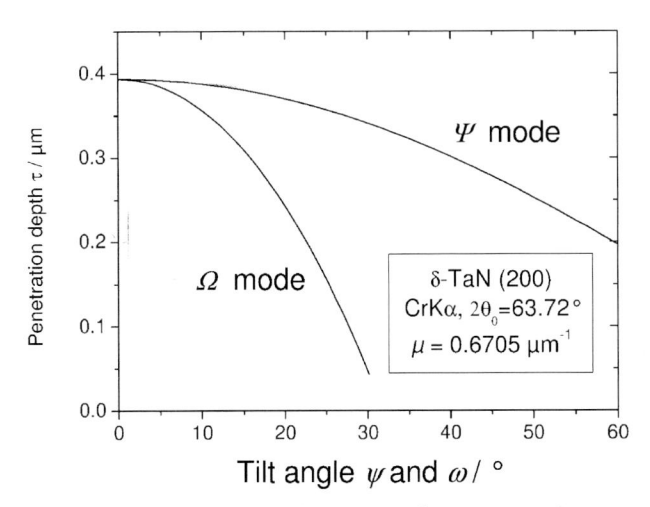

Figure 6.8 Penetration depths for TaN 200 reflection in Ω and Ψ mode recorded with Cr Kα radiation.

fits the measured data appropriately the derived in-plane stress may be considered a reliable result. The procedure might suffice to obtain useful input parameters for the optimization of a deposition process. Caution should be taken, however, when accurate stress values are desired or the interrelation between growth and intrinsic stresses is studied on a quantitative basis. Especially when highly textured films or layers with large intrinsic stresses are considered, underlying implicit assumptions may lead to large systematic errors in the stress values obtained from a linear fit of a d–$\sin^2\psi$ distribution.

Having arrived at this point one might have finished. Maybe one was able to measure a d distribution over a large ψ range and the data obtained could be fitted by linear model yielding reliable fit parameters. One might have even checked that the bowing of the film–substrate composite and the analysis by Stoney's formula yielded comparable in-plane stresses. Then one might stop here: one has a result. However, it may also happen that not all of the points mentioned above are obeyed, that – even worse and more probable – the $\sin^2\psi$ distribution exhibits no linear behavior, but is significantly bent or modulated. In this case it will be necessary to climb up the "ladder of complexity" (Fig. 6.1) of which we have just presented the bottom level. The following sections might then be helpful to learn about how the analysis has to be extended. In order to climb up the first rung we have to consider more thoroughly the elastic properties of polycrystalline aggregates.

Instrumental Box 7:

Detectors

Detectors convert the received x-ray photons into electrical pulses that are individually counted. The unit of the measured intensity scattered by the sample thus is in counts or counts per second (cps). Detectors may be distinguished by their dimensionality as point, line or area detectors (zero-, one- and two-dimensional systems). They may also be distinguished by their working principle, from which three types are most widely used. These are proportional counters, scintillation counters and electronic diode solid-state detectors (see Table i7.1 for a comparison).

Table i7.1 Properties of different detector types.

	Proportional counter	Scintillation counter	Solid-state detector
Working principle	Electron–ion pair cascade	Conversion to light and electrons and electron multiplication	Generation and separation of electron–hole pairs
Transducing medium	Xe	NaI:Tl	Si, Ge
Noise rate (cps)	1	0.1–10	<0.1
Maximum count rate (cps)	10^5–10^6	10^5–10^6	10^4
Resolution $\Delta E/E$ (%)	18–20	40–50	depends on type
Costs	Low	Medium	High

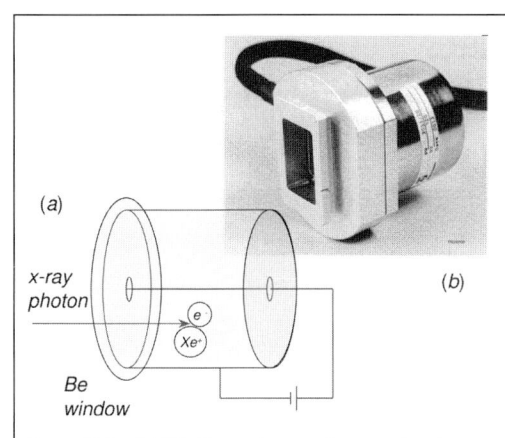

Figure i7.1 (a) Working scheme of proportional counter and (b) photograph of a commercial system (kindly provided by PANalytical).

The proportional counter basically consists of a metallic wire in a cavity filled with a working gas like Xe (Fig. i7.1). The wire is positively biased with voltages typically in the 1–2 kV range. X-rays enter via a Be window into the active gas volume, where they can cause the formation of an electron–ion pair by ionization. The electron is accelerated towards the wire and causes subsequent ionization along its path. The number of electron–ion pairs N produced in a cascade from one x-ray photon depends on the x-ray energy E and the ionization energy I of the active gas and maximally becomes $N = E/I$. One x-ray photon causes an electrical pulse in the wire and by an analysis scheme called pulse height discrimination the energy E can be determined. The energy resolution $\Delta E/E$ of proportional counters is in the 20% range.

The minimum count rate of a detector is determined by electronic noise and the monitoring of background radiation that stems from the hard component of cosmic radiation. It amounts to about 1 cps for the proportional counter. The maximum count rate follows from the time constants of the transduction processes. In a proportional counter the cascade is quenched by the addition of a quenching gas like CH_4 or CO_2. The time constants of one full detection process consisting of cascade ionization, quenching and an appropriate dead time for recovery of the system amounts to about 1 ms yielding a maximum count rate of about 10^6 cps for a proportional counter. Finally, the dynamic range of a detector is within the minimum and maximum count rate.

In a scintillation counter an x-ray photon is first converted to a light pulse; second, the light is converted to electrons and, third, the small number of generated electrons are multiplied to a large number in order to arrive at measurable current amplitudes (see Fig. i7.2). The scintillation counter has its basis in the optical detection of ionizing radiation by phosphorizing screens emitting light pulses whenever a high-energy particle or photon impinges. Although scintilla is the Greek word for spark, these light pulses are only of very low intensity. In a scintillation counter an additional photomultiplier tube that converts the light to electrons by the outer photoelectric effect performs the detection of the optical photons. These electrons are accelerated towards anodes of increas-

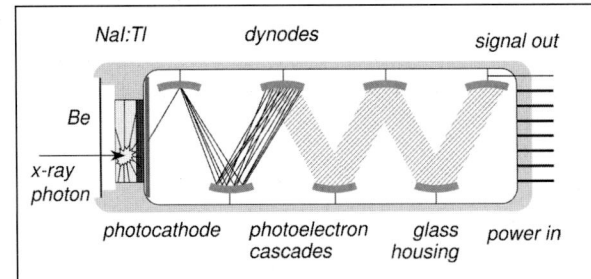

Figure i7.2 Working scheme of a scintillation counter.

ing potential (dynodes) causing the generation of further, secondary electrons and thus leading to a multiplication of the electron current. In a typical scintillation counter the phosphorizing material is made of Thallium-doped NaI of an optimized thickness to absorb as many x-ray photons as possible and transfer as many optical photons as possible to the second converter.

Whereas the multiplication of charges proceeds in the gas phase or in vacuum for proportional and scintillation counters, respectively, so-called solid-state detectors make use of the separation of electron–hole pairs generated in the active region of semiconductor devices. There are mainly two types in use, which are either based on p–i–n diodes made from Li-drifted Si or Ge or derive from charge-coupled devices (CCD). In Li-drifted p–i–n diodes impinging x-rays generate electron–hole pairs via the photoelectric effect. These charge carriers are subjected to the built-in electrical field in the intrinsic region of the diode from where the electrons or holes drift to the n- or p-type doped regions, respectively. CCDs basically are pixel arrays of transistors sensitive to visible light. The x-ray photon thus has to be converted into visible photons, which may be performed by a phosphorizing layer as in a scintillation counter.

Solid-state detectors are operated under cooling in order to improve the signal-to-noise ratio by suppressing thermally excited charge carriers. For a long time liquid nitrogen (LN_2) was used for this purpose, but recently the less complicated Peltier cooling has been introduced for various systems. The voluminous LN_2 tank necessary for the continuous cooling of solid-state detectors prohibited their use in x-ray diffraction, where the detector has to be moveable via the goniometer arm around the sample. Rather in x-ray spectroscopic applications stationary Si(Li) or Ge(Li) detectors were used. However, in the field of x-ray diffractometry proportional and scintillation counters were the detector workhorses for many years. With the introduction of Peltier cooling and new detector principles the use of solid-state detectors for x-ray diffraction applications received increased significance.

It can be seen from Table i7.1 that the energy resolution $\Delta E/E$ of the proportional and scintillation detector is about 20% or higher. Since the energy separation between α–β radiation amounts to 10% for Cu K, for instance, these detectors cannot perform α–β separation by adjustment of the energy window. It is this feature of x-ray detectors from which the occurrence of various elements in the diffractometer beam path like β filters and monochromators may be understood.

The counting statistics of all x-ray detectors may be assumed to follow a Poisson distribution, since individual counts are independent. It follows that the standard deviation for a large count number N is $\sqrt{N}$. Moreover, it can be concluded from the rules of error propagation that the absolute error amounts to $1/\sqrt{N}$. To be explicit: the "true" count number is found with a 68% probability within $[N - \sqrt{N}, N + \sqrt{N}]$ and 95% probability within $N \pm 2\sqrt{N}$, and so on. It has to be emphasized that the basic assumption of independent counts is only valid in the linear range of the detector.

For some measurements like in x-ray reflectometry (XRR) or high-resolution configurations a dynamical range of more than five or six orders of magnitude has to be covered. A practical approach to extend the dynamic range is by scanning the different count rate regions sequentially. The high count rate region – low 2θ in XRR – could either be measured with reduced generator power or by introducing an absorber like a thin metal plate into the beam path. The low count rate region is afterwards measured under conventional conditions and the full measurement range is obtained by the combination of both. Some diffractometer systems are equipped with an automatic beam attenuator that is activated when the count rate exceeds a specified intensity value.

In a conventional $\theta/2\theta$ scan the focusing condition would only be strictly obeyed for the detector at scattering angle $2\theta_0$ and inclination angle of the incident beam θ_0. In the vicinity of $2\theta_0$ the focusing condition is more or less violated the further the deviation from $2\theta_0$ becomes. However, the data collected around $2\theta_0$ are not useless, but just lead to an additional instrumental broadening of detected Bragg reflections. This principle is made use of by so-called position-sensitive detectors (PSDs) that extend over a certain 2θ range, which amounts for instance to 50 mm or 12° for a 240 mm radius goniometer in the case of the popular PSD from Braun. Essentially, this PSD operates like a proportional counter with the anode wire oriented parallel to the tangent of the goniometer circle. Position sensitivity is achieved by measuring the running time of the voltage pulse that is generated by one x-ray photon and received at both ends of the wire.

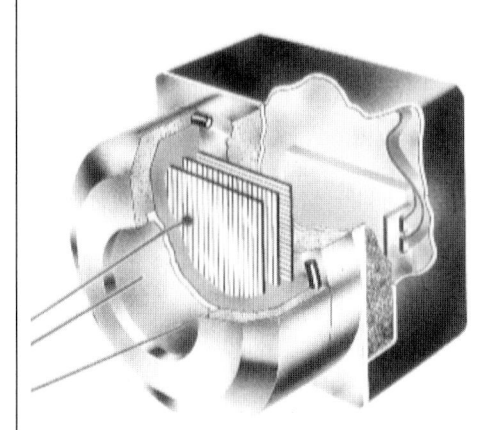

Figure i7.3 Working scheme of two-dimensional wire detector (kindly provided by Bruker AXS).

> PSDs may also be realized on the basis of electronic solid-state devices. Recently, such a detector of real-time multiple strip (RTMS) technology has been introduced having an active length of 9 mm in the scattering plane and consisting of about 100 parallel stripe-shaped solid-state detectors of 15 mm in width. For a goniometer radius of 240 mm a resolution of 0.0021° in 2θ is achieved. The use of one-dimensional detectors may significantly reduce the measurement time or improve the signal-to-noise (S/N) ratio for the same scan time. A measurement time reduction factor of about 10^2 could be demonstrated for instance for the RTMS detector to arrive at the same S/N ratio for a powder sample when compared to a conventional proportional counter scan. This point might be of utmost importance in thin-film work, where often only a small sample volume is available and any measure has to be taken to improve counting statistics.
>
> Wire detectors based on the principle of the proportional counter have not only been introduced for line detectors, but also for area detectors. A schematic representation of one commercially available two-dimensional multiwire proportional counter is shown in Fig. i7.3. It consists of two perpendicular sets of (2×512) or (2×1024) wires and exhibits an active circular area of 115 mm diameter.
>
> If one- and two-dimensional detectors are not bent and do not adapt to the goniometer radius R, the distance to the sample varies for different sensing elements. This has severe effects on the analysis of collected data, since measured intensities have to be corrected, line profiles are skewed and other corrections become necessary. One should be aware that in particular two-dimensional detectors require extended software tools for the analysis of the measured data when compared with constant-R input.

6.4
Diffraction Elastic Constants (DECs) s_1 and $1/2 s_2$

The fundamental equation we have derived in Eq. (6.9) would only be valid for isotropic matter. However, crystals are not elastically isotropic because of their crystalline structure. Imagine, for instance, a sphere or a ball that is made from a single crystal of an element or a compound. Although one might not recognize a crystallographic direction hkl by optical inspection of the sphere one might succeed by a mechanical test. For instance, by measuring the compressibility of the sphere for various directions the crystal reference frame $\{c_i\}$ in the sphere can be uncovered, since this property depends on crystallographic direction. In order to deal with elastic anisotropy, fourth-rank tensors are introduced in elasticity theory that mediate between stress and strain. The components s_{ijkl} of these tensors are either called elastic compliances

$$\varepsilon_{ij} = \sum_{k,l=1}^{3} s_{ijkl} \sigma_{kl} \tag{6.15}$$

if the strain is expressed by a linear combination of stresses, or elastic stiffnesses c_{ijkl}

$$\sigma_{ij} = \sum_{k,l=1}^{3} c_{ijkl} \varepsilon_{kl} \tag{6.16}$$

in the opposite case. Each fourth-rank tensor is the inverse of the other

$$s_{ijkl} = \left(c_{ijkl}\right)^{-1} \tag{6.17}$$

where the rules of matrix inversion have to be obeyed in the calculation. In consequence, the elastic properties of crystalline materials still depend linearly on the applied stress, but may exhibit pronounced anisotropies. This statement is even valid for materials from the high-symmetry cubic crystal classes, for which only three components of the full s_{ijkl} or c_{ijkl} tensor are independent. Use is made in the following of lower-case (s_{ijkl}) or upper-case (S_{ijkl}) letters to indicate the elastic constants of single crystals and polycrystals, respectively. Table 6.2 lists the single-crystal elastic compliances s_{ijkl} of some nitride compounds used in protective coating applications. Also the anisotropy constant s_0 is given in the table that is derived from the s_{ijkl} constants through

$$s_0 = s_{1111} - s_{1122} - 2s_{1212} \tag{6.18}$$

This constant is a measure for anisotropic behavior and vanishes for isotropic materials.

If we now assume that a polycrystalline aggregate may be modeled by a huge number of spherical crystallites it becomes evident that their mechanical response to an external load will depend on the orientation of crystallites towards the load and towards each other. In case of a random orientation distribution the macroscopic elastic behavior would become independent of direction due to an averaging over many crystallites of distinct orientation.

However, it has frequently been pointed out that x-ray diffraction is a highly selective probe and there exists an important distinction between a macroscopic mechanical test and the monitoring of a Bragg peak. Whereas in a load test all crystallites of the sample are subjected to testing, in a diffraction experiment only a subset of grains is probed for which the diffraction condition is fulfilled (compare Fig.

Table 6.2 Single crystal data of elastic compliances s_{ijkl} of nitride compounds used in protective coating applications (in units of 10^{-6} MPa^{-1}, transformation to Voigt notation via $s_{11} = s_{1111}$, $s_{12} = s_{1122}$, $s_{44} = 4s_{1212}$ and $s_0 = s_{11} - s_{12} - s_{44}/2$). The data are from the references as indicated. The macroscopic elastic constants E^{macro} (in GPa) and ν^{macro} were calculated on the basis of these data according to the model of Eshelby and Kröner.

Nitride	s_{1111}	s_{1122}	s_{1212}	s_0	Ref.	E^{macro}	ν^{macro}
c-BN	1.34	−0.25	0.52	0.55	[70]	907	0.12
CrN	1.86	−0.09	2.84	−3.73	[20]	330	0.22
TiN	2.17	−0.38	1.49	−0.43	[71]	428	0.20

1.9). Therefore, only those crystallites contribute to the Bragg reflection the lattice planes hkl of which have defined orientations. Also, as for the single-crystalline sphere mentioned above, the reflecting crystallites will exhibit a mechanical anisotropy, i.e. the shift of the Bragg peak will depend on Miller indices hkl albeit all crystallites are subjected to the same mechanical constraints from their surroundings. It is due to this selective reception of diffraction that elastic anisotropy enters into XSA.

In order to include these arguments in the formalism, an averaging has to be introduced into the fundamental equation of XSA, where the averaging is over the reflecting crystallites. Hooke's law as given in Eq. (6.2) may then be written in an averaged form for the ε_{33} stress component in the laboratory frame $\{l_i\}$

$$\left\langle \varepsilon_{33}^L \right\rangle^r = \left\langle \sum_{mn} s_{33mn}^L \sigma_{mn}^L \right\rangle^r \tag{6.19}$$

where the superscript r denotes the averaging over reflecting lattice planes. From the derivation of the fundamental equation it is evident that it is not modified with respect to its ϕ and ψ dependence. However, the prefactors in front of the trigonometric functions have to be adapted to take account of mechanical anisotropy. These factors depend on Young's modulus E and Poisson's ratio ν, and have to be substituted by hkl-dependent expressions

$$\frac{1+\nu}{E} \rightarrow \frac{1}{2} s_2(hkl)$$
$$-\frac{\nu}{E} \rightarrow s_1(hkl) \tag{6.20}$$

The new expressions are named diffraction elastic constants (DECs) and are abbreviated by s_1 and $1/2s_2$. Since the dimensions of these material constants are Pa^{-1} they are related to the group of elastic compliances, which is the reason for using the letter s for abbreviating them. In this notation the fundamental equation of the XSA for the biaxial stress case becomes

$$\varepsilon_\psi = \left[\frac{1}{2} s_2(hkl)\right] \sigma_\| \sin^2\psi + 2\left[s_1(hkl)\right] \sigma_\| \tag{6.21}$$

The d–$\sin^2\psi$ distribution thus becomes dependent on Miller indices hkl of the reflection under investigation. Again, this equation represents a way of expressing Hooke's law. A new expression for the stress-free tilt angle is obtained in the notation of DECs that reads

$$\sin\psi^*(hkl) = \sqrt{\frac{-2s_1(hkl)}{\frac{1}{2}s_2(hkl)}} \tag{6.22}$$

It will be outlined in the following section how the averaging over reflecting crystallites and their interaction with the surrounding volume has to be accomplished in order to derive the DECs s_1 and $1/2s_2$ from the components of the elasticity tensors.

6.5
Grain Interaction Models

As argued above, the Bragg equation acts as a selection rule deciding about which crystallites contribute to the actually measured reflection. The nonreflecting crystallites may be considered as a matrix into which the reflecting ones are embedded (see Fig. 1.9). The goal of an XSA investigation then becomes to derive from a strain measurement of a subset of crystallites the average stress to which the total volume of the sample is subjected. It is evident from this formulation that XSA is related to very fundamental and highly complex questions about the micromechanics and microstructure of polycrystalline matter. Answers to these questions have been given by various micromechanical models, from which the most relevant ones for application in XSA are briefly outlined.

From a basic point of view the strain–stress relation between a crystallite and the surrounding matrix depends on a large set of structural features like shape and dimensions, the interface structure between grain and matrix, their different elastic constants, and so on. It is assumed in most models that the elastomechanical properties of the matrix are isotropic. The elastic anisotropy is then related only to the crystalline nature of grains contributing to the measured Bragg reflection. These models are generally termed homogeneous elastic matrix (HEM) models. Evidently, they will not apply to samples that exhibit a preferential orientation of crystallites. In these cases the matrix also has to be considered as anisotropic, which is addressed in Section 6.6. The two most basic HEM models were put forward by Voigt in 1928 and Reuss in 1929, assuming that all crystallites in the sample exhibit the same strain (Voigt) or are all subjected to the same stress (Reuss) [6, 7].

For the Reuss approach the r-averaged stress–strain relation (Eq. (6.19)) then becomes

$$\left\langle \varepsilon_{33}^L \right\rangle^r = \sum_{mn} s_{33mn}^L \left\langle \sigma_{mn}^L \right\rangle^r \tag{6.23}$$

This expression has to be reformulated using of the elastic compliances in the crystallographic reference frame $\{c_i\}$ and the stress tensor in the sample reference frame $\{s_i\}$. Applying as many transformation matrices as required by the rank of the two tensors, one ends up with the lengthy, but straightforward expression

$$\left\langle \varepsilon_{33}^L \right\rangle^r = \sum_{i,j,k,l,m,n,o,p} a_{3i}^{LC} a_{3j}^{LC} a_{mk}^{LC} a_{nl}^{LC} a_{mo}^{LS} a_{np}^{LS} s_{ijkl}^C \left\langle \sigma_{op}^S \right\rangle^r \tag{6.24}$$

In this expression the single-crystal elastic compliances s_{ijkl}^C as for instance given in Table 6.2 have to be inserted. A comparable consideration can be made in the Voigt approach. It is shown in Ref. [8] that the development of Eq. (6.19) into a new d–$\sin^2\psi$ distribution may be performed in a unified manner for both the Voigt and Reuss approaches or any other HEM model. This leads to general expressions of the diffraction elastic constants s_1 and $1/2s_2$ that may be written as

$$s_1 = \frac{1}{2} \sum_{ijkl} a_i^{CL} a_j^{CL} \left(\delta_{kl} - a_k^{CL} a_l^{CL} \right) A_{ijkl}$$

$$\frac{1}{2} s_2 = \frac{1}{2} \sum_{ijkl} a_i^{CL} a_j^{CL} \left(3 a_k^{CL} a_l^{CL} - \delta_{kl} \right) A_{ijkl} \tag{6.25}$$

Here, the a_i^{CL} vectors are the ith column of the CL transformation matrix a_{ij}^{CL}. In addition, the choice of the A_{ijkl} tensor has to be made in accordance with the HEM model under consideration. For the Voigt and the Reuss approaches it may be inserted

$$A_{ijkl} = \begin{cases} \langle c_{ijkl}^{-1} \rangle & \text{Voigt} \\ s_{ijkl} & \text{Reuss} \end{cases} \tag{6.26}$$

In the Voigt approach of constant strain the A_{ijkl} derive from the inversion of the stiffness tensor, while in the Reuss approach the A_{ijkl} are identified with the compliance components. For the precise calculation of the DEC coefficients in arbitrary crystal symmetry the reader is referred to the literature [8–10]. For the cubic compounds, s_1 and $1/2 s_2$ simplify considerably. One obtains for the DECs in the Voigt model

$$s_1 = \frac{s_0 (s_{1111} + 2 s_{1122}) + 10 s_{1122} s_{1212}}{3 s_{1111} - 3 s_{1122} + 4 s_{1212}}$$

$$\frac{1}{2} s_2 = \frac{10 s_{1212} (s_{1111} - s_{1122})}{3 s_{1111} - 3 s_{1122} + 4 s_{1212}} \quad \text{(Voigt, cubic lattices)} \tag{6.27}$$

If the Reuss model is assumed to be valid, the DECs depend on the Miller indices hkl and the peak shift will thus depend on which reflection is actually under consideration

$$s_1 = s_{1122} + \Gamma s_0$$

$$\frac{1}{2} s_2 = s_{1111} - s_{1122} - 3 \Gamma s_0 \quad \text{(Reuss, cubic lattices)} \tag{6.28}$$

where the orientation of lattice planes is accounted for by the orientation parameter Γ we have already encountered in the discussion of microstresses in Chapter 3. It reads

$$\Gamma = \frac{h^2 k^2 + k^2 l^2 + h^2 l^2}{(h^2 + k^2 + l^2)^2} \tag{6.29}$$

and exhibits minimum and maximum values of 0 and $\frac{1}{3}$ for $h00$ and hhh Miller indices, respectively. Any arbitrary hkl will yield orientation parameters Γ between 0 and $\frac{1}{3}$. The DECs for some nitride compounds calculated in the Voigt (V) and Reuss (R) model as a function of Γ are shown in Fig. 6.9. The DECs are seen to be constant in the Voigt approach, but to depend on orientation in the Reuss model.

Since the underlying assumptions of both models are extreme cases, it can be assumed that neither the use of one nor the other will yield the correct stress-strain relation. It was in fact revealed by measurements on loaded steel samples as early as in the 1940s that more appropriate stress–strain relations were achieved when an average DEC value of the Voigt and the Reuss model was used [11]. Later, it was shown by Hill that the constant-strain (Voigt) and the constant-stress (Reuss) models represent limiting cases in HEM models [12]. The arithmetic mean DEC as calculated from both extreme cases became known as the Neerfeld–Hill approximation. This is widely used in practice, since it often yields DEC values of sufficient precision.

Another HEM model is due to the works of Eshelby, Kröner and Kneer (EKK), and is mentioned here because of its broad dissemination in the literature. The EKK model is based on the work of Eshelby, who calculated the strain on an ellipsoidal isotropic inclusion within an elastically isotropic matrix put to a stress test [13]. The matrix is assumed in this model to be infinitely extended and to act as a homogenous and isotropic elastic medium to the grain. Both the stress in the matrix and in the crystallites is below the yield stresses of the two different materials. Kröner applied the model to spherical crystallites of cubic symmetry within an isotropic matrix [14]. Finally, Kneer extended the model to include triclinic crystalline materials [15]. The DECs are calculated in a comparable manner as in Eq. (6.25) by inserting for $A_{ijkl} = S_{ijkl} + t_{ijkl}$ the sum of polycrystal compliances S_{ijmn} and the components of the elastic polarizability tensor t_{ijmn}, the latter of which describe the mechanical interaction between matrix and inclusion. The DECs of various one-elemental metals, alloys [8, 9] and selected ceramics [10] have been calculated in the framework of the EKK model. The use of these DEC values is recommended whenever thin-film materials are investigated for which s_1 and $1/2s_2$ are available. The appropriate values for nitride compounds are also shown in Fig. 6.9 as a function of orientation parameter Γ. If new materials with unknown DECs in the EEK model are investigated, we recommend calculating the DECs in the Neerfeld–Hill approximation, especially in those cases where only little is known about the microstructure and the grain interaction. These values will probably deviate from the EKK values by only a few percent and will yield sufficiently precise DECs for most applications.

We have now taken the second step on the "ladder of complexity" by taking into account the mechanical anisotropy of crystalline matter. The inclusion of DECs in the analysis of d–$\sin^2\psi$ distributions will yield more reliable results in many cases than using simply the macroscopic elastic constants E and ν. The DEC approach was developed in the area of x-ray stress analysis of steel and steel-derived alloys. Its introduction into the field closed a gap of 5–15% as observed between the load stress to which macroscopic steel work pieces were subjected and the result from x-ray stress analysis [1]. The difference might appear small when compared to other sources of systematic error in the XSA of thin films.

Initially, the HEM models were developed for the prediction of elastic constants of polycrystals from their single-crystal counterparts s_{ijkl} and the polycrystal was thereby assumed to be macroscopically isotropic with respect to its mechanical properties. Additionally, these models could also be applied to the calculation of DECs. The main problem with their application to thin polycrystalline films resides in the intrinsic structural anisotropy of thin films that manifests as typical

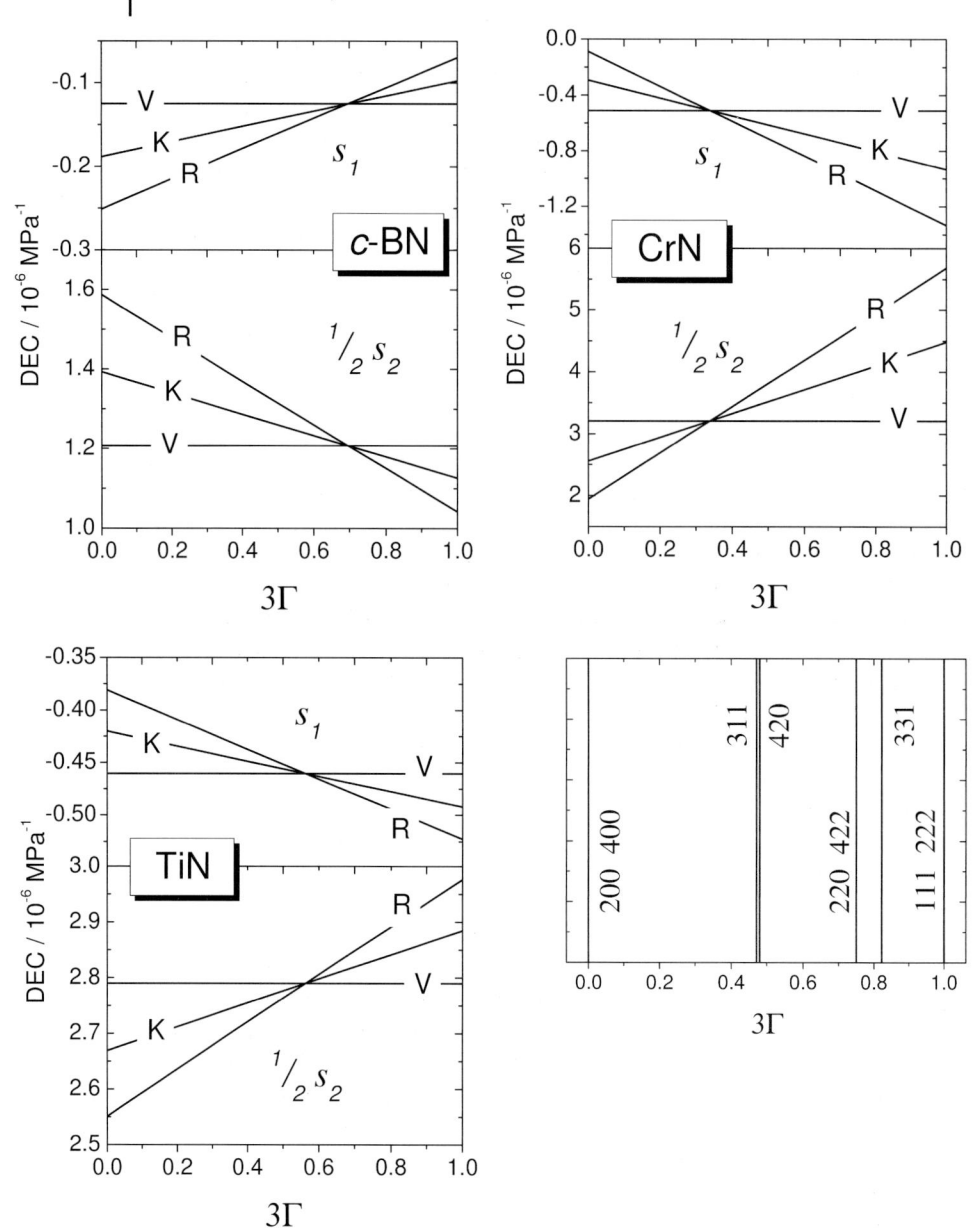

Figure 6.9 Diffraction elastic constants s_1 and $\tfrac{1}{2}s_2$ for some nitride compounds used for hard coating applications c-BN, CrN and TiN derived from the micromechanical models of Voigt (V), Reuss (R) and Eshelby-Kröner-Kneer (K). The lower right figure shows the position of Bragg peaks from NaCl-structured compounds on the 3Γ coordinate.

columnar growth morphology and which rather contradicts the assumption of macroscopic elastic isotropy. A model more suited was proposed by Vook and Witt in 1965 [16], who assumed a constant in-plane strain $\varepsilon_{11} = \varepsilon_{22} = \varepsilon_\parallel$ and a constant out-of-plane stress σ_{33} to exist in polycrystalline layers. The approach has recently been supported and extended by experimental and theoretical investigations leading to the introduction of so-called direction-dependent grain interaction models [17–19]. It seems that these models are well suited for thin-film investigations, but their introduction into the field has just begun.

6.6
The Effect of Texture

The occurrence of textures in thin films may further complicate the determination of residual stresses by analyzing the measured strain. For one reason, the nonrandom distribution of crystalline orientation may cause a Bragg reflection to be observable only for a restricted range of tilt angles. A reliable regression of the measured strain distribution may thus be impossible. The effect can be realized from the lower plot in Fig. 6.6, where the intensity of a TiN 422 reflection as a function of inclination angle ψ is displayed. Due to the pronounced 111 fiber texture in this layer the 422 intensity almost disappears around $\psi = 40°$. For such low-intensity peaks the determination of the interplanar spacing d might become largely erroneous or even impossible. The second obstacle to XSA posed by texture is a more fundamental one, since a textured polycrystal is elastically anisotropic even on a macroscopic scale and cannot be modeled by assuming the crystallites to be embedded in an elastically homogeneous matrix. The calculation of diffraction elastic constants as presented in Section 6.5 then becomes questionable or inadequate. Since textures or preferred orientation are such ubiquitous phenomena in thin-film growth one is confronted with a serious obstacle in residual stress analysis. Some approaches have been developed for textured bulk materials and in some cases these techniques have also been applied to the investigation of thin films. The evolving nature of the field should be remembered when the methods are outlined in the following for the evaluation of residual stress in the presence of texture.

The phenomenological appearance of texture in a strain distribution is illustrated in Fig. 6.10 as presented in a recent synchrotron study [20]. An inverse pole figure of a CrN layer signifies the film to exhibit a strong (110) fiber texture (Fig. 6.10(a)). A small cross-sectional cut of the specimen was probed in a transmission mode with the in- and out-going x-ray beam parallel to the surface. The height of the incident synchrotron beam was of the same order of magnitude as the film thickness of 9 µm. Also the strain distribution plotted versus $\sin^2\psi$ as obtained from the measurement of (311) lattice planes is displayed (Fig. 6.10(b)). Strains ε were calculated according to Eq. (6.4) by inserting measured d values and d_0 as determined from the analysis. The strain distribution is seen to exhibit a pronounced nonlinear behavior, which is caused by the preferred orientation of crystallites, as will be seen later in this section (Eq. (6.33)).

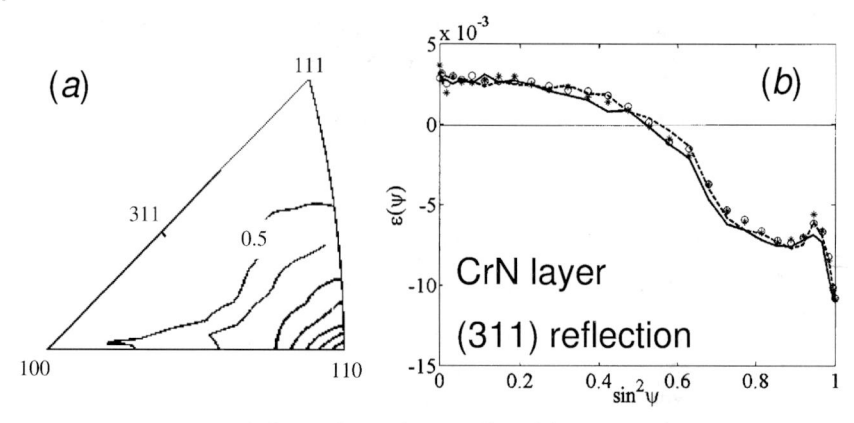

Figure 6.10 (a) Inverse pole figure of 9 µm thin CrN film exhibiting a pronounced [110] fiber texture. (b) Strain distributions of 311 reflection for different azimuth angles (from Ref. [20]).

If the $\sin^2\psi$ range in this plot had extended only to 0.6 instead of to 1.0, the influence of texture would have appeared as a bending of the curve. The hampered recognition of texture effects on the d distribution is a problem one may often encounter in practice. If the curve had been measured only up to $\sin^2\psi = 0.4$ one might have even deduced a linear behavior of the d distribution. This points again to the necessity of extending the tilt angle range as far as possible and to make use of parallel beam techniques. In the presented example both the preferred orientation and the residual stress state were found rotationally symmetric around the substrate normal and the methods presented in the following will also be restricted to these constraints.

A first approach to minimize the influence of texture on the results of XSA makes use of the weighted summation of d–$\sin^2\psi$ distributions from different Bragg reflections. This procedure might be applied in the case of weak textures when a set of different reflections can be measured and reliably analyzed within the whole accessible tilt angle range. If these presuppositions are fulfilled the distributions of lattice spacings $d(\psi)$ should be converted to unit cell edges $a_h(\psi) = (h^2 + k^2 + l^2)\, d(\psi)$. From the various a distributions the weighted sum should be calculated according to

$$\bar{a}(\psi) = \frac{\sum_h m_h a_h(\psi)}{\sum_h m_h} \tag{6.30}$$

with multiplicities m_h of Bragg reflections h as weight factors. The summation has the effect of averaging over different crystallographic directions and depending on how many reflections enter the sum a final a distribution with an improved linear behavior might be obtained. The determination of the stress σ from the slope of this curve has to be accomplished by the procedures outlined in Section 6.3 and by inserting the elastic constants E and ν of isotropic material. The diffraction elastic constants s_1 and $1/2s_2$ may not be applied, since the hkl dependence has been lost by the summation of different d distributions. This summation procedure might yield reliable results whenever the texture is not too strong and enough reflections over a large tilt range are measurable.

A further class of techniques to deal with texture in XSA may be termed orientation distribution function (ODF)-related techniques [20–24]. They may be applied whenever the ODF $f(g)$ has been determined from texture analysis. Most of these techniques rely on the introduction of so-called stress factors or anisotropic stress factors F_{ij}. The stress factors very generally represent the linear relationship between stress and strain, but which may depend on orientation angles ϕ and ψ and on Miller indices hkl in order to account for possible anisotropy

$$F_{ij}(\phi,\psi,hkl,s_{ijkl}) = \frac{\partial \varepsilon^L_{33}}{\partial \langle \sigma_{ij} \rangle} \tag{6.31}$$

The inclusion of texture is introduced by assuming that the stress factor is subjected to an orientation averaging that is described by the ODF. Since the ODF is generally known in the sample reference frame, ε^L_{33} has to be adapted to K_S by use of a_{ij}^{LS} matrices. Allowing all measured crystallites to exhibit a rotational degree of freedom λ around the measuring direction Q, one ends up with an ODF-weighted stress factor

$$F_{ij}(\phi,\psi,hkl,s_{ijkl}) = \sum_{m,n=1}^{3} \frac{\int \frac{\partial \varepsilon^S_{mn}}{\partial \langle \sigma_{ij} \rangle} a_{3m}^{LS} a_{3n}^{LS} f(\phi,\psi,\lambda) d\lambda}{\int f(\phi,\psi,\lambda) d\lambda} \tag{6.32}$$

For any general homogeneous elastic matrix model the A_{ijkl} coefficients discussed previously may be inserted for $\partial \varepsilon^S_{ij}/\partial \langle \sigma_{ij} \rangle$. The expression simplifies considerably when the symmetry constraints are applied. This particularly holds for axisymmetric texture (layer texture) and stress (biaxial stress model). For the most widely occurring 100, 111 and 110 fiber texture, for instance, it can be shown that stress–strain distribution becomes

$$\varepsilon(hkl) = \left(2s_{1122} + 2s_{1212} \sin^2 \psi + s_0 \, p^n \, \frac{h^2+k^2+ml^2}{h^2+k^2+l^2} \right) \sigma_{\parallel} \tag{6.33}$$

where the (p, n, m) triple adopts numerical values as $(p, 0, 0)$, $(½, 1, 2)$ and $(⅔, 1, 1)$ for 100, 110 and 111 fiber textures, respectively [25]. By means of Eq. (6.33) we are now able to explain the oscillatory behavior of the d_ψ or ε_ψ versus $\sin^2\psi$ distributions, which is frequently observed in the case of strongly textured materials. Due to the preferred orientation of the crystallites with respect to the fiber axis, the strains are no longer continuous functions of the pole angle ψ, which has to be expressed for any individual single-crystal orientation by means of the Miller indices hkl. Further, it is realized from the term containing hkl that, even for a uniform stress $\sigma_\parallel$, the plots of the strains in the distinct poles versus $\sin^2\psi$ generally do not fall on a straight line, because of the cyclic permutation of the Miller indices. The only exception for this elastically anisotropic behavior are reflections of the type $h00$ and hhh, which still should yield linear ε_ψ versus $\sin^2\psi$ plots, as well as the $\langle 111 \rangle$ texture, the anisotropy term of which in Eq. (6.33) does not contain an explicit dependence on hkl.

A different approach is offered by the crystallite group method, which has been developed to deal with very strong textures [1, 5, 26]. Ideally, this method is applied in thin-film work for samples with a single fiber texture component, although the method is capable of handling multiple components. The polycrystalline aggregate to be analyzed by the technique is imagined to be perfectly anisotropic which is in contrast to the model of a homogeneous elastic matrix. The set of all crystalline grains belonging to one fiber constitute one crystallite group and there are as many groups as fiber components. The environment of a grain from one of the groups is assumed to consist only of grains of the same orientation.

6.7
Classification of Stresses

6.7.1
Classification by Dimension

It is mentioned in the introduction of this chapter that the mechanical stresses considered in this book are so called "residual" or "material inherent" stresses. Much work has been done in the past century to describe, define and to classify these stresses. Heyn, one of the pioneers in this field, called them "hidden elastic stresses" [27] and used a simple model of springs embedded in a homogeneous matrix to explain the phenomenon. Probably the most general definition of residual stresses (RS) was given by Macherauch et al. [28]. According to this definition, residual stresses are those stresses that act in a closed body, which is free of any external load. The mechanical forces and moments related to them must compensate each other with respect to any macroscopic cross-section and axis through the body, respectively; otherwise the body would elongate or bend, which is obviously not the case.

This definition, however, up to now does not tell us anything about the linear dimension, on which the residual stresses act. Following Macherauch again, it seems reasonable to find a definition that defines these stresses on the basis of different length scales, for the polycrystalline aggregate in which we are interested, which in our case is a thin film consisting of many crystallites. Thus, being aware of the crystalline (film) structure, the residual stresses within the material are divided into first-, second- and third-type stresses, which are usually described by the symbols σ^I, σ^{II} and σ^{III}, respectively. The RS of type I, σ^I, which is also called macrostress, is the volume average over all crystallites and phases within a volume, which is large enough to represent the macroscopic material

$$\sigma^{\mathrm{I}} = \left(\int \sigma dV \Big/ \int dV\right)_{\text{many crystallites}} \tag{6.34}$$

A release of type I RS usually leads to macroscopic alternations of the body's shape, which is the basis of the mechanical stress analysis methods.

The residual stresses of types II and III are usually grouped under the general term microstresses. Type II micro-RS are homogeneous within an individual crystallite. They are defined as the mean deviation from the macro- or type I RS

$$\sigma^{\mathrm{II}} = \left(\int \sigma dV \Big/ \int dV\right)_{\text{one crystallite}} - \sigma^{\mathrm{I}} \tag{6.35}$$

Finally, type III micro-RS, σ^{III}, are inhomogeneous even on the atomic scale. They are defined as the position-dependent deviations from type II RS

$$\sigma^{\mathrm{III}}(x,y,z) = \sigma(x,y,z) - \sigma^{\mathrm{I}} - \sigma^{\mathrm{II}} \tag{6.36}$$

where $\sigma(x, y, z)$ denotes the actual residual stress at any point within the considered specimen.

From the thin-film point of view it is further important to note that the extension of the integration in Eq. (6.34) may be small with respect to the depth coordinate z. Steep gradients $\sigma^{\mathrm{I}}(z)$, the X-ray analysis of which is discussed in Section 6.10, thus are permitted. In practical work, this case always becomes relevant if rather soft (i.e. low-energy) x-rays providing only small penetration depths are used for the diffraction experiments. In these cases the experimentalist has to make sure that the lateral extension of the volume exposed to the x-ray beam is large enough to guarantee that the measured lattice strains are representative for the macro residual stress state within the film.

6.7.2
Residual Stresses in Multiphase Materials

The definition of the residual stresses by their dimension in the form given above does not take into account different crystalline phases of which the material under investigation may consist. Since x-ray diffraction methods are phase selective, we therefore have to extend our RS definition as regards this important point. The situation is shown in Fig. 6.11. We have to keep in mind that in the case of multiphase polycrystalline materials a large number of crystallites of the considered phase α within the exposed specimen volume always contribute to the measured signal. So it becomes obvious that any information obtained from the diffraction line profiles corresponds to average RS values rather than to the position-dependent stresses defined on different scales. For the phase-specific RS evaluated from the shift of the diffraction line profiles, for example by means of the $\sin^2\psi$ method, we therefore write

$$\langle \sigma \rangle^{\alpha} = \sigma^{\mathrm{I}} + \langle \sigma^{\mathrm{II}} \rangle^{\alpha} + \langle \sigma^{\mathrm{III}} \rangle^{\alpha} \tag{6.37}$$

Especially in the US literature, for example in the textbook of Noyan and Cohen, the $\langle\sigma\rangle^\alpha$ RS are often termed "pseudo macrostresses" (PMS), which takes into account their dualistic nature. On the one hand, the $\langle\sigma\rangle^\alpha$ have macroscopic character, because they are averages over a large crystallite ensemble. On the other hand, they also have microscopic character. This is due to the fact that the individual crystallites of each phase α may possess a special shape, size, texture and arrangement, which characterize the microstructure of the material. The sum of the volume fraction weighted phase-specific RS $\langle\sigma\rangle^\alpha$ yields the type I or macrostress

$$\sum_{\alpha=1}^{N} f_\alpha \langle\sigma\rangle^\alpha = \sigma^I \tag{6.38}$$

where f_α and N denote the volume fraction and the number of phases, respectively.

The microstresses $\langle\sigma^{II}\rangle^\alpha$ represent the average mutual constraints between the individual phases of the material, which have to compensate each other

$$\sum_{\alpha=1}^{N} f_\alpha \langle\sigma^{II}\rangle^\alpha = 0 \tag{6.39}$$

Concerning the existence and experimental determination of averaged type III micro-RS, $\langle\sigma^{III}\rangle^\alpha$, the interested reader is referred to particular literature [29, 30].

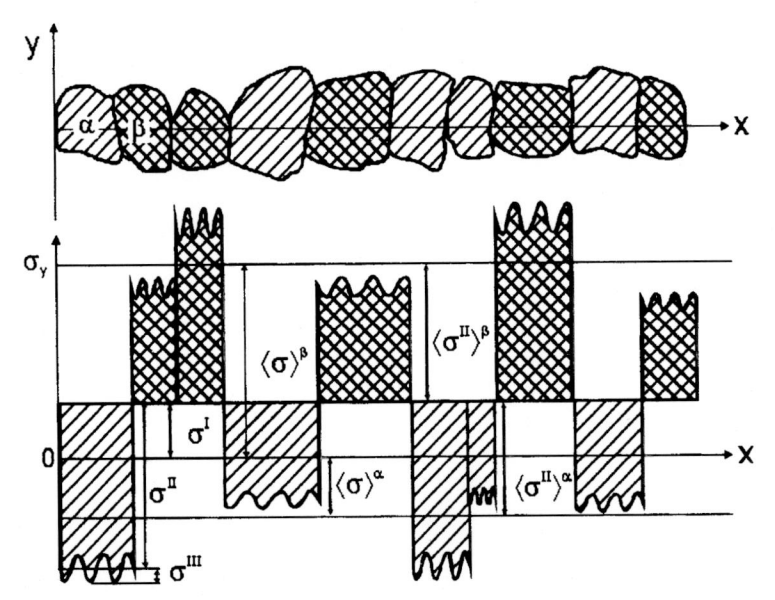

Figure 6.11 Definition of residual stresses in multiphase materials (after Ref. [1]).

6.7.3
Origin of Residual Stresses: Extrinsic and Intrinsic Stresses

In the previous sections we have introduced and classified the residual stresses rather formally, which is certainly useful from the experimental point of view, when we intend to analyze the residual stress state in our thin film by x-ray methods. So far so good! Materials scientists and thin-film engineers, however, who deal with the process of thin-film deposition, are of course more interested in learning more about the correlation between the residual stresses stored in the films and their origin in the deposition process. A general classification of residual stresses by means of their origin in the manufacturing and/or machining process of materials would go beyond the scope of this book. Here the reader is referred to the numerous textbooks, for instance Refs. [1, 4, 31], which deal with these problems as well as to the particular literature cited there.

Concerning the origin of residual stresses in thin films, however, a few remarks seem to be advisable, because they also touch the field of thin-film XSA in some respects. Long-range macrostress or phase-specific residual stress fields in polycrystalline materials usually arise from an inhomogeneous plastic deformation with respect to any sample cross-section or at least to any macroscopic sample area. Another important source for macroresidual stresses is the elastic strain between individual parts of technical compounds. The latter especially applies to thin films deposited on any kind of substrate (amorphous, single- or polycrystalline). According to their origin in the film evaporation process, it is convenient to distinguish between extrinsic and intrinsic residual stresses (see Fig. 6.12).

Extrinsic residual stresses have been generally defined as the thermal stresses that are due to the difference in the coefficients of thermal expansion between the film and the substrate, $\Delta\alpha$ (Fig. 6.12(a)). If T_{dep} is the deposition temperature, the thermal stress σ_T at any temperature T is proportional to [4]

$$\sigma_T \propto \left(T_{dep} - T\right)\Delta\alpha - \varepsilon^{pl} \tag{6.40}$$

Here, ε^{pl} is the plastic strain, which arises for large $\Delta\alpha$ or ΔT if the film stress exceeds the yield point at any temperature. Therefore, the total stress, which remains at room temperature within the film, is reduced by an amount proportional to ε^{pl}.

Intrinsic residual stresses are defined as those stresses having lattice inherent origins. In single-crystal heteroepitaxy, for example, where the lattice parameter of the growing film a_f often differs only by a small amount from that of the substrate a_s, the residual stress σ_M induced by this lattice misfit is proportional to the interfacial strain $\varepsilon_{\parallel}$ [32] (Fig. 6.12(b))

$$\sigma_M \propto \frac{a_{TF} - a_{Sub}}{\frac{1}{2}\left(a_{TF} + a_{Sub}\right)} = \varepsilon_{\parallel} \tag{6.41}$$

Equation (6.41) holds for pseudomorph film growth without taking into account bending effects of the substrate as well as misfit dislocations, which would reduce the interfacial strain by a certain amount. Other origins for intrinsic film stresses,

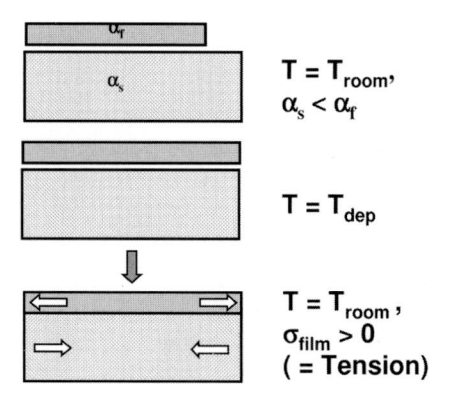

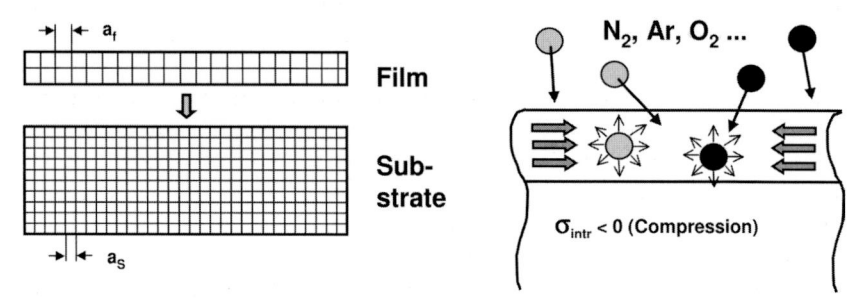

Figure 6.12 Possible origins for the formation of (a) extrinsic and (b, c) intrinsic residual stresses in thin films. (a) Thermal stress due to the difference between the coefficients of thermal expansion of the film and the substrate, α_{TF} and α_{Sub}, respectively. (b) Misfit-induced stress due to the mismatch between the lattice parameters of film and substrate, a_f and a_s, respectively. (c) Stresses due to defect incorporation during evaporation (after Ref. [31]).

which are of particular interest in the physical (PVD) or chemical vapor deposition (CVD) of polycrystalline films, are defect incorporation into the film during evaporation (Fig. 6.12(c)), the formation of nonequilibrium structures, recrystallization, grain boundary relaxation, ion bombardment, crystallite coalescence, etc., e.g. see Refs. [3, 33–36].

Compared with the thermal- or the misfit-induced stresses, which can be calculated (or at least estimated) by means of Eqs. (6.40) and (6.41), respectively, a theoretical evaluation of the residual stresses of the latter group is difficult, because their amount as well as their distribution within the growing film depend in a very complex way on the deposition kinetics. This, however, emphasizes the need for an experimental analysis of the thin-film stresses by means of rather sophisticated x-ray methods that have been developed to meet the special requirements with which one is confronted in thin-film diffraction.

6.8
Effect of Residual Stress Gradients

6.8.1
General Considerations

In Section 6.7 we learned that residual stresses in a polycrystalline material are not homogeneous but vary on different length scales, which were defined using a single grain or crystallite as "basic unit". However, because x-ray diffraction methods applied to a polycrystal always yield average information over a large number of some hundred or even up to some thousand crystallites within the exposed sample volume, we can restrict our considerations to residual stress gradients on the macroscopic scale. In other words, we are here interested in variations of macroresidual stress fields, which occur on a length scale that is comparable to the dimensions of the exposed sample volume. In the case of thin films having a thickness of a few micrometers or (much) less, the irradiated film volume usually extends over an area of some square millimeters parallel to the film plane and, in many cases, over the total film thickness in the z direction perpendicular to the film.

We now have to deal with the question as to which macroresidual stress states are possible in thin films. For the sake of simplicity, we make some assumptions, which are, however, justifiable for most practical problems with which the x-ray analyzer will be confronted. Concerning the sample geometry, the substrate is assumed in the following to be much thicker than the film, so that bending effects of the complete system due to internal moments [37] can be neglected. Furthermore, let us assume that the film consists of a single-phase material. In this case, the residual stresses analyzed by means of x-ray diffraction are of the first type or macrostresses, i.e. we have $\langle \sigma \rangle^\alpha = \sigma^I$.

In order to get an idea of the macro residual stress states, which may occur within thin films, we have to consider two of the fundamental equation systems in the theory of elasticity. One of them, the differential equilibrium conditions, has already been introduced in Section 6.2, but we remind the reader of the corresponding formulas, which read in compact form using the Einstein suffix notation

$$\sum_{j=1}^{3} \frac{\partial \sigma_{ij}}{\partial x_j} = 0 \quad (i = 1, 2, 3) \tag{6.42}$$

The second set of equations that we need for our considerations on thin-film stress states are the boundary conditions for any external body surface being free of external forces

$$\sum_{j=1}^{3} \sigma_{ij} n_j = 0 \tag{6.43}$$

where n_j is the exterior unit normal vector on the free surface of the body. It should be noted that the conditions of Eq. (6.42) must be fulfilled at any point, i.e. at the surface and in the interior of the body.

6.8.2
The Biaxial Stress State

A schematic survey of what may be possible with respect to the residual macrostress state in thin films is visualized in Fig. 6.13. Let us first consider the case of a closed thin and homogeneous film on a substrate. Defining the outer surface normal as the x_3 direction with x_1 and x_2 in the plane of the film, the boundary conditions of Eq. (6.43) take the simple form

$$\sigma_{13} = \sigma_{23} = \sigma_{33} = 0 \quad \text{at} \quad x_3 = 0 \tag{6.44}$$

Far from the edges of the film, Eqs. (6.42) and (6.43) do not contain any restrictions for the in-plane stress components σ_{11}, σ_{22} and σ_{12}. However, the existence of macrostress fields of the out-of-plane stress components σ_{13}, σ_{23} and σ_{33} within the film is associated with the variation of the in-plane stresses with respect to x_1 and x_2. If the in-plane residual stresses σ_{ij} ($i, j = 1, 2$) do not vary at least within the irradiated part of the film, the derivatives of the σ_{i3}-components with respect to x_3 are zero, and therefore the out-of-plane stresses have to vanish with respect to the total film thickness because they cannot change their values at the free surface given by the boundary condition of Eq. (6.43). Hence, the film stress state remains biaxial. At this point, it seems necessary to emphasize the difference between a biaxial and a plane stress state. Whereas a biaxial stress state allows for gradients of the

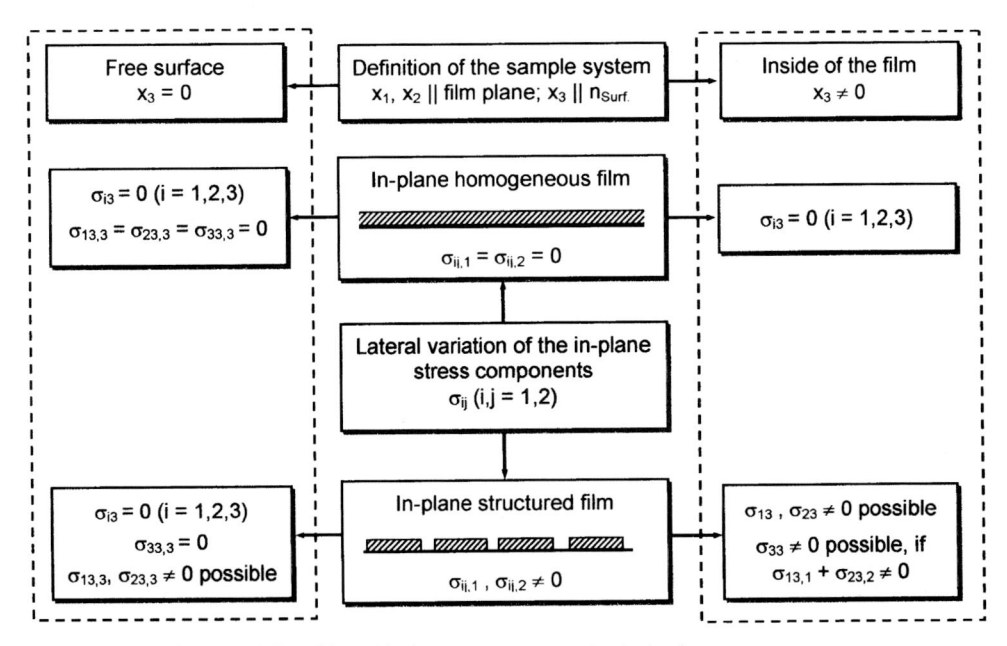

Figure 6.13 Possible residual macrostress states in single-phase thin films (from Ref. [31]).

in-plane stress components σ_{ij} ($i, j = 1, 2$) with respect to the film normal direction x_3, a plane stress state, which is used, for example, in the theory of elasticity to evaluate stress distribution in thin plates, is based on the assumption that the stress components parallel to the plate are functions of x_1 and x_2 only.

In real thin films grown on a substrate, e.g. by PVD or CVD, the in-plane residual stresses usually depend on the film thickness. It was already mentioned in Section 6.2 that the in-plane residual stress components in polycrystalline thin films in many practical cases reduce to one single component $\sigma_{11} = \sigma_{22} = \sigma_{\parallel}$. In the following we term this stress state "biaxial residual stress state of rotational symmetry". Gradients of $\sigma_{\parallel}$ with respect to the film thickness x_3 or z, $\sigma_{\parallel}(z)$, may occur in the films for many reasons. Think, for example, of growth parameters such as the process gas pressure in PVD, which are varied during film deposition. Doing this in a well-defined way, the mean free path length and therefore the kinetic energy of the particles striking the growing film surface can be adjusted. Increasing the strength of the particle bombardment, however, leads to an increase of the defect incorporation in the film lattice and therefore to higher compressive in-plane residual stresses. Thus, the film grower uses a tool to control "*in situ*" the residual stress development in the growing film, or, considering the final product, the in-plane residual stress gradient $\sigma_{\parallel}(z)$ over the film thickness.

In addition to the differential equilibrium conditions of Eq. (6.42) the macro- or first-type residual stresses have to obey macroscopic equilibrium conditions, which say that the forces and moments related to them have to compensate each other with respect to any body cross-section and axis, respectively. Let us consider a substrate of thickness t_{Sub} on which a thin film of thickness t is deposited. Assuming further the residual stress state of the system to be biaxial and of rotational symmetry, the integral of $\sigma_{\parallel}(z)$ over the film–substrate cross-section must vanish. Thus we have

$$\int_0^{t_{Sub}} \sigma_{\parallel}(z)\,dz + \int_{t_{Sub}}^{t} \sigma_{\parallel}(z)\,dz = 0 \qquad (6.45)$$

A hypothetical course of $\sigma_{\parallel}(z)$ for a film–substrate compound is shown in Fig. 6.14. In this case the steep (compressive) stress gradient in the film introduced by processing is seen to be compensated by rather small (tensile) stresses in the substrate, which extend, however, over a comparatively large depth zone in order to fulfill the condition of Eq. (6.45). Residual stresses, which are balanced between the film and its substrate, are termed "interlayer stresses" [4]. They form due to the mutual elastic constraint of the film and the substrate. If, However, the residual stress fields are balanced totally within the film, Eq. (6.45) becomes

$$\int_0^{t} \sigma_{\parallel}(z)\,dz = 0 \qquad (6.46)$$

These stresses are termed "intralayer stresses", because they arise from mutual elastic constraint of displacements within the film [4].

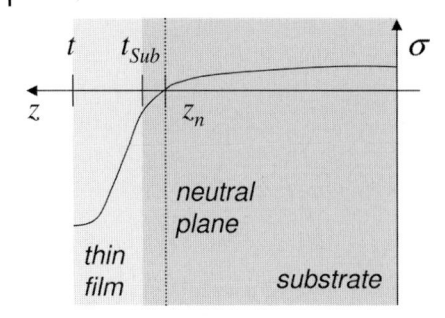

Figure 6.14 Hypothetical course of $\sigma_\parallel(z)$ for film–substrate compound. The stress in the film introduced by processing is seen to be compensated in the substrate.

Finally, it should be mentioned that complex triaxial residual stress states may occur in the case of structured films (see lower part of Fig. 6.13). Spatially limited deposit structures play an important role, for example, as integrated circuits in microelectronics [38]. Here the relaxation of the in-plane stresses near the free edges leads to complex triaxial residual stress states at least near the edges [39]. However, even individual crystallites, which form a columnar grain structure in closed polycrystalline films, may be considered as an assembly of freestanding columns. Assuming in this case a strong interaction with full elastic accommodation of the thermal misfit at the column–substrate interface, but a rather weak interaction between the individual columns, high shear stresses σ_{12} and normal stresses σ_{33} perpendicular to the interface may arise, which can stimulate film debonding.

6.9
Detection of Residual Stress Gradients in Thin Films

6.9.1
Basic Relations

In Section 6.8 it was shown that residual stresses in thin films may vary with respect to the film thickness for several reasons. In thin hard coatings based of TiN, for example, such gradients are even desired, because they may help to improve the mechanical properties of drills or cutting tools. Thus high compressive in-plane stresses $\sigma_\parallel$ at the top of the layer are known to contribute to the high hardness and to the excellent resistance against abrasive wear of such films, whereas rather small residual stresses near the film–substrate interface, however, guarantee a sufficient adhesion of the film.

We now have to answer the question as to how such (in many cases steep) residual stress gradients within a thin film can be detected and evaluated quantitatively. To start with, we reconsider the fundamental equation of the XSA in its biaxial form for the case of rotational in-plane symmetry, i.e. $\sigma_{11} = \sigma_{22} = \sigma_\parallel$ and $\sigma_{13} = \sigma_{23} =$

6.9 Detection of Residual Stress Gradients in Thin Films

$\sigma_{33} = \sigma_{12} = 0$. Assuming now the residual stress–strain state to depend on the depth z, we have to write

$$\varepsilon_\psi(hkl, z) = \left[\tfrac{1}{2}s_2(hkl)\sin^2\psi + 2s_1(hkl)\right]\sigma_\parallel(z) \tag{6.47}$$

Because of the exponential attenuation of the x-rays in the material, we have to distinguish between the actual or "real space" strain depth profiles $\varepsilon_\psi(hkl, z)$ on the one hand and the exponentially weighted or "Laplace space" depth profiles $\varepsilon_\psi(hkl, \tau)$, which are the experimentally accessible quantities, on the other hand. If the strain-free lattice spacing $d_0(hkl)$ does not depend on z (i.e. we exclude chemical gradients from further considerations), real and Laplace space strain profiles are correlated by (see Eq. (4.10) and Section 4.3)

$$\varepsilon_\psi(hkl, \tau) = \frac{d_\psi(hkl, \tau)}{d_0(hkl)} - 1 = \frac{1}{d_0(hkl)} \frac{\int_0^t d_\psi(hkl, z)\, e^{-z/\tau} dz}{\int_0^t e^{-z/\tau} dz} - 1 \tag{6.48}$$

Because of its finite upper limit the integral in the numerator of Eq. (6.48) is of the Fredholm type. For $t \gg \tau$, which holds for bulk materials and thick films, the upper limit can be replaced by infinity. Then the integral takes the form of a "true" Laplace transform with respect to $1/\tau$ [40]. Assuming further the DECs $s_1(hkl)$ and $1/2s_2(hkl)$ to be also independent of the depth z, the corresponding integration of the right-hand side of Eq. (6.47) can be restricted to the stress $\sigma_\parallel$

$$\sigma_\parallel(\tau) = \frac{\int_0^t \sigma_\parallel(z)\, e^{-z/\tau} dz}{\int_0^t e^{-z/\tau} dz} \tag{6.49}$$

With Eqs. (6.47)–(6.49) we now use the tools that are necessary for the evaluation of the residual stress depth distribution within the film from the experimentally determined lattice strain profiles. We may write

$$\sigma_\parallel(\tau) = \frac{\varepsilon_\psi(hkl, \tau)}{F_\parallel(hkl, \psi)} \tag{6.50}$$

where $F_\parallel(hkl, \psi)$ are the so-called stress factors. For quasi-isotropic materials having no preferred texture the $F_\parallel(hkl, \psi)$ are given by $F_\parallel(hkl, \psi) = 1/2s_2(hkl)\sin^2\psi + 2s_1(hkl)$, otherwise they have to be calculated using the single-crystal elastic constants and the ODF, which has to be determined beforehand by quantitative texture analysis.

Equation (6.50) yields the in-plane residual stress depth profile in the form of a discrete distribution of more or less noisy and improperly spaced data in Laplace space or τ space. In many thin-film development projects, however, one is of course more interested in real space quantities and therefore the data have to be transformed back into z space. However, it has already been outlined in Section 4.3 that the inverse numerical Laplace transform (INLT) is one of the most difficult problems in numerical mathematics, because especially in case of noisy data the systems of equations to be solved become ill-conditioned [41], which leads to physically meaningless results. For practical use, it has been proved more convenient to describe the real space stresses $\sigma_\parallel(z)$ by appropriate functions, which can be easily transformed into $\sigma_\parallel(\tau)$ by Eq. (6.49). The unknown coefficients in the transformed functions are then obtained by least squares fitting. Table 6.3 summarizes some frequently used approaches and their "true" Laplace transforms. In cases where the upper limit of the integrals in Eq. (6.49) may not be replaced by infinity, only simple polynomials, the transforms of which are then given by Eq. (4.14) are used in practice. The "finite" transforms of the other functions given in Table 6.3, are rather complicated and therefore generally not suited for least squares fitting.

Table 6.3 Some frequently used functions for describing residual stress depth profiles and their Laplace transforms.

Function	$\sigma(z)$	$\sigma(\tau)$
Polynomial	$\sum_{k=0}^{N} a_k z^k$	$\sum_{k=0}^{N} a_k k! \tau^k$
Exponentially damped	$a_0 + a_1 e^{-a_2 z}$	$a_0 + \dfrac{a_1}{a_2 \tau + 1}$
Exponentially damped polynomial	$\left(\sum_{k=0}^{N} a_k z^k\right) e^{-a_{N+1} z}$	$\sum_{k=0}^{N} a_k \dfrac{k! \tau^k}{(a_{N+1} \tau + 1)^{k+1}}$

6.9.2
X-ray Penetration Depth for the General Case of Asymmetric Diffraction

The above-mentioned considerations make clear that the detection of nonuniform residual stress fields $\sigma_\parallel(z)$ is based on the appropriate variation of the penetration depth τ of the x-rays in the material. In Chapter 1 τ was formally introduced as the reciprocal value of the linear absorption coefficient μ, i.e. $\tau = \mu^{-1}$. For normal incidence of the incoming x-ray beam it specifies the depth at which the beam intensity is reduced to $1/e$ of the primary intensity I_0. In Chapter 4 τ was specified for the case of grazing incidence diffraction and the configuration factor k_α was derived to account for the beam path within the investigated sample.

It will be seen in the next section that the detection of lattice strain depth profiles requires experiments performed in general asymmetric diffraction geometry. This geometry is defined by two conditions as visualized in Fig. 6.15: (a) the incoming and the diffracted beam form different angles α and β, respectively, with the sample surface and (b) the planes of refraction, which contain the incoming and the diffracted beam and the surface normal s_3, do not coincide as in the case of grazing incidence diffraction considered in Chapter 4. For the general diffraction experiment the configuration factor k takes the form

$$k_{\alpha\beta} = \frac{1}{\sin\alpha} + \frac{1}{\sin\beta} \qquad (6.51)$$

The angles α and β depend on the Bragg angle θ, the inclination angle ψ between the scattering vector Q and the surface normal s_3. In addition, a new angle of rotation has to be introduced that accounts for the rotation of the sample around the scattering vector itself [42] and is denoted by η in the following. The use of η allows for reformulation of α and β by

$$\sin\alpha = \sin\theta \cos\psi - \cos\theta \sin\psi \cos\eta \qquad (6.52)$$

$$\sin\beta = \sin\theta \cos\psi + \cos\theta \sin\psi \cos\eta$$

For $\eta = 0°$ the above equations yield $\alpha = (\theta - \psi)$ and $\beta = (\theta + \psi)$, respectively, which corresponds to the asymmetric Ω mode of XSA, where the sample tilt ψ is realized by an axis perpendicular to the diffraction plane. For $\eta = 90°$ the geometry changes into the symmetric Ψ mode, where the ψ tilt axis lies in the diffraction plane, and we have $\sin\alpha = \sin\beta = \sin\theta \cos\psi$. In the case of grazing diffraction, which is usually applied in thin-film stress analysis, we have $\theta < \psi$, and consequently there exits some angle η_{min} were parallel incidence (i.e. $\alpha = 0°$) is reached

$$\eta_{min} = \arcsin\left(\frac{\sqrt{\sin^2\psi - \sin^2\theta}}{\sin\psi \cos\theta}\right) \qquad (6.53)$$

The penetration depth τ_{63} defined in Section 4.2 by the condition that it contains 63% of the integrated intensity now takes the form

$$\tau_{63} = \frac{1}{\mu k_{\alpha\beta}} = \frac{1}{\mu} \frac{\sin\alpha \sin\beta}{\sin\alpha + \sin\beta} \qquad (6.54)$$

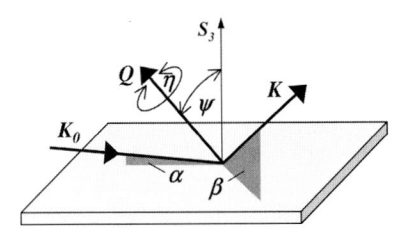

Figure 6.15 Schematic view of the general case of asymmetric diffraction. The angle η describes the rotation of the sample around the scattering vector Q.

By inserting Eqs. (6.52) into Eq. (6.54), we obtain a general formulation for the penetration depth [42]

$$\tau_{\psi\eta}(\theta) = \frac{\sin^2\theta - \sin^2\psi + \cos^2\theta \, \sin^2\psi \, \sin^2\eta}{2\mu \sin\theta \, \cos\psi} \tag{6.55}$$

The subscripts ψ and η in Eq. (6.55) should indicate that in angle dispersive diffraction τ can be adjusted by a sample tilt ψ as well as by a sample rotation η, whereas the variation $\Delta\theta$ of θ realized, for example, in a θ–θ scan to record a diffraction line profile, leads to an additional variation $\Delta\tau$ of the penetration depth.

The above equations define the x-ray penetration depth on the basis of pure geometrical considerations. Up to now the effects of refraction and total external reflection (Section 4.5), which become important for grazing diffraction geometry, have not been taken into account. Recalling the corresponding expression for the depth $\tau_{1/e}$ given by Eq. (4.30), we are easily able to allow for grazing diffraction conditions. We therefore rewrite Eq. (4.30) in a slightly different form that also includes larger incidence angles α

$$\tau_{1/e}(\alpha) = \frac{\lambda}{4\pi}\left(\sqrt{\sqrt{\left(\sin^2\alpha - \sin^2\alpha_c\right)^2 + \chi''^2} + \sin^2\alpha_c - \sin^2\alpha}\Big/2\right)^{-1} \tag{6.56}$$

where α_c and χ'' are the critical angle and the imaginary part of the refractive index. We now additionally take into account the path length of the diffracted beam within the material. From Fig. 6.16 it becomes obvious that the penetration depth $\tau_{1/e}(\alpha)$ is reduced by a certain amount corresponding to the attenuation of the diffracted beam, which forms an angle β with the sample surface. For the penetration depth $\tau_{63}(\alpha, \beta)$ we obtain

$$\tau_{63}(\alpha,\beta) = \frac{\sin\beta}{\sin\alpha + \sin\beta}\,\tau_{1/e}(\alpha) \tag{6.57}$$

Hence, Eq. (6.57) represents the most general formulation for the (kinematic) x-ray penetration depth valid for both grazing and nongrazing diffraction.

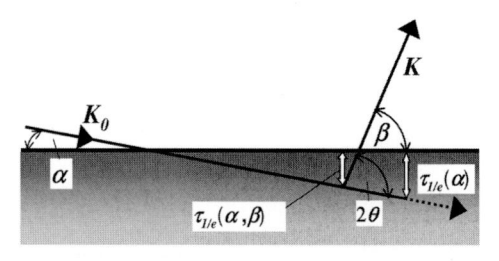

Figure 6.16 Derivation of penetration depth for the general case of asymmetric diffraction.

Finally, we give a general formulation for the absorption factor $A_{\alpha\beta}$ valid for any asymmetric diffraction geometry. The derivation is analogous to that given in Section 4.1 for the GIXRD absorption factor (cf. Eqs. (4.1)–(4.4)). Using the expressions for $\sin\alpha$ and $\sin\beta$ in Eqs. (6.52) we obtain

$$A_{\alpha\beta}(\theta) = \left(1 + \tan\psi \cot\theta \cos\eta\right)\left(1 - \exp\left(-\mu t k_{\alpha\beta}\right)\right) \tag{6.58}$$

with $k_{\alpha\beta}$ given by Eq. (6.51).

6.9.3
Special Methods for X-ray Stress Gradient Analysis

For the detection of the $d_\psi(hkl, \tau)$ depth profiles different concepts and methods have been developed, which mainly differ by the way used for the variation of the penetration depth τ. In Section 6.9.2 we derived a general formulation for τ (cf. Eqs. (6.55) and (6.57)) and showed that it depends on four parameters, namely the three angles θ, ψ and η and the linear absorption coefficient μ. In angle-dispersive x-ray diffraction considered in this book, μ and θ are determined by the wavelength λ and the reflection hkl, respectively, which are used for the diffraction experiment. Thus only the tilt angle ψ and the rotation angle η remain for the variation of τ. The crucial difference between these two possibilities is that the use of ψ is always connected with an additional variation of the measuring direction, whereas the variation of τ by means of η leaves the orientation of the scattering vector with respect to the sample system unchanged [43].

Table 6.4 summarizes some of the most interesting methods applied to residual stress depth profiling in thin films as well as the particular formulation of τ used for depth variation. It should be emphasized that no method solves all problems that may occur in thin-film stress gradient analysis at once. Thus the choice of the most appropriate measuring technique always depends on the particular question under consideration. In the case of untextured or at least weakly textured materials, for example, use can be made of those methods that are based on strain depth profiling with variable orientation (φ, ψ) of the diffraction vector with respect to the sample system. For strongly textured samples, however, strain depth profiling at fixed orientations should be advantageous. In other cases, if one is interested in the residual stress state in a certain depth below the surface, the application of those methods that are based on strain analysis in a predefined depth is recommended.

In the following we discuss some of the concepts listed above in more detail. We focus especially on those methods that enable the analysis of steep residual stress gradients in very thin films and films with strong texture, because these are the most difficult problems with which people are confronted when analyzing stresses in thin films by x-ray diffraction.

Table 6.4 Methods applied to depth-resolved XSA of thin films. The footnotes denote the most frequently used/possible modes from Ref. [31].

Method	Penetration depth	Remarks	Ref.
$\sin^2\psi$ method/Ψ mode	$\dfrac{1}{2\mu}\sin\theta\cos\psi$	Sample tilt around axis parallel to the diffraction plane[a, b]	[24, 44]
$\sin^2\psi$ method/Ω mode	$\dfrac{1}{2\mu}\dfrac{\sin^2\theta-\sin^2\psi}{\sin\theta\cos\psi}$	Sample tilt around axis perpendicular to the diffraction plane[a, b]	[33]
$\sin^2\psi$-method/mixed Ω–Ψ mode	$\dfrac{1}{\mu}\dfrac{\sin\alpha\sin(2\theta-\alpha)}{\sin\alpha+\sin(2\theta-\alpha)}\cos\psi_d$	Sample tilt in the Ω mode by $\lvert\alpha-\theta\rvert$ towards small incidence, followed by successive tilts in the Ψ mode by ψ_d with subsequent azimuth correction[a, b/d]	[45, 46]
Low incidence beam angle diffraction (LIBAD)	$\dfrac{1}{\mu}\dfrac{\sin\alpha\sin(2\theta-\alpha)}{\sin\alpha+\sin(2\theta-\alpha)}$	2θ scan for constant (small) incidence angles α between the primary beam and the surface[b, d]	[47, 48]
Scattering vector method	$\dfrac{\sin^2\theta-\sin^2\psi+\cos^2\theta\,\sin^2\psi\,\sin^2\eta}{2\mu\sin\theta\cos\psi}$	Lattice spacing depth profiling by sample rotation η around the diffraction vector for fixed orientation (φ,ψ)[a, c/d]	[49, 50]
Grazing incidence–grazing exit diffraction (GID)	$\dfrac{\lambda}{4\pi}\left\{\dfrac{1}{2}\left[\dfrac{\sqrt{(\alpha^2-\alpha_c^2)^2+\chi''^2}}{+\alpha^2-\alpha_c^2}\right]\right\}^{-\frac{1}{2}}$	Lattice spacing depth profiling at small incidence angles α near the critical angle of total reflection, α_c, either in symmetric ($\psi\approx 90°$) or asymmetric ($\psi\approx 0$) Bragg diffraction geometry[a, c]	[51, 52]

a Stress gradient analysis with direct conversion $\varepsilon(\tau)\to\sigma(\tau)$.
b Strain depth profiling with variable orientation (φ,ψ) of the diffraction vector.
c Strain depth profiling at fixed orientation (φ,ψ) of the diffraction vector.
d $\sin^2\psi$-based analysis at predefined penetration depths.

6.9.4
Grazing-Incidence Diffraction (GID)

Figure 6.17 shows the in-plane residual stress distribution in a very thin polycrystalline Si film with a thickness of $t = 110$ nm on quartz, which was prepared by laser crystallization of amorphous a-Si [53]. To keep the penetration depth τ as small as possible, lattice spacing depth profiling was performed in the symmetric Ψ mode at very large ψ angles between 87.5° and 89.5° in steps of $\Delta\psi = 0.1°$. However, in spite of the grazing diffraction conditions, τ remains larger than t until $\psi\approx 88.8°$, where α approaches the critical angle $\alpha_c = 0.22°$ of total external reflection (Fig.

6.17(a)). The strain-free lattice parameter d_0 needed for strain calculation was determined from measurements in the strain-free direction ψ^* of the biaxial stress state, which was assumed to be valid in the present case.

Figure 6.17(b) shows the plot of the evaluated residual in-plane stress distribution against τ_{eff}. The "data gap" between about 0.015 and 0.04 µm is due to the steep slope of the $\tau_{eff}(\psi)$ curve in the range between $\psi \approx 89°$ and 89.3°. The high tensile residual stresses > 1 GPa detected in the film are obviously of extrinsic nature and arise as a result of the difference between the thermal expansion coefficients of the film and the substrate, α_f and α_s, respectively, during cooling from preparation to room temperature. It is further interesting to note that the stress is nearly constant in the "bulk" of the film but decreases within a very small layer near the free surface. Using optical Raman spectroscopy, the average in-plane film stress was found to be 1.32 ± 0.14 GPa, which fits quite well into the evaluated residual stress profile [53].

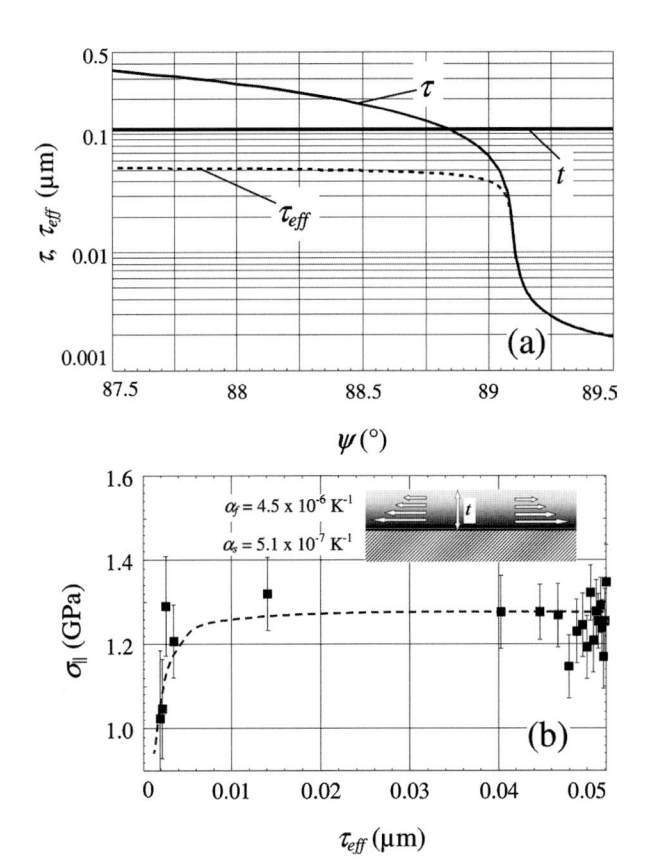

Figure 6.17 (a) Penetration depth τ and effective penetration depth τ_{eff}, according to Eq. (6.56) of Cu Kα radiation in a poly-Si film (thickness t = 110 nm) on quartz at grazing incidence, 111 reflection, 2θ = 28.3°. (b) In-plane residual stress distribution in the film. α_{TF} and α_{Sub} are the coefficients of thermal expansion of the film and the substrate, respectively [54].

The example demonstrates that the GID method is well suited for the detection of steep residual strain and stress gradients in very thin surface layers. It should be pointed out that the method can be applied even to films with strong texture. In those cases strain depth profiling has to be performed at lattice planes that are perpendicular to the film plane [52]. However, the evaluation of the in-plane residual stress depth profiles $\sigma_\parallel(\tau)$ from the measured lattice spacing profiles $d_\perp(hkl, \tau)$, according to Eq. (6.50), requires the knowledge of the exact value for the strain-free lattice parameter $d_0(hkl)$, because even a small uncertainty would lead to a considerable shift of the $\sigma_\parallel(\tau)$ curves on the absolute scale, in the same sense as is known from triaxial bulk stress analysis by means of neutron diffraction methods.

6.9.5
The Scattering Vector Method

Most of the x-ray methods applied to the analysis of residual stress gradients are based on the measuring principle of the $\sin^2\psi$ method introduced in Section 6.2. It requires untextured or at least only weakly textured materials, which provide diffracted intensity within a wide angular range of the inclination angle ψ. Thin films, however, are often highly textured, which restricts the available measuring directions to small angular ranges near the intensity poles. In such difficult cases the gradient techniques based on the $\sin^2\psi$ method often yield only qualitative results with respect to the residual stress distribution over the film thickness.

To overcome the difficulties related to this restriction, an alternative method has been proposed which uses the rotation η of the sample around the scattering vector Q in a fixed orientation (φ, ψ) with respect to the sample system in order to vary the x-ray penetration depth (see Fig. 6.15) [42]. Figure 6.18(a) illustrates that for some fixed position ψ_k of the inclination angle the rotation η around the scattering vector bridges the gap $\Delta\tau = |\tau_\Omega - \tau_\psi|$ between the Ω and the Ψ mode of the XSA [49]. Thus, recording diffraction profiles after stepwise rotation of the sample around Q (H-mode), τ is now varied by means of η (cf. Eq. (6.55)) and $d_{\varphi\psi}(hkl, \tau)$ profiles are obtained even for the same orientation (φ, ψ) of the scattering vector with respect to the sample system $\{s_i\}$. In the fundamental equation of the XSA the inclination angle ψ now becomes a usual (fixed) parameter in the same sense as the azimuth angle φ and, therefore, individual components $\sigma_{ij}(\tau)$ can be separately evaluated, if $d_{\varphi\psi}(hkl, \tau)$ profiles measured in different orientations (φ, ψ) are combined in an appropriate way [49, 50].

For a biaxial residual stress state of rotational symmetry the evaluation procedure is as follows (see Fig. 6.19). According to the equations given in Section 6.9.1 the in-plane residual stress distribution $\sigma_\parallel(\tau)$ can be calculated directly from one lattice parameter depth profile $d_\psi(hkl, \tau)$ obtained in the H mode at some inclination angle ψ_i by

$$\sigma_\parallel^{(\psi_i)}(\tau_\eta) = \left(\frac{d_\psi(hkl, \tau_\eta)}{d_0(hkl)} - 1\right) \frac{1}{F_\parallel(hkl, \psi_i)} = \frac{\varepsilon_\psi(hkl, \tau_\eta)}{F_\parallel(hkl, \psi_i)} \tag{6.59}$$

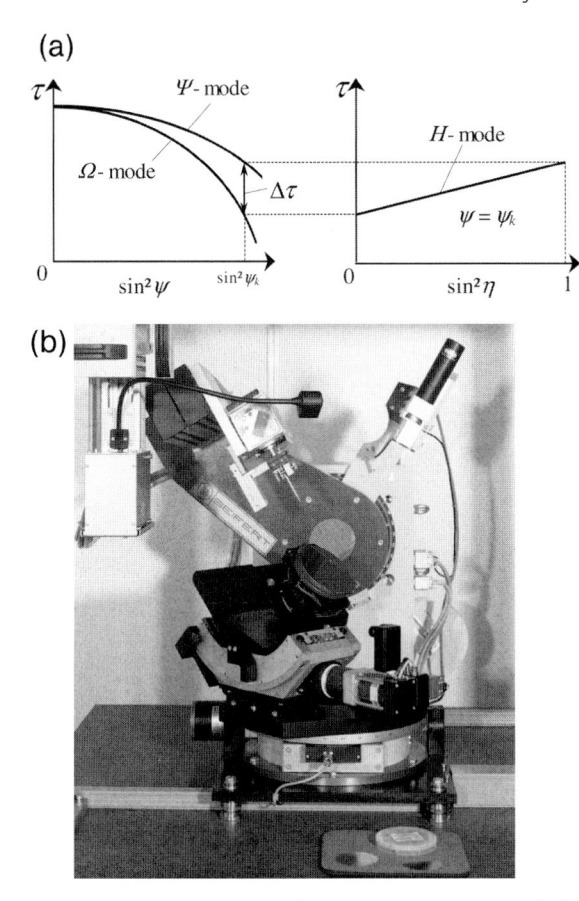

Figure 6.18 (a) Correlation of the penetration depths τ in Ω, Ψ and H (scattering vector) geometry; (b) Eta diffractometer for RS measurements in the eta mode (courtesy of Seifert).

The subscript η indicates that τ is varied by means of η. Because the expression defined by the above equation is very sensitive with respect to $d_0(hkl)$, it becomes obvious that two or more stress depth profiles $\sigma_{\parallel}^{(\psi_i)}(\tau)$ obtained for different ψ_i will only fit together in one plot without any offset if the correct $d_0(hkl)$ is used in the calculations. This criterion, however, can be used to determine the strain-free lattice parameter itself as well as the residual stress profile $\sigma_{\parallel}(\tau)$ in a self-consistent way, i.e. without a stress-free reference sample.

The example shown in Fig. 6.20 demonstrates the merits of the scattering vector method compared to the "conventional" $\sin^2\psi$ analysis in the case of strongly textured films. The investigations refer to a TiCrN/TiN hard coating system deposited by PVD on a high-speed steel substrate [55]. By texture analysis the coating was found to have a very strong $\langle 111 \rangle$ fiber texture and therefore the $\sin^2\psi$ analysis is restricted to ψ angles near the intensity poles (Fig. 6.20(a)). This lack of experi-

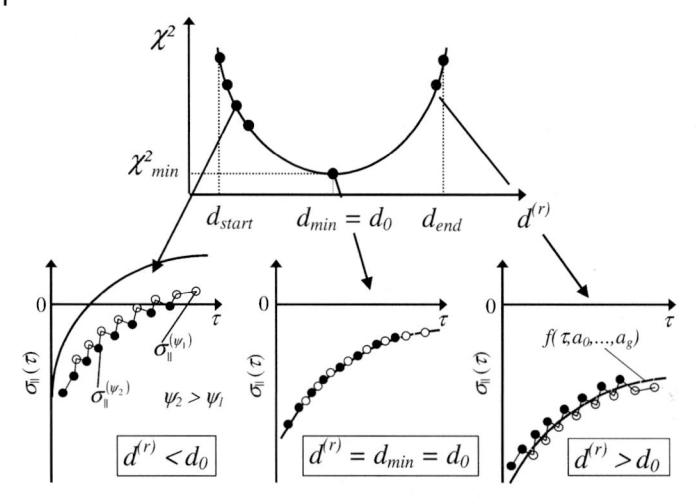

Figure 6.19 Schematic view of the residual stress and $d_0(hkl)$ evaluation by means of the scattering vector method. The stress profiles obtained for the individual inclinations ψ_i only fit together without any offset if the correct $d_0(hkl)$ is used. Note that different values of $d^{(r)}(hkl)$ taken for $d_0(hkl)$ in the calculations lead additionally to a shift of the residual stress profiles on the absolute scale [43].

mental data prevents a full quantitative residual stress gradient analysis on the basis of the methods summarized, for example, in Ref. [31]. However, some qualitative conclusions may be drawn from the deviations of the d–$\sin^2\psi$ curve from the Reuss straight line, which yields the average in-plane residual stress $\langle \sigma_\| \rangle$ over the coating thickness. Assuming nonlinearities in the d–$\sin^2\psi$ plot due to elastic anisotropy (texture) can be neglected in the case of TiN (anisotropy ratio $A = 2c_{44}/(c_{11} - c_{12}) \approx 0.9$), the convex curvature of the plot indicates an increase of the compressive in-plane stress with depth.

In such difficult cases the scattering vector method yields more detailed and reliable information. Lattice spacing depth profiles $d(311, \tau)$ were recorded in the H mode by stepwise sample rotation around the 311 intensity poles of the $\langle 111 \rangle$ film texture ($\psi = 58.5°, 80°$), and the formalism shown in Fig. 6.19 was applied to the data to evaluate $\sigma_\|(\tau)$ (Fig. 6.20(b)). It can be seen that the course of the residual stress depth profile agrees well within the error margins with the result of the $\sin^2\psi$ stress analysis as well as with the qualitative predictions for the residual stress depth distribution made on the basis of the curved d–$\sin^2\psi$ plot.

6.9.6
Realization of H Mode on a Four-Circle Diffractometer

The rotation of the sample around the scattering vector Q having an orientation (φ_P, ψ_P) with respect to the sample system $\{s_i\}$ (Fig. 6.15) can be performed in different ways. A straightforward solution, which has been realized, for example, by the concept of the ETA diffractometer (Fig. 6.18(b) and 6.21), would be to adjust the angle ψ_P between Q and the surface normal s_3 by means of some one-circle goniometer or X-cradle segment, which has to be mounted on the Φ axis of the diffractometer. In this case, a rotation φ is identical with the rotation η around Q because the scattering vector is parallel to the Φ axis.

In principle however, the rotation η can be realized on any diffractometer that is equipped with three axes that are perpendicular to each other. In the following, we give without derivation some useful equations that describe the rotation η on a four-circle standard diffractometer such as a θ–θ diffractometer with a Euler cradle. We refer to the experimental setup as shown in Fig. 6.21. In our notation ω denotes the sample rotation around the Ω axis perpendicular to the diffraction plane, whereas φ and ψ have their usual meanings. Further, we assume the orientation of Q to be restricted to the first octant of the sample system $\{s_i\}$, i.e. $0 \leq \varphi_P \leq 90°, 0 \leq \psi_P$

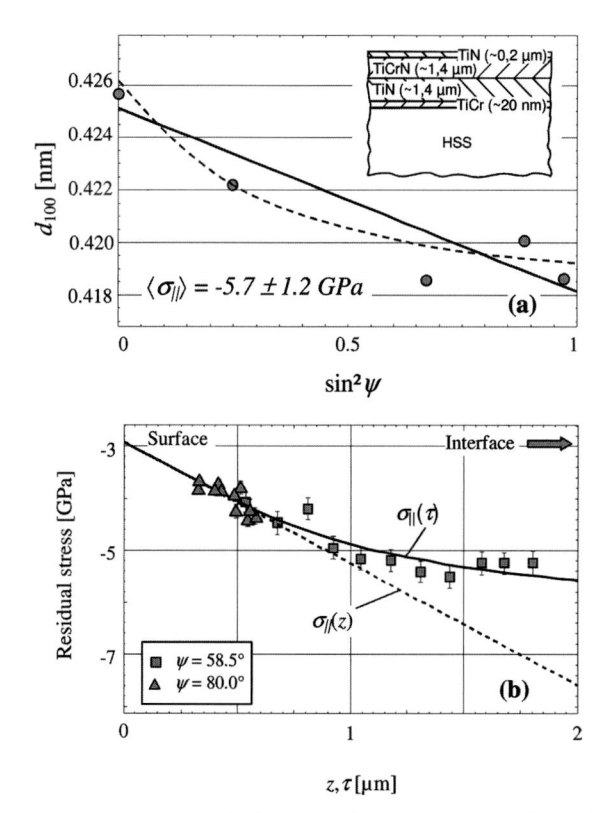

Figure 6.20 X-ray residual stress analysis performed on a thin hard coating system. (a) d–$\sin^2\psi$ plot, Cu Kα, (111) reflection (lattice spacings normalized to $d(100)$). The solid lines correspond to the Reuss straight line the slope of which yields the average in-plane coating stress $\langle\sigma_\|\rangle$. The dashed lines mark the nonlinearity due to the stress gradient $\sigma_\|(z)$. (b) In-plane residual stress distribution calculated by the variation procedure outlined in Fig. 6.19 from the lattice spacing depth profiles $d(311, \tau)$ measured in the scattering vector mode. $\sigma_\|$ is the Laplace-stress (fit curve using the approach $\sigma_\|(\tau) = c_0 + c_1[\tau - te^{-t/\tau}/(1 - e^{-t/\tau})]$; $\sigma_\|(z) = c_0 + c_1 z$ are real space stresses [55].

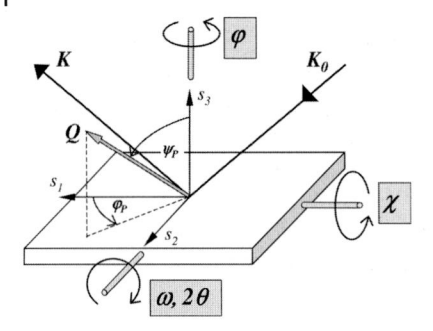

Figure 6.21 Realization of the sample rotation around the scattering vector Q on a standard four-circle diffractometer. The diffractometer axes are set up with respect to the sample reference system $\{s_i\}$.

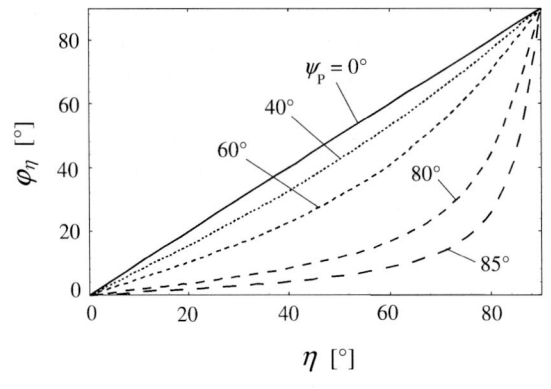

Figure 6.22 Correlation between the rotations φ_η and η for different inclination angles ψ_P according to Eq. (6.60) [56].

$\leq 90°$ and that η lies between 0° (Ω mode with $\psi < 0°$) and 90° (Ψ mode of the XSA). Then the (virtual) rotation η of the sample around Q is realized by three successive rotations (φ) and tilts (χ and ω) [56]

$$\varphi = -\left(\varphi_P + \pi\right) + \arccos\left(\frac{\cos\eta}{\sqrt{1-\sin^2\psi_P \sin^2\eta}}\right) = -\left(\varphi_P + \pi\right) + \varphi_\eta \qquad (6.60)$$

$$\chi = \arcsin\left(\frac{\sin\psi_P \sin\varphi_\eta}{\sqrt{1-\sin^2\psi_P \cos^2\varphi_\eta}}\right) \qquad (6.61)$$

$$\omega = \theta - \arcsin(\sin\psi_P \cos\varphi_\eta) \qquad (6.62)$$

The φ rotation in Eq. (6.60) may be understood to be composed of two virtual rotations. The first one, $\varphi^{(1)} = -(\varphi_P + \pi)$, is to bring Q into the diffraction plane defined by K_0 and K, whereas the second one, $\varphi^{(2)} = \varphi_\eta$, is directly correlated to the η rotation. Figure 6.22 shows that $\psi_P = 0$ corresponds to the trivial case $\varphi_\eta = \eta$. For $\psi_P > 0$ the χ tilt given by Eq. (6.61), which tilts Q back into the diffraction plane, "undertakes" a part of the η rotation. This part becomes larger as ψ_P approaches 90°. The final ω tilt given by Eq. (6.62) is to adjust the diffraction condition for Q, which is fulfilled if Q is the bisector of the angle $(\pi - 2\theta)$.

6.10
Applications

Residual stress and preferred orientation are two structural properties that often occur closely related in thin films. If a deeper insight into the growth mechanisms is desired or one intends to tailor one of these properties it will be a valuable approach to investigate both properties in common [5, 57, 58]. Particularly for plasma- or ion beam-assisted deposition processes the ion bombardment onto the surface of the growing film may have severe effects on the formation of certain textures or stress states (see Ref. [59] for a recent study of Ti adhesion layers). It has been revealed for various nitride compounds like TiN [60–62], TaN [63], epitaxial GaN [64] and other material systems like Si [65, 66] that the texture type may be adjusted by including an ion beam component into the process. However, too large an ion energy or momentum may cause severe resputtering of deposited atoms or may induce large stresses into the film that can even cause its detachment from the substrate.

In layer systems with a sufficient number of Bragg reflections the combined investigation of texture and stress state may be accomplished by virtue of the whole-pattern fitting techniques that record various 2θ scans for varying tilt angles ψ. These methods have already been introduced in Sections 3.5.4 and 5.5.2, where the emphasis was on microstructure and texture, but some of the evaluation techniques presented there may also be applied for an evaluation of residual stresses, in particular those presented in Refs. [22, 24, 67]. The advantage of these techniques lies in the many microstructural parameters like lattice parameter, grain size, microstress, texture, residual stress, etc., that may be determined simultaneously. They require, however, sufficient diffraction information, i.e. a minimum number of Bragg peaks in the accessible 2θ range that can be evaluated and, therefore, noncubic materials will probably be more suited than the cubic nitride compounds presented in this chapter.

An interesting question of thin-film elasticity is related to the applicability of bulk elastic constants to specimens that are structured on the nanometer scale. The question has already been briefly addressed in Section 3.6. Here, it should be pointed to a study of W/Cu multilayers in which the intragranular elastic constants of the W layers were investigated [68]. The superlattice period Λ was varied between 3.1 and 24 nm and the multilayers were put under tensile stress, while concomitantly a $\sin^2\psi$ distribution of a W Bragg reflection was recorded. The elastic constants were found to differ significantly from the known bulk values, while the deviation was more severe in 3.1 nm period samples. This investigation emphasizes how doubtful the assumption may be that macroscopic elastic constants are also valid in nanostructured thin films.

This example highlights again the complications that might be associated with the investigation of residual stress in thin films. Although we have climbed up to the highest level of the ladder of complexity, it should not be forgotten that even further heights may have to be climbed. A question unconsidered in the previous sections is that of the chemical composition of the investigated sample. In the special case of hard nitride coatings, for instance, stoichiometry deviations are well known to occur and in general the stoichiometry of thin polycrystalline films is often undefined on a level of a few at%. These deviations would directly translate into variations of the elastic constants and therefore imprecise diffraction elastic constants when calculated from the single-crystal data of the stoichiometric compound. Also the occurrence of a porosity network as often observed in thin films or the subsequent in-diffusion of air and postoxidation of internal surfaces may have the effect of a second chemical phase. These effects may all be associated with systematic errors in the results of determined stresses. A further complication might be related to the evolution with time of both stress and texture state. It has recently been revealed that the $h00$ texture of a 196 nm thin TiN layer prepared by magnetron sputtering changed to an hhh texture on a timescale of about 1.5 hours after finishing the deposition process [69].

However, things may not always be so complicated and rather often a more practical approach must be applied, for instance, when the mechanical or tribochemical properties of the layers have to be optimized. We therefore recommend to start any investigation of thin-film residual stress by recording the d–$\sin^2\psi$ curve and testing the application of Eqs. (6.11) or (6.21) by assuming either the isotropic or the quasi-isotropic case to apply. For many material systems these models may already lead to reliable results. If the d–$\sin^2\psi$ curve, however, displays nonlinearities as shown in some of the figures given in this chapter one must probably climb a step higher on the ladder of complexity according to how deep one's understanding of the structural film properties have to be developed. For practical cases it might be helpful to optimize intended properties by comparing the d–$\sin^2\psi$ distributions of layers prepared under different deposition conditions even if they show pronounced nonlinearities. For other cases, it might be unavoidable to develop a thorough understanding of the stress and texture state as a function of layer depth.

Exercises

6.1 Demonstrate the usefulness of the generalized Hooke's Eq. (6.2) for isotropic matter by calculating the strains caused by a hydrostatic pressure ($\sigma_{ij} = p\delta_{ij}$). Compare with the situation of uniaxial compression in Eq. (6.1) and realize the effect of the Poisson ratio.

6.2 What is the "ideally isotropic" value of Poisson's ratio v for volume conservation under uniaxial loading?

6.3 Derive the inverse of the generalized Hooke's Eq. (6.2), i.e. stress tensor components σ_{ij} as a function of strain tensor components ε_{ij}.

6.4 Derive the matrix a_{ij}^{LS} for the transformation of laboratory to specimen coordinates from the rotation matrix of ϕ around the z axis and ψ around the x axis.

6.5 Derive the $\sin^2\psi$ relation in the framework of the biaxial stress model of Eq. (6.11) by inserting Eq. (6.10) into Eq. (6.9).

6.6 What is the error introduced by the rules of error propagation into the slope $(1+v)\sigma_{\parallel}/E$ of a $\sin^2\psi$ distribution from a systematic error Δd in d?

6.7 List the differences between a rocking curve scan of a fiber reflection in texture analysis and the measurement of a d–$\sin^2\omega$ curve in XSA. Comment on the tilting of x-ray reflecting lattice planes with respect to s_3 and compare to Fig. 1.9.

6.8 Calculate the a_{ij}^{SC} matrix that transforms the s_3 direction in $\{s_i\}$ to $h = 111$ and $h = 331$ in $\{c_i\}$.

6.9 Consider a single-crystalline sphere made from rocksalt-structured TiN. Calculate the compressibility along 100, 110 and 111 by using the elastic constants of Table 6.2.

6.10 Calculate the DEC in the Neerfeld–Hill approximation of c-BN for $h = 200$. Calculate, firstly, the orientation parameter Γ and, secondly, the DEC according to Voigt and Reuss by applying the elastic constants c_{ij} given in Table 6.2.

6.11 What is the systematic error $\Delta\varepsilon$ caused by thermally induced variations of the unit cell dimensions? What is the maximum limit for ΔT if $\Delta\varepsilon < 5 \times 10^{-5}$ is intended for a strain measurement of TiN (thermal expansion coefficient $\alpha = 9.3 \times 10^{-6}\ K^{-1}$)?

References

Monographs and conference proceedings relevant to this chapter

C. Genzel, *Problems related to x-ray stress analysis in thin films in the presence of gradients and texture*, in: *Diffraction Analysis of the Microstructure of Materials*, Springer, Berlin, 2004, p. 473.

V. Hauk, *Structural and Residual Stress Analysis by Nondestructive Methods*, contributions by H. Behnken, C. Genzel, W. Pfeiffer, L. Pintschovius, W. Reimers, E. Schneider, B. Scholtes and W. A. Theiner, Elsevier, Amsterdam, 1997.

I. C. Noyan, J. B. Cohen, *Residual Stress Measurement by Diffraction and Interpretation*, Springer, Berlin, 1987.

International Conference on Residual Stresses VII: ICRS-7, Xi'an, China, 2004, S. Denis et al. (eds), Mat. Science Forum **490/491** (2005).

International Conference on Residual Stresses VI: ICRS-6, Oxford, UK, 2000, IOM Communications, London.

International Conference on Residual Stresses V: ICRS-5, Linköping, Sweden, 1997, Proc. edited by &. Ericsson, M. Odén, A. Andersson.

European Conferences on Residual Stresses, ECRS-5, Delft, The Netherlands, 1999, Mat. Science Forum **347–349** (2000).

Thin Film Stresses and Mechanical Properties IX, MRS Symp. Proc. **695** (2000).

Thin Film Stresses and Mechanical Properties VIII, MRS Symp. Proc. **594** (1999).

Special papers

[1] V. Hauk: *Structural and Residual Stress Analysis by Nondestructive Methods*, Contributions by H. Behnken, C. Genzel, W. Pfeiffer, L. Pintschovius, W. Reimers, E. Schneider, B. Scholtes and W. A. Theiner, Elsevier, Amsterdam, 1997.

[2] I. Szabó, *Geschichte der mechanischen Prinzipien*, Birkhäuser Verlag, Basel, 1976, p. 356.

[3] J. Perry, A further study of the state of residual stress in TiN films made by physical vapor deposition methods, *J. Vac. Sci. Technol.* A **8** (1990) 3186.

[4] I. C. Noyan, T. C. Huang, B. R. York, Residual Stress/Strain Analysis in Thin Films by X-ray Diffraction, *Crit. Rev. Sol. St. Mater. Sc.* **20** (1995) 125.

[5] A. Saerens, P. Van Houtte, B. Meert, C. Quaeyhaegens, Assessment of different X-ray stress measuring techniques for thin titanium nitride coatings, *J. Appl. Cryst.* **33** (2000) 312.

[6] W. Voigt, *Lehrbuch der Kristallphysik*, B. G. Teubner, Leipzig, 1928.

[7] A. Reuss, Berechnung der Fließgrenze von Mischkirstallen auf Grund der Plastizitätsbedingungen für Einkristalle, *Z. angew. Math. Mech.* **9** (1929) 49.

[8] H. Behnken, V. Hauk, Berechnung der röntgenographischen Elastizitätskonstanten (REK) des Vielkristalls aus den Einkristalldaten für beliebige Kristallsymmetrie, *Z. Metallkde.* **77** (1986) 620.

[9] F. Bollenrath, V. Hauk, E. H. Müller, Zur Berechnung der vielkristallinen Elastizitätskonstanten aus den Werten der Einkristalle, *Z. Metallkde.* **58** (1967) 76.

[10] H. Behnken, V. Hauk, Die röntgenographischen Elastizitätskonstanten keramischer Werkstoffe zur Ermittlung der Spannungen aus Gitterdehnungsmessungen, *Z. Metallkde.* **81** (1990) 891.

[11] H. Neerfeld, Zur Spannungsberechnung aus röntgenographischen Dehnungsmessungen, *Mitt. K. W. I. Eisenforsch.* **24** (1942) 61.

[12] R. Hill, The Elastic Behaviour of Crystalline Aggregates, *Proc. Phys. Soc. (London)* **A65** (1952) 349.

[13] J. D. Eshelby, The determination of the elastic field of an ellipsoidal inclusion and related problems, *Proc. Roy. Soc. (London)* **A241** (1957) 376.

[14] E. Kröner, Berechnung der elastischen Konstanten des Vielkristalls aus den Konstanten des Einkristalls, *Z. Physik* **151** (1958) 504.

[15] G. Kneer, Über die Berechnung der Elastizitätsmoduln vielkristalliner Aggregate mit Textur, *Phys. Stat. Sol.* **9** (1965) 825.

[16] R. W. Vook, F. Witt, Thermally Induced Strains in Evaporated Films, *J. Appl. Phys.* **36** (1965) 2169.

[17] M. van Leeuwen, J.-D. Kamminga, E. J. Mittemeijer, Diffraction stress analysis of thin films: Modeling and experimental evaluation of elastic constants and grain interaction, *J. Appl. Phys.* **86** (1999) 1904.

[18] M. Leoni, U. Welzel, P. Lamparter, E. J. Mittemeijer, J.-D. Kamminga, Diffraction analysis of internal strain-stress fields in textured, transversely isotropic thin films: theoretical basis and simulation, *Phil. Mag. A* **81** (2001) 597.

[19] U. Welzel, M. Leoni, E. J. Mittemeijer, Diffraction Elastic Constants and Stress Factors; Grain Interaction and Stress in Macrocsopically Elastically Anisotropic Solids; The Case of Thin Films, in: *Diffraction Analysis of the Microstructure of Materials*, p. 363.

[20] J. Almer, U. Lienert, R. L. Peng, C. Schlauer, M. Odén, Strain and texture analysis of coatings using high-energy x-rays, *J. Appl. Phys.* **94** (2003) 697.

[21] H. Behnken, V. Hauk, Berechnung der röntgenographischen Spannungsfaktoren texturierter Werkstoffe - Vergleich mit experimentellen Ergebnissen, *Z. Metallkde.* **82** (1991) 151.

[22] M. Ferrari, L. Lutterotti, Method for the simultaneous determination of anisotropic residual stresses and texture by x-ray diffraction, *J. Appl. Phys.* **76** (1994) 7246.

[23] C. Genzel, A Study of X-Ray Residual Stress Gradient Analysis in Thin Layers with Strong Fibre Texture. I. Evaluation of the Stress Factors F_{ij}, *phys. stat. sol. (a)* **165** (1998) 347.

[24] P. Scardi, Y. H. Dong, Residual stress in fiber-textured thin films of cubic materials, *J. Mater. Res.* **16** (2001) 233.

[25] C. Genzel, X-Ray Stress Gradients Analysis in Thin Layers – Problems and Attemps at Their Solution, *phys. stat. sol. (a)* **159** (1997) 283.

[26] H. U. Baron, V. Hauk, Röntgenografische Ermittlung der Eigenspanungen in Kristallitgruppen von fasertexturierten Werkstoffen, *Z. Metallkde.* **79** (1988) 127.

[27] E. Heyn, Eine Theorie der „Verfestigung" von metallischen Stoffen infolge des Kaltstreckens, *Festschr. Kaiser-Wilhelm-Ges.* (1921) 121.

[28] E. Macherauch, H. Wohlfahrt, U. Wolfstieg, Zur zweckmäßigen Definition von Eigenspannungen, *Härterei Techn. Mitt.* **28** (1973) 203.

[29] H. Mughrabi, Dislocation wall and cell structures and long-range internal stresses in deformed metal crystals, *Acta Metall.* **31** (1983) 1367.

[30] H. Mughrabi, T. Ungar, M. Wilkens, Gitterparameteränderungen durch weitreichende innere Spannungen in verformten Metallkristallen, *Z. Metallkde.* **77** (1986) 571.

[31] C. Genzel, Problems Related to X-Ray Stress Analysis in Thin Films in the Presence of Gradients and Texture, in: *Diffraction Analysis of the Microstructure of Materials*, p. 473.

[32] J. H. van der Merwe, The theory of two-phase interfacial structure, *J. Microscopy* **102** (1974) 261.

[33] D. S. Rickerby, A. M. Jones, B. A. Bellamy, Structure of Titanium Nitride Coatings Deposited by Physical Vapour Deposition: A Unified Structure Model, *Surf. Coat. Technol.* **36** (1988) 631.

[34] R. W. Hoffmann, The Mechanical Properties of Non-Metallic Thin Films, in: *Physics of Non-Metallic Thin Films*, NATO Avanced Study Institute Series **B14** (1976) 273.

[35] H. Windischman, Intrinsic Stress in Sputter-Deposited Thin Films, *Crit. Rev. Sol. Stat. Mat. Sc.* **17** (1992) 547.

[36] W. D. Nix, B. M. Clemens, Crystallite coalescence: A mechanism for intrinsic tensile stresses in thin films, *J. Mater. Res.* **14** (1999) 3467.

[37] S. N. G. Chu, A. T. Macrander, K. E. Strege, W. D. Johnston, Misfit stress in InGaAs/InP heteroepitaxial structures grown by vapor-phase epitaxy, *J. Appl. Phys.* **57** (1985) 249.

[38] W. D. Nix, Mechanical Properties of Thin Films, *Metall. Trans. A* **20** (1989) 2217.

[39] W. G. Sloof, B. J. Kooi, R. Delhez, T. H. d. Keijser, E. J. Mittemeijer, Diffraction analysis of nonuniform stresses in surface layers: Application to cracked TiN coatings chemically vapor deposited on Mo, *J. Mater. Res.* **11** (1996) 1440.

[40] H. Dölle, V. Hauk, Der theoretische Einfluß mehrachsiger tiefenabhängiger Eigenspannungszustände auf die röntgenographische Spannungsermittlung, *Härterei Techn. Mitteil.* **34** (1979) 272.

[41] I. J. Craig, A. M. Thompson, Why Laplace Transformation are Difficult to Invert Numerically, *Computers in Physics* **8** (1994) 648.

[42] C. Genzel, Formalism for the Evaluation of Strongly Non-Linear Surface Stress Fields by X-Ray Diffraction in the Scattering Vector Mode, *phys. stat. sol. (a)* **146** (1994) 629.

[43] C. Genzel, A self-consistent method for X-ray diffraction analysis of multiaxial residual-stress fields in the near-surface region of polycrystalline materials. I. Theoretical concept, *J. Appl. Cryst.* **32** (1997) 770.

[44] M. Leoni, Y. H. Dong, P. Scardi, Strain-Texture Correlation in r.f. Magnetron Sputtered Thin Films, *Mater. Sci. Forum* **321–324** (2000) 439.

[45] T. Dümmer, B. Eigenmann, M. Stüber, H. Leiste, H. Müller, O. Vöhringer, *Z. Metallkde.* **90** (1999) 780.

[46] T. Dümmer, B. Eigenmann, D. Löhe, X-Ray Analysis of Residual Stresses and Stress-Free Lattice Parameters in Thin Gradient Coatings, *Mater. Sci. Forum* **321–324** (2000) 81.

[47] K. von Acker, L. De Buyser, J.-P. Celis, P. van Houtte, Characterization of thin nickel electrocoatings by the low-incident-beam-angle diffraction method, *J. Appl. Cryst.* **27** (1994) 56.

[48] H. Mohrbacher, K. van Acker, B. Blanpain, P. van Houtte, J.-P. Celis, Comparative measurement of residual stress in diamond coatings by low-incident-beam-angle-diffraction and micro-Raman spectroscopy, *J. Mater. Res.* **11** (1996) 1776.

[49] C. Genzel, X-ray stress analysis in presence of gradients and texture, *Adv. X-ray Analysis* **44** (2001) 247.

[50] C. Genzel, M. Broda, D. Dantz, W. Reimers, A self-consistent method for X-ray diffraction analysis of multiaxial residual stress fields in the near-surface region of polycrystalline materials. II. Examples, *J. Appl. Cryst.* **32** (1999) 779.

[51] M. F. Doerner, S. Brennan, Strain distribution in thin aluminium films using x-ray depth profiling, *J. Appl. Phys.* **63** (1988) 126.

[52] C. J. Shute, J. B. Cohen, Strain gradients in Al-2at% Cu thin films, *J. Appl. Phys.* **70** (1991) 2104.

[53] P. Lengsfeld, N. H. Nickel, C. Genzel, W. Fuhs, Stress in undoped and doped laser crystallized poly-Si, *J. Appl. Phys.* **91** (2002) 9128.

[54] C. Genzel, X-ray residual stress analysis in thin films under grazing incidence – basic aspects and applications, *Mat. Sci. Technol.* **21** (2005) 10.

[55] C. Genzel, W. Reimers, Depth-resolved X-ray residual stress analysis in PVD (Ti,Cr)N hard coatings, *Z. Metallkde.* **94** (2003) 655.

[56] C. Genzel, Entwicklung eines Mess- und Auswerteverfahrens zur röntgenographischen Analyse des Eigenspannungszustandes im Oberflächenbereich vielkristalliner Werkstoffe, Habilitation thesis, Humboldt Universität, Berlin, (2000).

[57] G. Abadias, Y. Y. Tse, Diffraction stress analysis in fiber-textured TiN thin films grown by ion-beam sputtering: Application to (001) and mixed (001)+(111) texture, *J. Appl. Phys.* **95** (2004) 2414.

[58] B. Okolo, P. Lamparter, U. Welzel, E. J. Mittemeijer, Stress, texture, and microstructure in niobium thin films sputter deposited onto amorphous substrates, *J. Appl. Phys.* **95** (2004) 466.

[59] M. Birkholz, C. Genzel, T. Jung, X-ray diffraction study on residual stress and preferred orientation in thin titanium films subjected to a high ion flux during deposition, *J. Appl. Phys.* **96** (2004) 7202.

[60] A. Bendavid, P. J. Martin, X. Wang, M. Wittling, T. J. Kinder, Deposition and modification of titanium nitride by ion assisted arc deposition, *J. Vac. Sci. Technol.* **A13** (1995) 1658.

[61] J. E. Greene, J.-E. Sundgren, L. Hultman, I. Petrov, D. B. Bergstrom, Development of preferred orientation in polycrystalline TiN layers grown by ultrahigh vacuum reactive magnetron sputtering, *Appl. Phys. Lett.* **67** (1995) 2928.

[62] J. W. Gerlach, U. Preckwinkel, H. Wengenmair, T. Kraus, B. Rauschenbach, Biaxial alignment of TiN films prepared by ion beam assisted deposition, *Appl. Phys. Lett.* **68** (1996) 2360.

[63] C.-S. Shin, D. Gall, Y.-W. Kim, N. Hellgren, I. Petrov, E. Greene, Development of preferred orientation in polycrystalline NaCl-structure δ-TaN layers grown by reactive magnetron sputtering: Role of low-energy ion surface interaction, *J. Appl. Phys.* **92** (2002) 5084.

[64] B. Rauschenbach, S. Sienz, S. Six, J. W. Gerlach, Synthesis of metal nitrides by low-energy assisted film growth, *Surf. Coat. Technol.* **142–144** (2001) 371.

[65] M. Birkholz, E. Conrad, W. Fuhs, Crystallinity of Thin Silicon Films Deposited at Low Temperatures: Combined Effect of Biasing and Structuring the Substrate, *Jpn. J. Appl. Phys.* **40** (2001) 4176 .

[66] P. Reinig, F. Fenske, W. Fuhs, V. Alex, M. Birkholz, Highly (100)-Oriented Growth of Polycrystalline Silicon Films on Glass by Pulsed DC Magnetron Sputtering, *J. Vac. Sc. Technol. A* **20** (2002) 2004.

[67] N. C. Popa, D. Balzar, Elastic strain and stress determination by Rietveld refinement: generalized treatment for textured polycrystals for all Laue classes, *J. Appl. Cryst.* **34** (2001) 187.

[68] P. Villain, P. Goudeau, P.-O. Renault, K. F. Badawi, Size effect on intragranular elastic constants in thin tungsten films, *Appl. Phys. Lett.* **81** (2002) 4365.

[69] N. Schell, J. Bottiger, W. Matz, J. Chevallier, Growth mode and texture development in TiN films during magnetron sputtering – An in situ synchrotron radiation study, *Nucl. Instr. Meth.* **B 199** (2003) 133.

[70] P. B. Mirkarimi, K. F. McCarty and D. L. Medlin, Review of Advances in Cubic Boron Nitride Film Synthesis, *Mater. Sci. Eng. Rep.* **21** (1997) 47.

[71] W. Kress, P. Roedhammer, H. Bilz, W. Teuchert, A. N. Christensen, Phonon anomalies in transition-metal nitrides: TiN, *Phys. Rev.* **B 17** (1978) 111.

7
High-Resolution X-ray Diffraction

Mario Birkholz and Paul F. Fewster

Epitaxial thin films are in the focus of this chapter after the various structural characteristics of polycrystalline thin films have been presented in previous ones. In epitaxial layers the crystal lattice of the underlying single-crystal substrate is intended to be continued into the thin film. These material systems thus exhibit a high degree of crystalline perfection. However, some deviations from the perfect crystal lattice do occur in epilayers and it is the measurement of these deviations with which this chapter is concerned. As before, the structural information is derived from the position, intensity and line shape of Bragg reflections. However, in addition, the high degree of crystallinity necessitates refining the methodology with respect to both experimental equipment and theoretical understanding.

Special experimental equipment has to be introduced in order to resolve the very closely spaced peaks that occur in the diffraction pattern of epilayer–substrate material systems. This is the reason for naming the field high-resolution x-ray diffraction (HRXRD). It often happens that observable peaks are as close as 0.001° apart and that the structural information to be derived deserves the precise measurement of this spacing. This cannot be obtained with the diffractometer settings presented so far for the investigation of polycrystalline material. The precision $\Delta d/d$ by which interplanar spacings may be determined is given by

$$\frac{\Delta d}{d} = \frac{\Delta \lambda}{\lambda} + \frac{\delta}{\tan \theta} \quad (7.1)$$

where $\Delta \lambda$ is the spectral width or bandwidth of the radiation used and δ is the beam divergence. In order to arrive at the low $\Delta d/d$ values mentioned above crystal monochromators and analyzers have to be introduced into the beam path; these select only those x-rays that are Bragg reflected from a single crystal or a whole set of them. These systems are presented in Instrumental Box 8. An overview of $\Delta d/d$ ratios relevant to XRD and HRXRD is given in Fig. 7.1 that has been taken from the seminal paper of Bartels [1]. The diagram clearly shows the necessity of achieving an instrumental resolution in the region of seconds of arc in order to investigate thin-film devices produced in the electronic industry.

Moreover, the theoretical basis has to be extended in order to interpret the Bragg peaks of epitaxial films. Use has so far been made of the kinematical theory of x-ray diffraction in this book, but a proper understanding of epitaxial film–substrate diffraction patterns requires the use of the dynamical theory. The basic principles of the latter will shortly be outlined in this chapter. The reader is referred to the sources mentioned at the end of the chapter for a comprehensive introduction into the field.

The extension of both experimental equipment and theoretical understanding enables us to perform the measurement of rocking curves, so-called reciprocal space maps and their analysis in terms of HRXRD. The basis of various techniques to be presented here lies in the investigation of single crystals, which will frequently become evident. A closer look into this field of XRD is also of interest to the thin-film analyst, since it enables a better understanding of the microstructure and x-ray reflectivity of single-crystalline substrates and monochromators. Moreover, single-crystal techniques turn out to yield results that are of interest in the analysis of epitaxial thin films.

The last point is illustrated in this chapter by presenting examples from the area of optoelectronics and microelectronics, by which we mean III–V compound layers like cubic $Ga_{1-x}Al_xAs$ or hexagonal $Ga_{1-x}In_xN$ and group IV alloys like $Si_{1-x}Ge_x$. The functionality in these compounds is related to the engineering of their electronic band structure by tuning the composition x and its spatial variation on a nanometer scale. One may thereby change the wavelength of emitted light from light-emitting diodes (LEDs) and laser diodes (LDs) or the mobility of electronic charge carriers in electronic devices. Optoelectronic and microelectronic devices prepared from these materials are genuine thin-film systems that cannot be realized with bulk materials. Some background information on their structural properties and technological applications are found in Structure Box 8.

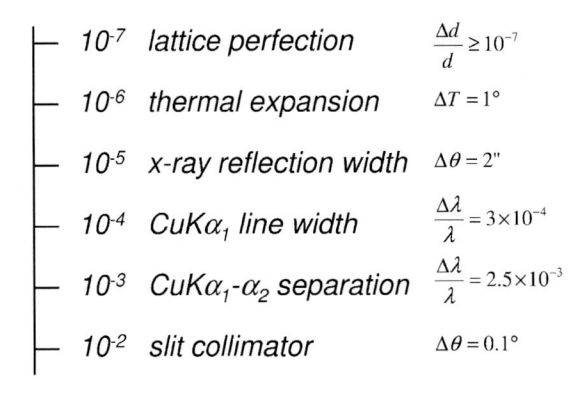

10^{-7}	lattice perfection	$\frac{\Delta d}{d} \geq 10^{-7}$
10^{-6}	thermal expansion	$\Delta T = 1°$
10^{-5}	x-ray reflection width	$\Delta\theta = 2''$
10^{-4}	$CuK\alpha_1$ line width	$\frac{\Delta\lambda}{\lambda} = 3\times 10^{-4}$
10^{-3}	$CuK\alpha_1$-α_2 separation	$\frac{\Delta\lambda}{\lambda} = 2.5\times 10^{-3}$
10^{-2}	slit collimator	$\Delta\theta = 0.1°$

Figure 7.1 Relative d spacings and precision according to Ref. [1].

Structure Box 8:

Epitaxial Thin Films for Electronic Applications

The structures of semiconducting compounds to be considered here have already been introduced in Structure Box 2. They are the diamond and the sphalerite structure in the case of cubic compounds and the wurtzite structure in the case of hexagonal ones. As underlying single-crystalline substrates either Si or GaAs wafers are typically used for cubic compounds, while sapphire Al_2O_3 crystals are mostly used for hexagonal ones. Evidently, the unit cell edges of the substrate should be as close as possible to those of the thin film to be grown. In most cases the crystallographic surfaces used for epitaxy are of the (001) and the (00.1) type, respectively. Deposition temperatures in epitaxial processes are typically in excess of 500 °C in order to activate a high surface diffusivity and grain boundary mobility and to achieve a high degree of crystalline perfection. When choosing a substrate for an epitaxial film the match of thermal expansion coefficients also has to be considered to avoid the occurrence of overly strong strains during cooling from the process temperature to room temperature.

Cubic Group IV Alloys

Switching frequencies in microelectronic silicon devices are restricted due to the maximum mobility of electronic charge carriers which are 1450 and 370 cm^2 V^{-1} s^{-1} for electrons and holes, respectively, at room temperature in Si. The general tendency of effective mass of charge carriers in the periodic system is to decrease with increasing period. Higher mobilities in Si devices can thus be obtained by randomly substituting Si atoms in the lattice by Ge and $Si_{1-x}Ge_x$ devices have paved the way for high-frequency applications. They are prepared in molecular beam epitaxy (MBE) or chemical vapor deposition (CVD) processes by epitaxy of thin $Si_{1-x}Ge_x$ films on Si (001) wafers with x ranging between 10 and 30%. A prominent device example is the heterojunction bipolar transistor (HBT), where use is made of the $Si_{1-x}Ge_x$ layer in the active base region of the transistor. This fast switching electronic device can be integrated into conventional CMOS/BiCMOS technology (see Fig. s8.1 for an example of device architecture [1]).

It is evident from the larger atomic diameter of Ge compared to Si that $Si_{1-x}Ge_x$ films on Si substrates experience a strong compressive stress. The stress might be accommodated via mechanical relaxation through the formation of misfit dislocations. If the dislocation density exceeds a critical value it may cause significant scatter of charge carriers and a reduction of mobility. One thus deals with a typical optimization problem of thin-film technology where a physical property is to be improved, but may be recursively diminished by the measures taken. One approach of optimization makes use of alloying additional carbon into the layer and to prepare the ternary compound $Si_{1-x-y}Ge_xC_y$ with small carbon content y of about 1%. The inclusion of carbon into SiGe turns out to have further beneficial effects like the suppression of plastic relaxation by dislocation pinning and the reduction of B out-diffusion of p-doped layers in succeeding processing steps. The grading of the Ge content in the HBT base as realized from Fig. s8.1 is directly relevant to device performance, since it is used to tailor the current amplification by the transistor.

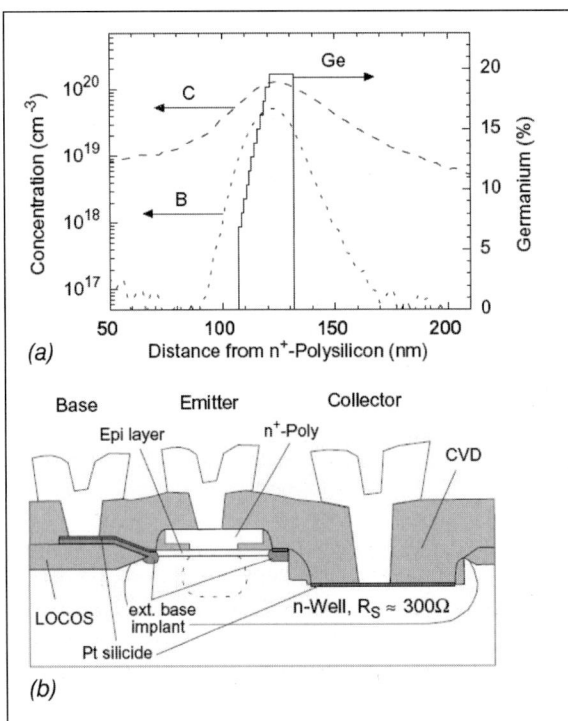

Figure s8.1 Architecture of silicon-germanium heterojunction bipolar transistor. (a) B, C and Ge concentration profiles with Ge profile determined by HRXRD. The spread in B and C profiles is artificially enhanced by the SIMS measurement process. (b) Schematical cross section of HBT structure [1].

Cubic III–V Compounds

This group comprises the set of binary $A^{III}B^V$ and ternary compounds that derive from the combination of (Al, Ga, In) with (P, As, Sb). The cubic binaries adopt the sphalerite structure and their unit cell edges a and electronic band gaps E_g are both displayed in Fig. s8.2. The ternary compounds $A^{III}_{1-x}\widetilde{A}^{III}_{x}B^V$ or $A^{III}B^V_{1-x}\widetilde{B}^V_{x}$ are found along the alloy tie lines of this plot. These material systems show an almost perfect miscibility having either group III elements randomly distributed over Zn sites or group V elements over S sites of the sphalerite structure. This property enables band gap engineering by controlling the materials composition x. The most prominent example is $Ga_{1-x}Al_xAs$ the band gap of which can be adjusted between 1.35 and 2.2 eV. This ternary is exceptionally attractive for optoelectronic applications, because the band gap variation is associated with a cell edge expansion of only 0.15% over the whole range of composition. Further degrees of freedom can be introduced by variation of both the III and V component leading to quaternary compounds.

The heteroepitaxy of ternary and quaternary compounds from the III–V system is mostly performed on single-crystal GaAs or InP (001) substrates that are commercially available with a high degree of crystalline perfection. The deposition process is carried

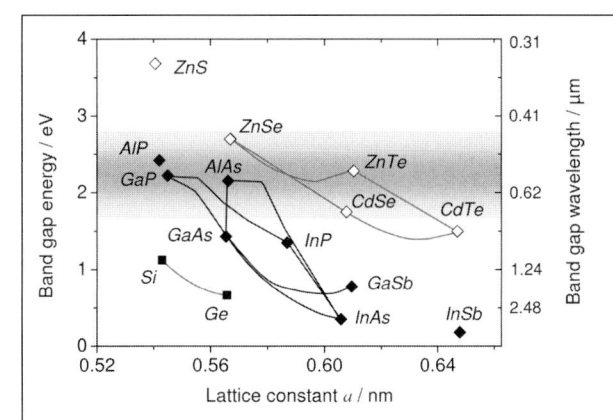

Figure s8.2 Variations of cubic cell edge a and electronic band gap E_g in some III–V and II–VI compounds and along some alloy tie lines. The shaded area between 0.44 and 0.75 μm indicates that region to which the human eye is sensitive.

out by either metal-organic chemical vapor deposition (MOCVD) where precursor gases like $A^{III}(CH)_3$ and $B^V H_3$ are typically used or by MBE which operates by evaporating the elements A^{III} and B^V from so-called effusion cells and directing the molecular beams towards the substrate. The growth of single atomic layers on a reproducible basis may be performed by both techniques enabling the production of highly abrupt compositional profiles.

Epitaxial thin films from the III–V system became the materials basis of optoelectronic devices due to their direct electronic band gap and the high mobilities of charge carriers. The success of these compounds derives from the fact that their band gaps can be tailored to the minimum loss windows in optical fiber communication around 1.3 and 1.55 μm, which is achieved by the preparation of $In_{1-x}Ga_xAs_yP_{1-y}$ films on InP substrates. In this case the band gap is tailored by the variation of the As content, $E_g = (1.35 - 0.72y + 0.12y^2)$ eV, for fixed x, while the lattice match to the substrate is obtained by varying the Ga content for which $x = 0.47y$ is required. Other prominent optoelectronic devices are light-emitting diodes (LEDs) and laser diodes (LDs) which are found, for instance, in compact disk drives. Basically, the function of these devices is enabled by the stacking of layers of different band gap as represented, for example, by a vertical-cavity surface-emitting laser (VCSEL) that may comprise some hundred individual layers. The major shortcoming of the cubic III–V compounds is that their range of emittable wavelengths is restricted to wavelengths higher than 500 nm, prohibiting the generation of blue light by these devices. The long lasting search for an adequate material system for this purpose was decided in the 1990s by the preparation of first blue light emitting GaN-based devices.

Hexagonal Group III Nitrides

Nitrogen III–V compounds like $Ga_{1-x}In_xN$ are well suited as blue light and UV emitters according to the band gap energies of 3.5 and 1.9 eV of GaN and InN, but their epitaxy is complicated by the crystallographic wurtzite structure. Additionally, their introduction into optoelectronic applications was impeded for a long time by the inability of p-doping them and a missing source for adequately reactive nitrogen. These problems were overcome, although complications remained when compared to the cubic III–V compounds that are related to the mismatch to single-crystal substrates. Sapphire Al_2O_3 is the most widely used substrate, but the system exhibits a rather large lattice mismatch. Figure s8.3 shows the epitaxial relation for GaN on a (00.1) sapphire cleavage plane with the layer unit cell rotated by 30° with respect to the substrate one. A relative mismatch of $\Delta a/a = (2a_L\cos30° - a_S)/a_S = 16.1\%$ can be realized from $a_{GaN} = 0.3189$ nm and $a_{Al2O3} = 0.4759$ nm. Moreover, the mismatch in thermal expansion coefficients is also high, and both effects can induce the occurrence of high dislocation densities during processing to the order of 10^{10} cm^{-2} causing twist and tilt of individual grains in the film.

A special structural feature of the noncentrosymmetric wurtzite structure is related to piezoelectrical fields that develop upon stress or strain. These fields are held responsible for affecting device performance, e.g. they may cause an increase of laser diode threshold currents or a reduction in LED emission efficiency. The determination of residual strains may therefore become a very relevant issue in hexagonal III–V-based devices in order to estimate the strength of built-in piezoelectric fields.

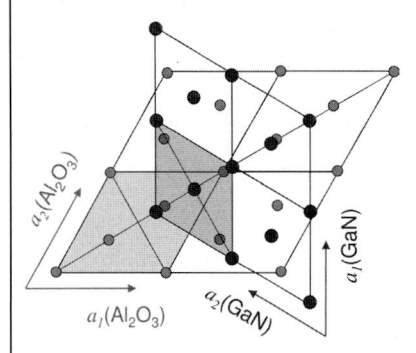

Figure s8.3 Epitaxial relation between (00.1) basal planes of GaN and Al_2O_3 sapphire.

References

Monographs and review articles relevant to this box

M. Grundmann (ed.), *Nano-Optoelectronics*, Springer, Berlin, 2002.
O. Madelung (ed.), Landolt-Börnstein III/22: *Semiconductors: Intrinsic Properties of Group IV Elements and III–V, II–VI and I–VIII Compounds*, Springer, Berlin, 1996.

Special paper

[1] D. Knoll, B. Heinemann, H. J. Osten et al., Si/SiGe:C heterojunction bipolar transistor in an epi-free well, single-polycrystalline technology, IEDM 1998, IEEE 1998.

7.1
Strain, Strain Relaxation and Composition in Epitaxial Layers

Ideally, epitaxy proceeds as so-called pseudomorphic growth with the substrate lattice continuing into the thin film (see Fig. 7.2). In the majority of epitaxy processes the compound to be deposited is distinct from the substrate – and the process is called heteroepitaxy. It is evident that the continuation of the substrate crystal lattice will be associated with the incorporation of strain into the layer if the layer and substrate unit cell edges a_L and a_S are distinct or even the crystallographic structures differ. A reliable measure for the strain to be expected is the difference in cell edges, which is called mismatch and reads $\Delta a = a_L - a_S$ in the case that both layer and substrate belong to the cubic crystal class. Sometimes, mismatch may also denote the relative mismatch $m = (a_L - a_S)/a_S$. It has been observed that epitaxy occurs if cell edges match with a relative mismatch of less than about 14%. Irrespective of the precise mismatch value, heteroepitaxy will inevitably introduce strain as is visualized on the left of Fig. 7.2 for an epitaxial process of cubic substances on a (001) surface. The adoption of the layer's lattice to that of the substrate causes a tetragonal distortion in the film unit cells. As can be seen from the figure the unit cells extend or shrink from their original value a_L to $a_\perp$ along the out-of-plane direction according to whether a_L is larger or smaller than a_S. The in-plane cell parameter is then denoted by $a_\parallel$.

According to the formalism presented in Chapter 6, the strain in the out-of-plane direction $\varepsilon_{33} = (a_\perp - a_L)/a_L$ is related for a cubic substrate–layer system via

$$\varepsilon_{33} = -\frac{2c_{1122}}{c_{1111}}\varepsilon_{11} \tag{7.2}$$

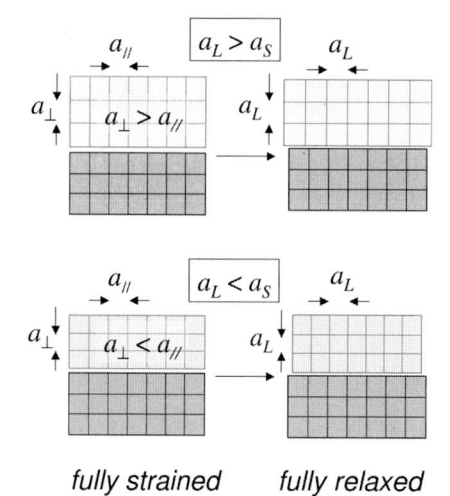

Figure 7.2 Strained (left) and relaxed (right) layers for $\Delta a > 0$ (top) and $\Delta a < 0$ (bottom).

to the in-plane strain $\varepsilon_{11} = (a_\| - a_L)/a_L$, where c_{1122} and c_{1111} are the single-crystal elastic stiffness constants of the layer. If the layer is fully strained, the in-plane cell edge perfectly coincides with the substrate cell and $a_\| = a_S$ holds. It may then be derived that the layer out-of-plane edge may be represented by the sum

$$a_\perp = a_L + (\kappa - 1)\Delta a \qquad (7.3)$$

where the elastic stiffness constants have been combined into the constant κ

$$\kappa = 1 + \frac{2c_{1122}}{c_{1111}} \qquad (7.4)$$

For other orientations (hkl) of the epitaxial system distinct κ values have to be inserted that are functions of the elastic constants only [2]. For vanishing elastic anisotropy, i.e. c_0 close or equal to zero, κ can be represented as a function of the Poisson ratio v, like $\kappa = 1 + v/(1 - v)$ in the case of (001) oriented growth of a cubic material system.

In deriving Eq. (7.3) use has been made of the fact that only the thin film is subjected to the tetragonal distortion, while the substrate lattice parameter is assumed to remain constant in the in-plane and in the out-of-plane direction. This assumption is justified when the thickness of the substrate greatly exceeds the thickness of the layer. According to Eq. (7.3) the difference in the lattice spacing causes a splitting of a Bragg reflection hkl into one from the substrate and – shifted from that – one from the film. A common substrate orientation in the epitaxy of cubic compounds is (001) and the two 00l reflections are separated in this case by the scattering angle difference

$$\Delta 2\theta_{00l} = 2\arcsin\left(l\frac{\lambda}{a_S}\right) - 2\arcsin\left(l\frac{\lambda}{a_\perp}\right) \qquad (7.5)$$

into which Eq. (7.3) for the out-of-plane cell edge of the layer would have to be inserted. Hence, the splitting depends on the elastic constants and the mismatch. The investigated reflection typically is the 004 for diamond-structure materials like $Si_{1-x}Ge_x$/Si, or, in addition, the 002 for sphalerite-structure materials.

It has been realized in many investigations that mechanical relaxation phenomena occur above a certain threshold of strain and that the in-plane layer edge may then deviate from that of the substrate, $a_\| \neq a_S$. A critical thickness t_c is identified for various epitaxial film–substrate systems above which a transition from the fully strained state to a relaxed state is observed. The degree of relaxation can be described by the relaxation parameter R

$$R = \frac{a_\| - a_S}{a_L - a_S} \qquad (7.6)$$

R takes values between 0 and 1 according to whether the film in-plane lattice constant $a_\|$ equals that of the substrate a_S or that of the unstrained layer a_L. The extrema of the fully strained state, $R = 0$, and the fully relaxed state, $R = 1$, are shown schematically in Fig. 7.2 for the cases that the unstrained lattice constant of the layer a_L is larger or smaller than that of the substrate a_S. Similar to the derivation above it is found that the out-of-plane cell parameter as a function of R is given by

$$a_\perp(R) = a_L + (1-R)(\kappa - 1)\Delta a \tag{7.7}$$

Again, a Bragg reflection of an epitaxial film–substrate systems will split into two reflections that are due to the layer and the substrate. Moreover, and this is the interesting point, their separation scales with the degree of relaxation R. In the most relevant case of a symmetric 00l reflection for epitaxial growth on a (001) cubic substrate the angular separation in units of the scattering angle $\Delta 2\theta_{00l}(R)$ becomes

$$\Delta 2\theta_{00l}(R) = 2\arcsin\left(l\frac{\lambda}{a_S}\right) - 2\arcsin\left(l\frac{\lambda}{a_\perp(R)}\right) \tag{7.8}$$

The phenomenon of relaxation depends on many factors. A very important parameter, of course, is the lattice mismatch Δa. Other influential factors are the elastic constants, and sometimes the characteristics of the deposition process. On a microscopic scale the elastic relaxation of the film is accompanied by the generation of dislocations. Since they are all introduced by the difference in the film and the substrate cell edge they are summarized as misfit dislocations; these can occur by bending over threading dislocations at the interface or from dislocation loops forming at the growing surface and gliding to the interface. Interestingly, the degree of relaxation R is directly related to the dislocation density and Eq. (7.8) thus offers a technique for its determination. Often too little consideration is given to the question of contaminants introduced during the growth process that might have substantial effects, especially for concentrations $>10^{18}$cm^{-3}. According to the concept of the critical thickness a breakdown of the fully strained pseudomorphic growth would occur for $t > t_c$ and from there on the film would grow in a partially or fully relaxed state. In many cases, however, this single-value model is not sufficient to account for the structural evolution within the film, which is the case, for instance, when the relaxation becomes a depth-dependent function $R(t)$.

It should be noted that the accuracy of determined relaxation values is coupled to the accuracy in elastic constants and therefore the constant κ and that tabulated elastic constants are often of less precision than expected. This holds, in particular, when ternary or quaternary compounds are under investigation. It should be kept in mind, moreover, that thin-film material parameters may significantly deviate from tabulated values that were often determined from bulk samples.

The degree of strain can become an important property in epitaxial films, since fully strained layers may be associated with low dislocation densities. Dislocations turn out in various materials systems to act as effective recombination centers and strain, therefore, is directly coupled to the layer electronic properties. In some cases, strain may even affect the band structure, as occurs for instance in InGaAs, where the valence band heavy hole/light hole degeneracy is lifted by strain and causes a variation of carrier mobility. Thus controlling the strain state is a highly relevant issue for tailoring of electronic properties.

A further important parameter in epitaxial electronic films like $Ga_{1-x}Al_xAs$ or $Si_{1-x}Ge_x$ is the composition x. The film composition affects both the position and the intensity of the Bragg reflection. The unit cell lengths of ternary III–V compounds may be approximated by a linear relationship called Vegard's rule that has already been introduced in Chapter 2. A linear relation is also assumed to account for the composition dependence of the κ constant. Both quantities are thus described by

$$a_L(x) = a_L^0(1-x) + a_L^1 x \qquad (7.9)$$
$$\kappa(x) = \kappa_L^0(1-x) + \kappa_L^1 x$$

where superscripts 0 and 1 refer to the two phases involved. For some material systems, like $Si_{1-x}Ge_x$ for instance, the lattice parameter may accurately be described by a quadratic function of composition x. In any case, an accurate $a(x)$ relation should be known when the composition is to be derived from the peak splitting. In generalizing Eq. (7.3) the out-of-plane cell parameter of a fully strained layer is then given by

$$a_\perp(x) = a_L(x) + [\kappa(x) - 1]\Delta a(x) \qquad (7.10)$$

where either the linear Eqs. (7.9) or higher polynomial expressions have to be inserted. For a $00l$ reflection the film composition may directly be deduced from the peak separation according to

$$\Delta 2\theta_{00l}(x) = 2\arcsin\left(l\frac{\lambda}{a_S}\right) - 2\arcsin\left(l\frac{\lambda}{a_\perp(x)}\right) \qquad (7.11)$$

HRXRD thus presents a simple technique for the determination of chemical composition. It should be remembered, however, that the reflection position may also be affected by the degree of relaxation R and we would have to evaluate a more complicated case, $\Delta 2\theta_{00l}(R, x)$. Techniques have been developed that allow for the concomitant determination of both composition x and relaxation parameter R as is shown in Section 7.2.

7.2
High-Resolution Rocking Curves

Assume that we now want to perform a high-resolution measurement. We have a sample comprising a Ge layer on a GaAs 001 substrate. The aim of the investigation is to determine the degree of relaxation R by measuring a single reflection with the rocking curve technique. This method has already been introduced in Chapter 5 and for its application the detector is set to the constant scattering angle $2\theta_B$ of the reflection under investigation while the sample is rotated around ω in the vicinity of the Bragg angle θ_B. Recall that the sample's axis of rotation is denoted by ω instead of θ if its rotation is uncoupled from the detector movement. The relaxation R is determined by measuring the 004 reflection which should split into a film peak and a substrate peak. Before starting the experiment the precision afforded in scattering angle resolution $\Delta 2\theta$ is estimated in order to set up the experimental constraints. The measurement of the 004 should be associated with a peak separation of a quarter of the mismatch, i.e. $\Delta a/4$, on the d axis which translates to a difference of $66.046° - 66.007° = 0.039°$ in 2θ units when Cu Kα radiation is applied. If the precision in the relaxation R to be determined should amount to 10%, i.e. $\Delta R = 0.1$ on an absolute scale, we would require a resolution of about $0.004°$ in 2θ for the fully strained layer. However, the relaxation of the layer strain would be even smaller than the $0.039°$ separation and would cause a further reduction in the afforded experimental resolution $\Delta 2\theta$.

The example clearly demonstrates that the resolution required for the investigation of epitaxial layers exceeds by far the slit-controlled instrumental resolution of the diffractometer configurations presented so far. A low beam divergence in the $0.01°$ and $0.001°$ range may only be obtained by virtue of double-crystal or multiple-crystal diffractometers with crystals in the incident beam path and the diffracted beam path. Such diffractometer systems are presented in Instrumental Box 8. Figure 7.3 shows a commercial multiple-crystal diffractometer.

In consequence of the much smaller angular spread on which the scattering intensity varies in high-resolution work the angular units of minutes of arc and sec-

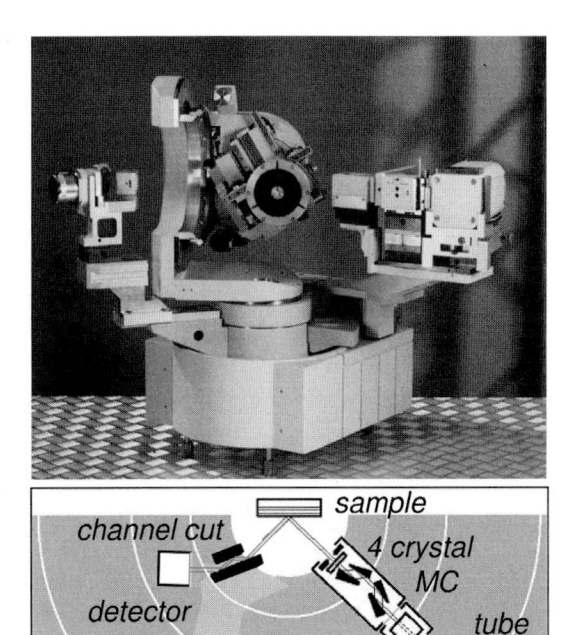

Figure 7.3 Setup of a multiple-crystal diffractometer (picture kindly provided by PANalytical).

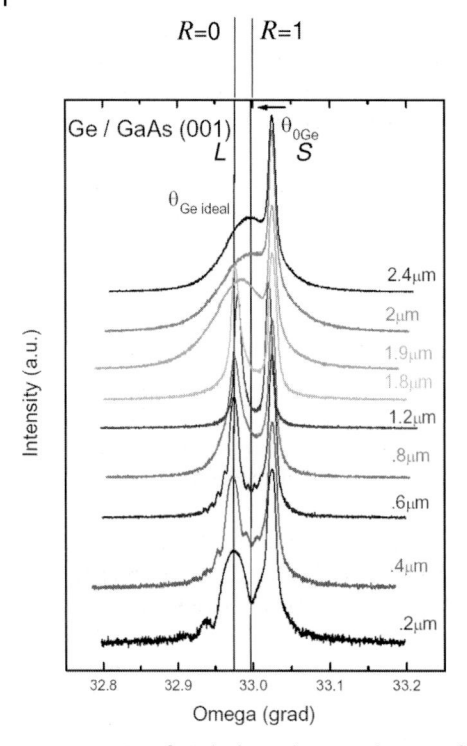

Figure 7.4 Set of 004 high-resolution rocking curves of Ge epilayers on GaAs substrate. The layer thickness is indicated in μm for each curve and increases from the bottom to the top (from Ref. [3]).

onds of arc are often used instead of degrees. One arc minute (1′) and one arc second (1″) are the 1/60 and 1/3600 part of 1°, respectively, and as a rule of thumb one might remember the approximate relation 3″ ≈ 0.001°. A further distinction between diffraction experiments with epitaxial and polycrystalline films derives from the higher sensitivity in the former compared to the latter. This is caused by the fact that all lattice planes of the sample contribute to the Bragg peak of an epitaxial film, whereas in a polycrystalline film a large set of grains is inadequately oriented to scatter into the detector. Therefore, also very thin epitaxial films may be measured which holds in particular if the film has grown perfectly pseudomorphic and is in a fully strained state.

The following example demonstrates the wealth of information that can be deduced from high-resolution rocking curves. The measurements were performed with thin Ge films on GaAs 001 substrates, which is associated with a relative mismatch $(a_{Ge} - a_{GaAs})/a_{GaAs}$ of only 0.08%. Figure 7.4 shows a set of symmetric 004 rocking curves that were measured for Ge films of varying thickness [3]. For small thicknesses one first sees the splitting of the reflection into a substrate peak S on the high-angle side and a layer peak L on the low-angle side. This arrangement is evidently caused by $a_L > a_S$ and would be reversed in the opposite case. The peak separation is due to the elongation of Ge unit cells associated with the tetragonal splitting and from the measurement of the peak position difference the out-of-plane unit cell edge of the layer $a_\perp$ is deduced (see the top part of Fig. 7.2). The layer's in-plane cell edge remains unresolved from this measurement and can only indirect be determined by use of Eqs. (7.6) or (7.7). Interestingly, the relevant features of the patterns are contained in a 0.2° wide range of rocking angle ω (or 0.4° in scattering angle 2θ) which amounts to 720″. The half widths of the tallest observable peaks are only of the order of 70″ emphasizing the necessity to perform the measurements with a highly collimated and monochromatic x-ray beam. The plot also indicates the theoretical position of fully strained and fully relaxed Ge layers by vertical lines and it is seen that the relaxation process starts to occur for layer thicknesses in excess of 600 nm. For thicker films the Ge peak shifts towards the fully relaxed position and broadens due to the increasing density of built-in crystallographic lattice faults. There is a significant overlap between the film and the substrate peak for the thickest film underlining again the necessity to make use of high-resolution equipment.

Interesting conclusions may also be drawn from the width of the layer peak. In general, the interpretation of the peak width proceeds along the same line of arguments as already outlined in Chapter 3, i.e. the peak width is broadened by (a) small dimensions of coherently scattering domains and (b) crystal lattice faults. In application of this argument we would expect the peak width to account solely for the film thickness in the case that a perfect pseudomorphically grown sample is investigated. And as with the determination of grain size by the formulas of Scherrer and others the peak width inversely scales with the height of unit cell columns, which equals the total film thickness when an epilayer is considered. There is no argument that would contradict the application of the derivation outlined in Section 3.4 to the case of epilayers, provided the scattering is weak (i.e. follows the kinematical approximation) and we thus end up with the formula

$$t = \frac{K'\lambda}{\beta\cos\theta} \tag{7.12}$$

relating the peak width β on the 2θ scale and thickness t of an epitaxial film that is free of dislocations or any other lattice faults. However, the analogue also applies to the microstructure, i.e. the peak broadening might also be due to dislocations or other crystalline lattice faults. Both phenomena are indeed observed in the Ge layer 004 rocking curve. For the thinnest samples, the peak width is seen to decrease with increasing thickness while it becomes broader again after t exceeded a thickness of 600 nm due to the increasing incorporation of crystalline lattice faults into the film. This is exactly the point for the onset of relaxation as derived from peak separation. These observations may consistently be interpreted by assuming that an increase of plastic deformation is introduced into the films after exceeding a thickness of 600 nm.

There is a further method to derive the film thickness from epilayer rocking curves other than the layer peak width. This is due to the fringing observed in the vicinity of the layer peak. It can be seen in the example figures that the Ge peaks of the thinnest films exhibit small interference fringes to the left and to the right of the main peak. The fringing has the same cause as the Kiessig fringes observed in x-ray reflectometry (XRR) and is due to interferences of x-rays that are reflected from the surface of the film and from the substrate–film interface. Fringing is a general phenomenon to be observed from the scattering of parallelepiped-shaped samples that has already been met in Chapters 1, 3 and 4. However, whereas the XRR technique may be considered as scanning the vicinity of the central 000 peak, the rocking curve technique elucidates Bragg reflections hkl with at least one of the Miller indices differing from zero. The distance $\Delta\theta_m$ between the fringes of order m can again be analyzed in terms of a modified Bragg equation with the film thickness substituting for the interplanar spacing. The formula

$$t = \frac{\lambda}{\Delta\theta_m} \frac{1}{2\cos\theta} \tag{7.13}$$

describes how the film thickness may be obtained from the fringe spacing for planes parallel to the surface. The observation of the fringing thus provides a valuable tool to determine the thickness of epitaxial layers. As can be seen from the figure, the fringing can only be analyzed if the density of crystalline disorder associated with dislocations, mosaicity, surface roughness, etc., is not too strong. In the case that disorder exceeds a critical level the fringes are washed out and may not be observed.

We conclude from the discussion of this first example that three types of peaks can be observed in epilayer rocking curves, caused by the substrate, the layer and the thickness fringes of the layer. S, L and F will abbreviate these different groups of peaks in the following.

In the next example it is considered how the composition x and the degree of relaxation R may be determined in common from a binary or ternary compound layer like $Si_{1-x}Ge_x$ or $Ga_{1-x}Al_xAs$. Thin-film alloys from the SiGe system have been developed for high-frequency microelectronic devices that have found application in the area of wireless communication, telecommunications and other fields. Heterojunction bipolar transistors (HBTs) using epitaxial $Si_{1-x-y}Ge_xC_y$ base layers with x and y around 20% and 0.5%, respectively, demonstrated high maximum oscillation frequencies that were increased from 60 GHz in 1998 to values in excess of 240 GHz in 2005 [4, 5]. The process development of appropriate $Si_{1-x}Ge_x$ layers required accurate knowledge of the Ge concentration substitutionally incorporated into the Si lattice, which could be gained from HRXRD rocking curves [6–10].

In order to derive the degree of relaxation R and the composition x in common it is evident from the discussion above that more structural information is needed than might be obtained from the measurement of one Bragg reflection alone even if it divides into separate peaks from the substrate, the film and some thickness fringes. The generalization of the formulas for a fully or partially relaxed layer yields for the $00l$ peak splitting of (001) oriented cubic materials

$$\Delta 2\theta_{00l}(R,x) = 2\arcsin\left(l\frac{\lambda}{a_S}\right) - 2\arcsin\left(l\frac{\lambda}{a_\perp(R,x)}\right) \tag{7.14}$$

Also, Eq. (7.7) for the out-of-plane cell edge $a_\perp$ of the layer dependent on both quantities R and x is simply found by inserting x-dependent functions for a_L and κ

$$a_\perp(R,x) = a_L(x) + (1-R)\big[\kappa(x) - 1\big]\Delta a(x) \tag{7.15}$$

depending on two unknown quantities with degree of relaxation R and composition x.

The required additional information can be derived from asymmetric reflections, which are measured with the scattering vector Q tilted with respect to the substrate surface normal s_3. By measuring one of the asymmetric reflections structural information is achieved about the out-of-plane cell edge, i.e. $a_\perp$ and the in-plane cell length $a_\parallel$ simultaneously. The latter quantity may be written in its (R, x)-dependent variant by making use of the definition equation of relaxation, Eq. (7.6), and it is obtained

$$a_\parallel(R,x) = a_L(x) - (1-R)\Delta a(x) \tag{7.16}$$

The common determination of R and x thus proceeds by first measuring a symmetric reflection $00l'$ and secondly scanning an asymmetric one hkl. The two scans yield the peak separation $\Delta 2\theta_{00l}(R, x)$ and $\Delta\Psi_{hkl}$. Zaumseil worked out this formalism for $Si_{1-x}Ge_x$ on Si (001) [6], with 004 and 113 serving as symmetric and asymmetric reflections, respectively. He argued straightforwardly that the measured angular difference $\Delta\Psi_{hkl}$ of the asymmetric reflection is the difference of inclination angles of hkl lattice planes between the layer and the substrate. The geometry is shown in Fig. 7.5(a), from which it can be seen that the angles of inclination of (hkl) lattice planes with respect to (001) planes are given by

$$\Psi_S = \arccos\left(\frac{l}{\sqrt{h^2 + k^2 + l^2}}\right)$$

$$\Psi_L = \arccos\left(\frac{l}{a_\perp\sqrt{(h^2+k^2)/a_\parallel^2 + l^2/a_\perp^2}}\right) \tag{7.17}$$

In addition to Eq. (7.14) one thus arrives at a second equation

$$\Delta\Psi_{hkl} = \Psi_L - \Psi_S \tag{7.18}$$

depending on relaxation R and Ge content x via the included $a_\perp$ and $a_\parallel$. The solution may be obtained by a numerical regression of R and x to Eqs. (7.14) and (7.18). Alternatively, a composition–relaxation (x, R) net may be drawn for the material system under investigation. Such a net is shown in Fig. 7.5(b) for the SiGe system, which has been calculated for the angular peak differences of 004 and 113 reflections. The developed formalism does not only apply to SiGe, but to any composition- and relaxation-dependent cubic epitaxial system of (001) orientation [6]. Also the x dependence of κ may be included into the formalism, which could be neg-

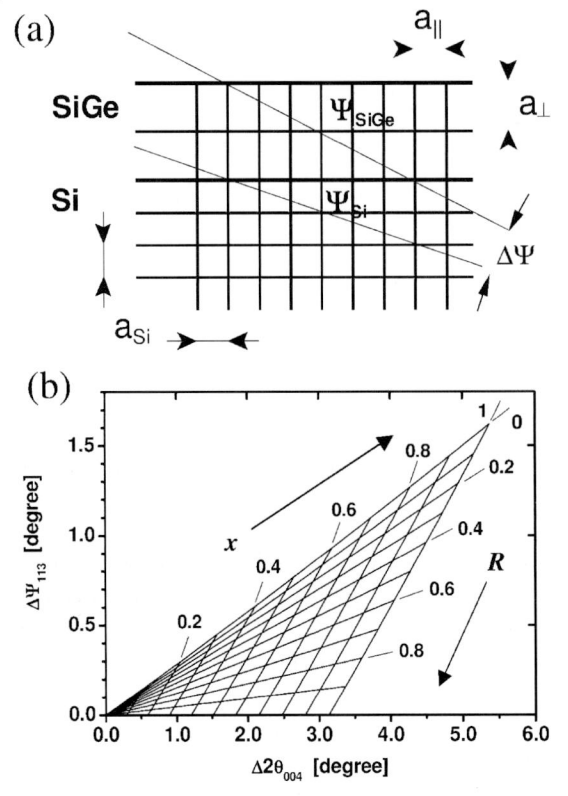

Figure 7.5 (a) (113) lattice plane inclination in Si substrate and a pseudomorphic SiGe layer on top. (b) Composition–relaxation (x, R) net corresponding to L–S peak position differences for symmetric 004 and asymmetric 113 reflection (compare Ref. [6]).

lected in the case of $Si_{1-x}Ge_x$, because the κ values of the endpoints Si and Ge differ by only 1% (according to the data given in Ref. [2]).

Table 7.1 lists the structural properties that may be determined from the different features of high-resolution rocking curves. It can be seen that highly relevant structural features of epitaxial films can be obtained. The development of HRXRD techniques as an analytical tool and the preparation of ever more complex optoelectronic devices have proceeded hand-in-hand. This progress has been accompanied by the increasing availability of software code that enables the regression of rocking curve scans with the dynamical theory of x-ray diffraction. After a short introduction into the basics of this theory we will take up again the issue of rocking curves and their quantitative interpretation in the case of the measurement of multilayer and superlattice systems.

Most HRXRD work is performed in the coplanar geometry, i.e. the sample normal is in the same plane as the wave vectors of the incoming and exiting beam K_0 and K_h. The line focus of an x-ray tube may thus be applied enabling a large x-ray

Table 7.1 Structural properties of thin films to be derived from high-resolution rocking curves. The features in the last two rows are discussed in later sections of this chapter.

HRRC feature	Structural property
Peak separation	Composition, strain
Peak widths	Layer thickness (Scherrer), crystalline perfection of either substrate or layer, i.e. mosaicity, dislocation density etc.
Fringing	Layer thickness
Integral intensity	Layer thickness, composition
Diffuse background	Structural defects like dislocation density, etc.
Satellite separation	Superlattice period

spot on the sample surface and an averaging over sample areas of the order of 10 mm². In contrast to polycrystalline XRD work the observation of a Bragg reflection from the layer is concentrated into a narrow ω range, because of the high degree of preferred orientation. With the extremely narrow, but intense Bragg reflection obtained HRXRD is at the opposite extreme of polycrystalline films that have broad reflections of small intensity. One should thus carefully distinguish between the rocking curves or Ω scans of polycrystalline materials (Chapter 5) and high-resolution rocking curves as required for the analysis of electronic device structures.

The group of structural parameters mentioned above like strain, degree of relaxation, dislocation density and composition and their depth profiles may be considered as microscopic structural features. In addition, macroscopic structural properties may significantly influence device performance by which we summarize phenomena like substrate bending, inclined growth due to substrate surface orientation and thickness inhomogeneity. One should be aware of the possibility that the substrate–film system may deviate from an ideal material system in these respects. These deviations can also affect the appearance of the rocking curve: the bowing of the sample, for instance, and an associated line broadening may already be introduced by unsuitable installation of the sample on the sample holder. Fringing might vanish due to bowing of the sample, which can be dealt with by reducing the illuminated area to a focal spot size with less bowing.

Consider the example in Fig. 7.6 of a SiGe HBT structure that has begun to relax at the SiGe–Si substrate interface, creating a disrupted interface. The interference fringes have been lost in the case where the Bartels monochromator is used due to the lateral inhomogeneity due to local scattering plane rotations from the dislocations. The inclusion of the analyzer crystal has reduced the range of rotations accepted into the detector and the fringes are recovered. For mosaic crystals a similar approach can be used. An alternative approach is to reduce the area analyzed on the sample, or a combination of both to isolate a homogeneous region for analysis. Further measurement problems that occur for macroscopic imperfections are given in the exercise section.

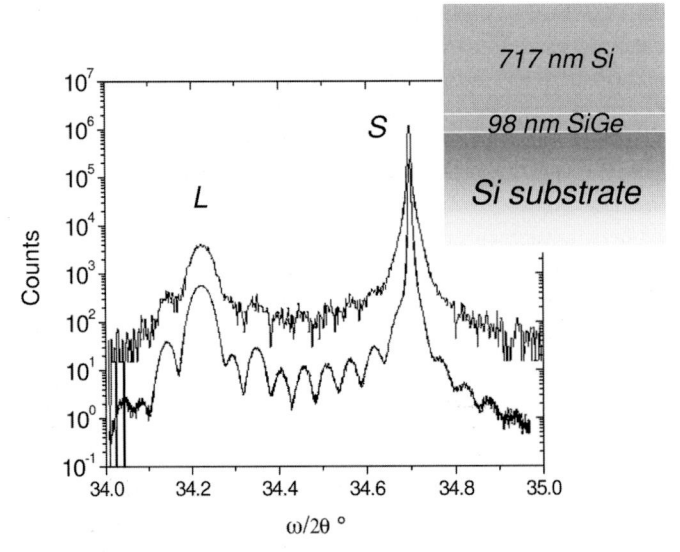

Figure 7.6 004 high-resolution rocking curves of a SiGe HBT structure with a small number of misfit dislocations at the SiGe–Si interface. The data were collected with (lower curve) and without (upper curve) an analyzer crystal. The profile without the analyzer is offset upwards by an order of magnitude for clarity.

7.3
Mosaicity and Extinction

The kinematic diffraction theory that has been made use of until now suffers from fundamental inadequacies in the quantitative description of the scattering process. These shortcomings become evident when the diffraction of highly perfect crystals is considered and one has to proceed to the use of dynamical theory. In a simplified picture of the scientific process it may be stated that the dynamical theory has its roots in two different approaches for which the bases were laid by C.G. Darwin and P.P. Ewald soon after the discovery of x-ray diffraction. Although these two approaches yield comparable results, their background is somewhat different. In the Darwin treatment the notion of mosaicity and extinction is central and is based on a geometrical interpretation, which will be outlined in this section together with its application in thin-film analysis. In contrast, the theory of Ewald is founded on a wave equation for the dielectric displacement field as will be seen in Section 7.4.

Both approaches are based on the diagnosis that the incoming beam suffers from a continuous intensity loss caused by reflection from lattice planes. This fact remains unconsidered in the kinematical theory where the attenuation of the x-ray within the crystal is almost exclusively attributed to photoelectron generation and quantified by the linear attenuation coefficient μ. Consequently, the integral intensity calculated by the kinematic theory will overestimate the scattered intensity and it was pointed out by Darwin in 1914 that Eq. (1.44) is incompatible with the conservation of energy principle [11].

In order to account correctly for the measured intensities the concept of mosaicity was introduced by Darwin to describe the microstructure in single crystals [12]. From the viewpoint of the mosaicity concept the crystal lattice does not extend in perfect periodicity from one crystal face to the opposite one. Instead, a single crystal is regarded to be typically comprised of small building blocks exhibiting small variations in orientation with respect to the sample reference frame $\{s_i\}$. The geometry compares with the two-dimensional analogue of a ceramic mosaic and this justified the naming of the phenomenon. Each mosaic block is connected to its neighbors via small-angle grain boundaries such that there are no voids but a series of dislocations are left that accommodate the misorientation. As in the description of texture the structural variations can be accounted for by an orientation distribution function and quite often the relative misorientation of the mosaic blocks in a single crystal is given by a Gaussian distribution with standard deviation Δ of the orientation distribution. A second important parameter to describe comprehensively mosaicity is evidently the average dimensions D of the mosaic blocks. A single crystal is regarded as an ideally perfect crystal if $\Delta = 0$. However, real crystals typically exhibit a certain degree of mosaicity. For GaAs wafers, for instance, values of about 10″ for Δ and ~2 µm for D have been reported [13], albeit these parameters are subject to constant technological improvement.

The notion of mosaicity is also used to describe the misorientation in epitaxial thin films. Especially in film–substrate systems with a large mismatch Δa the relaxation might become so strong that individual grains start to separate although the whole film remains connected and all the grains exhibit the same crystallographic orientation. Such films indeed mark the transition to the highly textured growth of polycrystalline films. A widely used approach is visualized in Fig. 7.7, where the misorientation among individual grains is characterized by twist and tilt. Here, tilt means orientational deviation from the out-of-plane direction, while twist indicates the rotation of grains in the film plane. Both orientation distributions are often accounted for by Gauss functions of the azimuth ϕ and the rocking curve angle ω as outlined in Chapter 5 (named ϕ and ψ there)

$$\text{Twist:} \quad W(\phi) = \frac{1}{\sqrt{2\pi}\Delta_\phi} \exp\left(-\frac{\phi^2}{2\Delta_\phi^2}\right)$$

$$\text{Tilt:} \quad W(\omega) = \frac{1}{\sqrt{2\pi}\Delta_\omega} \exp\left(-\frac{\omega^2}{2\Delta_\omega^2}\right) \quad (7.19)$$

where $\Delta\phi$ and $\Delta\omega$ are the width of the distributions. It has to be emphasized that the distribution widths in epitaxial films are much smaller than in highly textured polycrystalline films although the phenomenon is sometimes described in the same formalism. In depositions of GaN on sapphire – where the mismatch is as large as 16% – FWHM values $\Delta\phi$ between 0.1° and 1° and $\Delta\omega$ between 0.05° and 2° are typically observed. Twist is associated with edge threading dislocations and tilt with mixed threading dislocations [14].

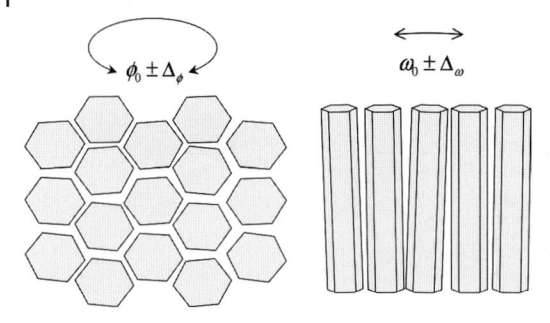

Figure 7.7 Schematics of twist and tilt in a columnar honeycomb structure viewed from the top (left) and from the side (right).

It should be realized that the use of the mosaicity concept frequently reaches beyond its original intention. For instance, in epilayers from group III nitrides the twist and tilt is sometimes so large that grain boundaries and voids between individual crystallites are already observed. In such cases the phenomenon should be distinguished from mosaicity and would better be named texture. It is evident that there exists a transition zone between mosaicity and texture (see Fig. 7.8 for the schematics of the different categories), and it sometimes might be ambiguous to give the right naming. In order to make appropriate use of the concept it is recommended to indicate the numerical value for the degree of misorientation whenever mosaicity is discussed.

The surplus damping of the x-ray beam within the crystal that was mentioned at the beginning of this section is now considered. This phenomenon is termed extinction and two types of extinction have to be distinguished that are named primary extinction and secondary extinction (Fig. 7.9). The damping of the x-ray intensity within a single mosaic block is termed primary extinction, which is caused by the successive scattering of the beam as it passes the atomic planes along its way through the block. As has been shown by Zachariasen, the kinematic intensity I_{kin}, given in Eq. (1.44), has to be multiplied by a prefactor $f(p)$ to arrive at the measurable integral intensity I_{pe}.

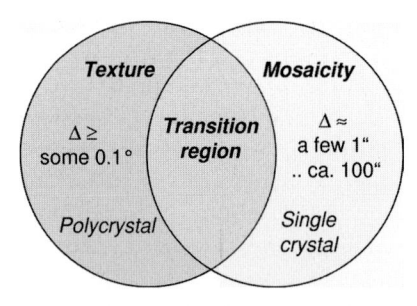

Figure 7.8 Schematics to distinguish mosaicity from texture according to the magnitude of the orientation standard deviation Δ. The distinction is not unequivocal, and there exists a transition region.

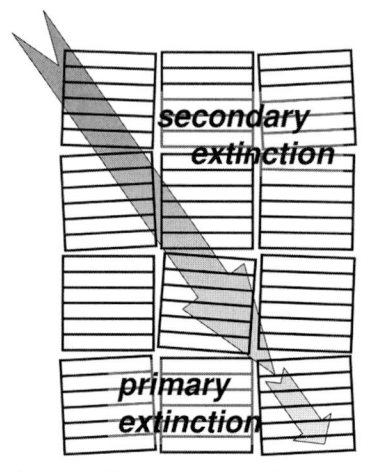

Figure 7.9 Primary and secondary extinction in a mosaic crystal.

$$I_{pe} = f(p)I_{kin} \tag{7.20}$$

The prefactor is called the primary extinction correction and reads for an unpolarized beam

$$f(p) = \frac{\tanh p + \cos 2\theta \tanh|p\cos 2\theta|}{2p\overline{C^2}} \tag{7.21}$$

where the parameter p is given by

$$p = \frac{r_e \lambda D \overline{C}}{\sin \theta} \frac{|F|}{V_{uc}} \tag{7.22}$$

$\overline{C}$ and $\overline{C^2}$ are the averaged polarization factors as introduced in Chapter 1 ($\overline{C} = (1 + \cos 2\theta)/2$ and $\overline{C^2} = (1 + \cos^2 2\theta)/2$), and also the other symbols have their usual meaning. The parameter p is seen to scale with the mosaic block size D.

Secondary extinction is caused by intensity loss by other mosaic blocks that are traversed by the x-ray beam along its path through the crystal. One thus has to extend the meaning of the known absorption coefficient μ by considering it only as one share of the more general extinction coefficient $\mu \to \tilde{\mu}$. If the deviation of the blocks from their average orientation can be assumed to be Gauss-like distributed with standard deviation Δ, secondary extinction might be dealt with by adding a further component to the linear attenuation coefficient

$$\tilde{\mu} = \mu + \frac{1}{\sqrt{2\pi}\Delta} \frac{r_e^2 \lambda^3 \overline{C^2}}{\sin 2\theta} \frac{|F|^2}{V_{uc}^2} \tag{7.23}$$

In the quantitative application of the mosaic model the extinction coefficient $\tilde{\mu}$ has to substitute the usual attenuation coefficient μ whenever it appears in the formulas used.

With the extinction corrections at hand we are now in the position to outline a procedure by which the film thickness may be deduced from the integral intensity of a substrate peak if this is a highly perfect single crystal. The procedure was introduced by Chaudhuri and Shah, who applied the mosaicity model to a film–substrate system [13]. In their study thin $Ga_{1-x}Al_xAs$ films on GaAs (001) substrates were investigated. They assumed that the epilayer and the substrate could both be modeled by mosaic crystals having Gaussian orientation distributions and standard deviations Δ_S and Δ_L, with the subscripts accounting again for substrate and layer. By applying the formulas given above, it could be shown that the substrate integral intensity I_S depends as

$$\frac{I_S}{I_0} = \frac{r_e^2 \lambda^3 \overline{C^2}}{\sin\omega \sin 2\theta} \frac{f(p_S)|F_S|^2}{V_{S,uc}^2} \frac{\exp(-\tilde{\mu}_L t k_\omega)}{\tilde{\mu}_S k_\omega} \tag{7.24}$$

on film thickness t [13]. k_ω is the geometry factor in the rocking curve configuration (see Chapter 5), $k_\omega = [1/\sin\omega + 1/\sin(2\theta - \omega)]^{-1}$. A high-resolution setup has definitely to be used in order to measure reliable single-crystal substrate intensities. A further technical point of concern is that a possible bending of the substrate has to be taken into account and corrected. Chaudhuri and Shah directly applied Eq. (7.24) by inserting a value for I_0 as determined by measurement of the direct beam. In addition, they determined the extinction coefficient $\tilde{\mu}$ and the primary extinction correction $f(p_S)$ from a measurement of an uncoated substrate. Two types of thickness results were derived from the measured substrate peaks: (a) with extinction correction and (b) without extinction correction (compare to Section 2.3). By applying the first method, Eq. (7.24), close agreement between the layer thicknesses determined compared with independent measurements was achieved. Best results were obtained when the substrate peak was calibrated versus a sample of known thickness. The second method yielded large deviations of up to 130% from the otherwise determined thickness values. The authors concluded that a precision within 4% uncertainty is achievable by application of the extinction correction according to Eq. (7.24).

Similar to the measurement of a substrate peak from a polycrystalline substrate, the intensity ratio of Eq. (7.24) may yield the film thickness independent of the crystalline, epitaxial or amorphous nature of the film. The formula also suggests that the $\tilde{\mu}_L t$ product of the layer might be determined from the intensity ratio $I_{S,after}/I_{S,prior}$ of the substrate peak after and prior to the deposition. Compared to the μt product considered so far, the product of extinction coefficient and thickness, $\tilde{\mu}_L t$, would not only comprise information on the chemical composition and density, but also on the microstructure, i.e. mosaicity of the film which might be highly valuable in optimizing the deposition process. However, an investigation applying this approach is not known to us so far. It is concluded that the approach of Chaudhuri and Shah might always be used for the film thickness determination, when no fringing can be observed. The technique has so far only be applied in very few cases and it would appear worthwhile to dispose of a larger number of experimental results.

The extinction coefficient $\tilde{\mu}$ provides us with a rough criterion, when the kinematic or dynamical theory might usefully be applied. For this purpose the extinction length has to be calculated, which is the inverse of the extinction coefficient, $\tau_{ex} = 1/\tilde{\mu}$. This length accounts for the distance within the sample up to which $1/e$ of the x-ray intensity has been swapped from the incoming to the diffracted beam. Extinction lengths are typically of the order of 10 to 100 µm and extinction is thus only observed in highly perfect single crystals. Due to this reason the usage of the kinematical theory is justified for most polycrystalline materials where the average grain size or the dimensions of the mosaic blocks are smaller than τ_{ex}. Samples that exhibit this property are named ideally imperfect crystals ($D < \tau_{ex}$) to distinguish them from ideally perfect crystals ($\Delta = 0$) mentioned above. In practice, all materials lie in between the two extremes. We have thus not made use of the wrong formulas in the previous chapters, but for the case when highly crystalline material is considered, dynamical theory should be applied.

7.4
Dynamical Theory of Ewald and Extensions

The starting point of Ewald's formulation of the dynamical theory is the concept of wave fields that are excited within the crystal when an x-ray beam enters from the outside [15]. In order to account for the interaction of wave fields with the electrical charges of the crystal the assumption of a spatially dependent polarizability $\chi(r)$ is made. The polarizability has already been dealt with in this book in conjunction with XRR, albeit it has been denoted there as deviation of the refractive index from unity, $n = 1 - \delta - i\beta$. In the dynamical theory the polarizability is accounted for by its spatially dependent form $\chi(r)$ that is related – for small χ values – to the index of refraction via $n(r) = 1 + \chi(r)/2$. Use is made of the Fourier expansion of the polarizability in basis functions $\exp(ihr)$, where h is a lattice vector defined by the crystal lattice geometry and accounted for by the set of Miller indices $h = hkl$

$$\chi(r) = \sum_h \chi_h \exp(ihr) \tag{7.25}$$

It has been shown in Chapter 4 that the zeroth order of polarizability is related to the average electron density $\rho_e = Z/V_{uc}$ via

$$\chi_0 = -\frac{\lambda^2 r_e}{\pi} \frac{Z}{V_{uc}} \tag{7.26}$$

when anomalous dispersion is neglected. From the comparison of Eq. (7.25) with the definition of the structure factor, Eq. (1.23), it can concluded that the structure factor F_h and the hth order polarizability component χ_h are related via

$$\chi_h = -\frac{\lambda^2 r_e}{\pi V_{uc}} \sum_i f_i(h) \exp(ihr_i) = -\frac{\lambda^2 r_e}{\pi V_{uc}} F_h \tag{7.27}$$

with the sum extending over all scattering centers in the unit cell. Absorption may be included into this formalism by the addition of anomalous dispersion corrections f_i' and f_i'' to the atomic form factors f_i, but this will be neglected here.

As in the kinematical theory, one is interested in the position and the intensity of interference reflections from the crystal lattice. In order to obtain the intensity, the dynamical theory starts from the field of dielectric displacement vectors $\boldsymbol{D}$ that are excited within the crystal by the incoming x-ray beam. Any other electromagnetic quantities like the electric field $\boldsymbol{E}$ – from which the intensity can be calculated – or the field of magnetization $\boldsymbol{H}$ derive from $\boldsymbol{D}$. In its most simple form the theory makes use only of planar wave fields, but generalizations for spherical wave fields were developed. By the application of Maxwell's equation and assuming the material to be nonconductive and nonmagnetic a wave equation for $\boldsymbol{D}$ is derived which reads

$$\Delta \boldsymbol{D}(\boldsymbol{r}) + K^2 \boldsymbol{D}(\boldsymbol{r}) + \nabla \times \nabla \times (\chi(\boldsymbol{r})\boldsymbol{D}(\boldsymbol{r})) = 0 \quad (7.28)$$

For the solution of the equation $\boldsymbol{D}$ can be considered as a sum of plane waves

$$\boldsymbol{D}(\boldsymbol{r}) = \sum_h \boldsymbol{D}_h \exp(-i\boldsymbol{k}_h \boldsymbol{r}) \quad (7.29)$$

where $\boldsymbol{k}_h$ is the wave vector. As in Chapter 4, K and k are used according to whether external wave vectors or those inside the crystal are to be accounted for. Inserting Eq. (7.29) into Eq. (7.28) yields after some manipulation

$$\frac{k_h^2 - K^2}{k_h^2} \boldsymbol{D}_h = \sum_n \chi_{n-h} \boldsymbol{D}_{n[h]} \quad (7.30)$$

which is termed the fundamental equation of dynamical theory [15]. The index n runs over all Miller index triples and the solution thus predicts the incoming x-ray beam to split into an infinite number of beams within the crystal. The symbol $\boldsymbol{D}_{n[h]}$ indicates the component of $\boldsymbol{D}_n$ perpendicular to $\boldsymbol{k}_h$. A thorough analysis reveals that in most experimental X-ray diffraction experiments and definitely in all cases of epilayer systems considered here, only two wave fields are sufficiently strong and of interest, which are due to the incident (forward diffracted or refracted) and the reflected (diffracted) beam. Indices 0 and h are generally used for physical quantities to characterize them and their wave vectors are related through the equation

$$\boldsymbol{k}_h = \boldsymbol{k}_0 + \boldsymbol{h} \quad (7.31)$$

which is the vector form of the Bragg equation. Only this will yield wave field solutions that are not exponentially damped to very small values within a distance of a few unit cells. This approach was introduced in 1917 by Ewald [15] and was later re-invented by Bloch for the description of the electron states in crystal lattices.

Equation (7.30) simplifies considerably for this two-beam solution and then reads

$$\{K^2(1+\chi_0) - k_0 k_0\} D_0 + K^2 C \chi_{\bar{h}} D_h = 0$$
$$K^2 C \chi_h D_0 + \{K^2(1+\chi_0) - k_h k_h\} D_h = 0 \qquad (7.32)$$

Here, C is the polarization factor (see Eq. (1.2) to discriminate between the **D** fields to lie either in ($C_\pi = |\cos 2\theta|$) or perpendicular the scattering plane ($C_\sigma = 1$)). The quantities

$$X_0 = \frac{1}{2K^2}\{k_0 k_0 - K^2(1+\chi_0)\}$$
$$X_h = \frac{1}{2K^2}\{k_h k_h - K^2(1+\chi_0)\} \qquad (7.33)$$

were denoted excitation errors (Anregungsfehler) by Ewald and describe the deviation of the refracted and the reflected wave vectors from the kinematic analogue, where $|k_0| = |k_h| = K(1+\chi_0)^{1/2}$ would hold and both X_0 and X_h would vanish. It is derived from the nonzero solution of the system of Eqs. (7.31) that the intensities of the two waves are related via

$$\left(\frac{D_h}{D_0}\right)^2 = \frac{\chi_h X_0}{\chi_{\bar{h}} X_h} \qquad (7.34)$$

and it is from this relation that the intensity of the diffracted beam can be calculated in the two-beam case of dynamical scattering. We have seen before that asymmetric reflections may add important contributions to the investigation of epilayers. In the dynamical theory asymmetric reflections are dealt with by introducing an asymmetry parameter γ being the ratio of direction cosines of incoming to outgoing beam. γ may be expressed with the help of the incidence angle ω by

$$\gamma = \frac{\gamma_0}{\gamma_h} = \frac{\sin(\omega)}{\sin(2\theta_B - \omega)} \qquad (7.35)$$

Evidently, γ would be unity for symmetric $\theta/2\theta$ scans, but deviates more from unity in rocking curve scans the more ω differs from θ_B.

A shift of Bragg reflections due to the refraction of the x-ray beam is observed for single crystals and epitaxial layers. If the peak centroid occurs at θ_B according to Bragg's equation, the peak is shifted by an amount $\Delta\theta$ to a value $\theta_0 = \theta_B + \Delta\theta$ and it holds

$$\Delta\theta = -\frac{\chi_0}{2}(\cot\omega + \tan\theta_B) \qquad (7.36)$$

Because of the subunity of the refractive index n (and $-\chi_0 > 0$) the shift of the peak is to larger diffraction angles θ (see Fig. 7.10).

Also a strong asymmetry is typically observed for Bragg peaks from highly perfect single crystals. The effect is due to the wave field nature of the x-ray beam within the sample and cannot be accounted for by the kinematical theory, where absorption solely due to photoelectron generation would cause a symmetric line profile. Peak shift and asymmetry are demonstrated in Fig. 7.10 for 220 and 440 reflections from highly perfect Ge single crystals as are used in x-ray monochromators, for instance. The line profiles were simulated by virtue of the rocking curve simulation program RCRefSim [16]. In the dynamical theory line profiles are described by the expression

$$I(\eta) = \left|\eta + \sqrt{\eta^2 - 1}\right|^2 \tag{7.37}$$

where η is the so-called deviation parameter, which is

$$\eta = \frac{2\Delta\theta \sin\theta_B - \chi_0(1-\gamma)}{C\sqrt{|\gamma|\chi_h\chi_{\bar{h}}}} \tag{7.38}$$

The deviation parameter describes the deviation from the scattering condition. Clearly as $\Delta\theta$ coincides with the Bragg condition the deviation is not zero, indicating the displacement of the scattering peak due to the refractive index. The above equation breaks down for large mismatches or significant deviations from the Bragg condition.

Often, measured peak intensities are given as a function of η or plotted where η serves as abscissa. The intrinsic width of the profile is given by

$$w = \frac{2C\sqrt{|\gamma|\chi_h\chi_{\bar{h}}}}{\sin 2\theta_B} \tag{7.39}$$

The intrinsic width of single-crystal material according to this formula is in the range of only a few arc seconds, which could not been measured for a long time after the development of the theory of XRD. The effect was caused by mosaicity and, therefore, the mosaicity-broadened peak width rather than the intrinsic width of x-rays diffracted by single crystals was often measured. The development of crystal

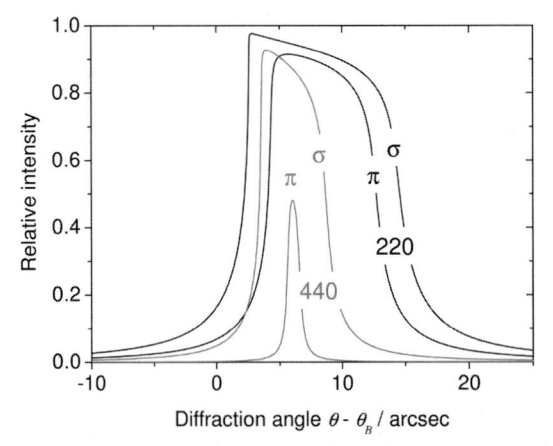

Figure 7.10 220 (black) and 440 (gray) Bragg reflection profiles from a highly perfect Ge single crystal for σ- and π-polarized plane-wave x-ray beam.

growth methods, HRXRD techniques and diffraction theory proceeded hand-in-hand. Meanwhile, both dislocation-free single crystals and high-resolution diffractometers are available, the latter of which allow for the measurement of the intrinsic half-width of such perfect crystalline matter.

The integrated intensity becomes

$$I = \frac{8\bar{C}}{3\sin 2\theta}\sqrt{|\gamma|\chi_h\chi_{\bar{h}}} \tag{7.40}$$

which is proportional to F instead of F^2 as in kinematical theory. And, finally, the extinction length as introduced in Section 7.3, is expressed by

$$\tau_{ex} = \frac{V_{uc}\sqrt{\gamma_0\gamma_h}}{r_e\lambda|F_h|} \tag{7.41}$$

The ratio of the extinction length over x-ray path ℓ in the sample gives a valuable indication for the kind of theory to be used: if $\ell/\tau_{ex} \ll 1$ then the kinematical theory should suffice, while for $\ell/\tau_{ex} \gtrsim 1$ the dynamical theory has to be applied. Since the extinction length, which is of the order of several micrometers, might often exceed the film thickness the use of kinematical theory might be adequate in these cases. However, when the film is fully strained, the wave fields excited in the film and in the substrate cannot be considered as independent, but are coupled and then dynamical theory has to be applied. These are – in short – the basic equations describing the x-ray peak position and profile within the dynamical theory. They evidently differ significantly from those of the kinematical theory.

A general approach to deal with imperfections in crystals in the framework of the dynamical theory was developed by Takagi and Taupin [17, 18]. In this formalism the crystal's imperfections are accounted for by a deformation field $u(r)$ for which any spatially dependent deviation of the lattice constituents from their ideal position may be inserted. The deformation field is incorporated into the Fourier expansion of the polarizability by rewriting it as

$$\chi(r) = \sum_h \chi_h \exp\bigl(i h(r + u(r))\bigr) \tag{7.42}$$

If the derivation of the fundamental Eqs. (7.30) is re-done using of this modified polarizability one finally ends up with a new set of equations to describe the two-beam case

$$\begin{aligned}\frac{i\lambda\gamma_h}{\pi}\frac{\partial D_h}{\partial z} &= \chi_0 D_h + C\chi_h D_0 + X_h D_h \\ \frac{i\lambda\gamma_0}{\pi}\frac{\partial D_0}{\partial z} &= \chi_0 D_0 + C\chi_{\bar{h}} D_h\end{aligned} \tag{7.43}$$

These are the Takagi–Taupin equations, which formally represent a set of differential equations in D_0 and D_h. The deformation fields $u(r)$ are buried in the polarizability Fourier components. The approach is much more general than the Ewald theory since distortions can be included. Various types of deformations have been considered since Eq. (7.42) was established, from the variety of which some will be discussed in the next section that are of relevance in the analysis of thin films, multilayers and superlattices.

7.5
High-Resolution Rocking Curves and Profiles from Layer Structures

Dynamical diffraction may have a significant influence on x-ray diffraction patterns from epilayer–substrate material systems. The magnitude of the effect depends on how strong the x-ray wave fields in the film and in the substrate are coupled. If the film is relaxed the wave fields in the film may be restricted to this region and only a weak coupling with wave fields in the substrate occurs. In this case the film's diffraction pattern might be simulated in a sufficiently precise fashion by a kinematical or a semikinematical model. If the film, however, is highly strained such that the in-plane lattice constants of film and substrate are nearly equal a strong coupling of wave fields will cause the structural information of the film to manifest in a diffraction pattern that is properly accounted for only by simulating with dynamical theory.

The equations of Takagi and Taupin contain the ratio of the D wave fields of the diffracted over the incident beam and the gradient of this quantity, which is the rate of change along any arbitrary direction of the sample. The application to problems of thin films and surfaces is evident and was developed soon after the equations were established. The development of a solution started with the investigation of compositional and stress variations as introduced into semiconductors by doping with ion implantation techniques [19, 20]. Later, the formalism was also applied to the simulation of the diffraction pattern from epilayer systems [21]. In the adaptation of the Takagi–Taupin equations to the thin-film problem the sample is considered as a finite set of lamellas where within the jth lamella the same deformation field $u_j(r)$ is considered constant. Depth profiling is achieved by introducing as many different lamellas with individual deformation fields as needed to simulate appropriately the sample. Use has been made of this conceptual approach in the simulation of XRR. It turns out that the reflectivity of each lamella is of the form

$$\mathfrak{R}_t = \eta + \sqrt{\eta^2 - 1}\frac{S_1 + S_2}{S_1 - S_2} \tag{7.44}$$

where the quantities S_1 and S_2 are given by

$$S_{1,2} = \left(\mathfrak{R}_0 - \eta \pm \sqrt{\eta^2 - 1}\right)\exp\left(\mp iT\sqrt{\eta^2 - 1}\right) \tag{7.45}$$

and the other variables have their usual meaning. The recursion usually starts with the reflectivity from an infinitely thick assumed substrate ($t \to \infty$)

$$\mathfrak{R}_\infty = \eta - \mathrm{sign}(\mathrm{Re}(\eta))\sqrt{\eta^2 - 1} \tag{7.46}$$

7.5 High-Resolution Rocking Curves and Profiles from Layer Structures

Instrumental Box 8:

Crystals as X-ray Beam Conditioners

Single crystals or highly textured polycrystals can be used as effective means for beam conditioning by manipulating the x-ray beam divergence δ and bandwidth $\Delta\lambda/\lambda$. Because beam conditioning is inevitably accompanied by intensity losses the general rule is obeyed that the beam divergence and bandwidth should compare to the angular spread of the structural features to be resolved by the experiment.

For the investigation of polycrystalline samples two of the highly textured mosaic crystals have already been encountered: the HOPG graphite and LiF monochromators. Whereas the graphite monochromator is mostly applied in a bent shape in the focusing geometry of the diffractometer (identification of chemical phases), the LiF device is mainly used as a flat plate in parallel beam configurations (texture and residual stress analysis) and grazing-incidence experiments. Both diffracted-beam monochromators are also distinct with respect to mosaicity, which is the degree of twist and tilt of individual coherently scattering domains that make up the crystal as outlined in Section 7.3.

In the field of high-resolution x-ray diffraction specially fabricated Si and Ge single crystals are used to achieve the extremely low divergence and bandwidth necessary for the investigation of epitaxial films. It is usual to denote these systems as monochromators if installed on the incident beam side and as analyzers on the diffracted beam side. These beam conditioners became possible with the advent of semiconductor single crystals of high crystalline perfection in the 1960s. Today, commercially available single crystalline silicon is the material having by far the largest ratio of crystalline perfection over costs.

One diffractometer setup for high-resolution rocking curves is that of double-crystal diffraction. In this configuration a single-crystal monochromator is inserted such that a certain Bragg reflection $(hkl)_{mc}$ is selected from the incident beam by setting the monochromator to the appropriate Bragg angle (see Fig. i8.1). Its alternative naming as double-crystal spectrometer stems from applications in elemental analysis. One step in the alignment of the monochromator for maximum intensity is the rotation of the sample on the Θ or Ω axis in combination with the rotation of the monochromator on its Θ axis and from this procedure stems the frequently used term double-axis diffraction. However, the monochromator is neither rotated nor moved throughout the rocking curve recording. To remove the influence of wavelength dispersion the first and second crystals should operate under the same Bragg angle.

Two configurations are possible in double-crystal diffraction that are distinct by the relative geometrical arrangement of monochromator and sample. The first configuration is the parallel (+ −) setting and the second is the nonparallel (+ +) one. In this nomenclature each sign stands for the sense of rotation of one reflection, when looking from above on the scattering plane and the first one always having a + sign. If the sense of rotation is the same for the second reflection a further + sign is added, while a − sign is used to indicate the change from a clockwise to anticlockwise rotation of the reflected beam or vice versa (see Fig. i8.1). The symbols n and m are in use if the order (hkl) of

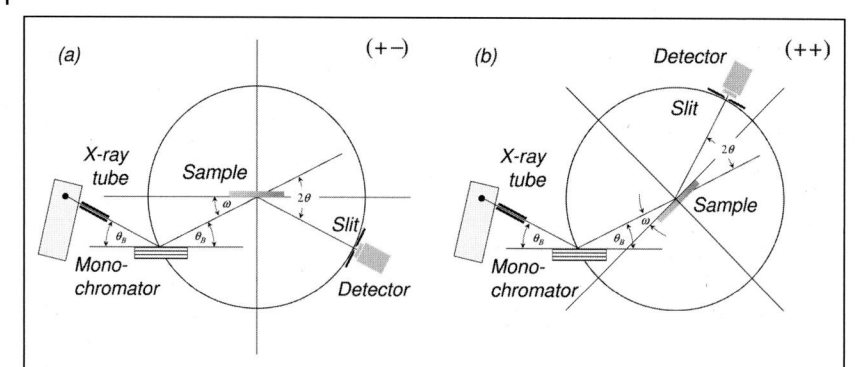

Figure i8.1 Double-crystal diffraction in (a) the parallel (+ −) and (b) the nonparallel (+ +) setting.

reflections is different and the possible configurations are then abbreviated by e.g. (+n, +m), (+n, −m).

The double-crystal configuration may not only be used for the investigation of the properties of one of the crystals, but as a full monochromating system when the exiting beam is directed on the sample. The divergence and bandwidth of the x-ray beam may substantially be tailored by the use of such combined monochromator devices. For instance, the (+n, +n) or (+ +) setting can be adjusted such that it results in the selection of a smaller wavelength band $\Delta\lambda$ compared to the primary beam, from which its terminology as dispersive setting is justified. The (+ −) setting, however, is a nondispersive setting.

The interaction between $\Delta\lambda$ and δ in any process of monochromation can be visualized in a condensed form by use of the DuMond diagram (see Fig. i8.2). In such a diagram the quantity $\lambda/(2d)$ versus θ of each reflecting crystal is represented by a single sin curve indicating the Bragg equation to relate abscissa and ordinate. Bandwidth and divergence are accounted for by ranges $\Delta\lambda$ and δ around the central wavelength λ_0 and the position θ to which the monochromator crystal is set. In the DuMond diagram of double-crystal diffraction two sin curves have to be drawn and their separation is given by the angle between their reflecting planes. This angle amounts to 0° in the parallel setting and to $2\theta_B$ in the nonparallel one. Figure i8.2 displays the DuMond diagram for two Ge single crystals acting as monochromator in the (+ −) configuration and in the (+ +) configuration. The Bragg reflection used is the Ge (220) the scattering angle of which amounts to 45.299° for Cu $K\alpha_1$ and 45.417° for Cu $K\alpha_2$ (according to ICDD 000-04-0545 $a_{Ge} = 0.56576$ nm).

It can be seen in Fig. i8.2 that both sin curves coincide in the (+ −) configuration, while they cross for (+ +). The Bragg angles and $\lambda/(2d)$ positions of the Ge 220 reflection are shown for Cu $K\alpha_1$ and Cu $K\alpha_2$ radiation in the insets of enlarged parts of the plot. It can be deduced from the right-hand that by setting both crystals to Bragg angle of 2θ(Ge220, Cu $K\alpha_1$) and restricting the divergence of the outgoing beam by appropriate slits a selection of bandwidth around λ_0 Cu $K\alpha_1$ can be achieved, which is not possible with the (+ −) setting. The comparison of both DuMond diagrams convincingly demonstrates the

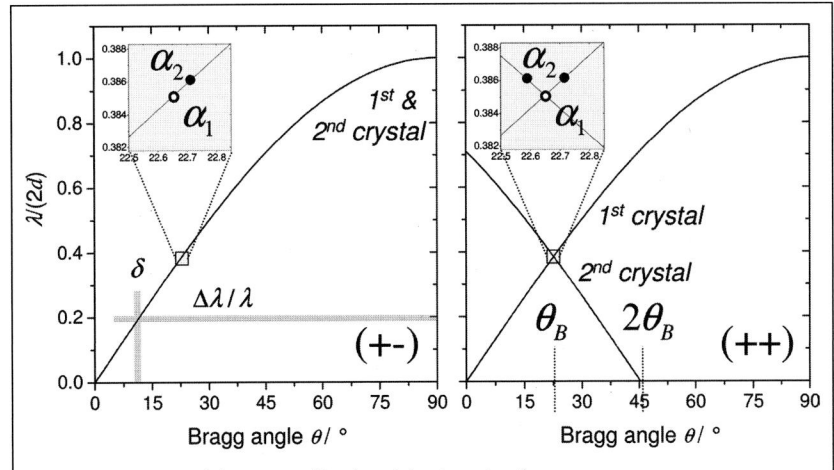

7.5 High-Resolution Rocking Curves and Profiles from Layer Structures

Figure i8.2 DuMond diagrams of (+ −) and (+ +) setting for Ge double-crystal monochromator operating with 220 Bragg reflections under Cu Kα radiation.

ability of the (+ +) setting to act as dispersive monochromator combination, while the (+ −) setting cannot.

It appears trivial to mention that two technical presuppositions had to be fulfilled before these monochromator methods could be established: (i) the precision of the gear settings rotating the crystal reproducible had to be in the 0.001° range and (ii) the mosaicity of the two crystals has to be substantially smaller than the intended bandwidth $\Delta\lambda$, being for most laboratory set-ups the difference of the α_1-α_2 splitting of the operating Bragg reflection. These arguments may help to understand why such techniques became routinely available only some decades after the invention of x-ray diffraction.

As a further step in reducing the wavelength dispersion $\Delta\lambda/\lambda$ Bonse and Hart introduced a monolithically grooved single-crystal monochromator in 1965 that also became known as the channel-cut crystal (CCC) monochromator (see Fig. i8.3). The channel is limited by crystallographic lattice planes of the same Miller indices (hkl) and the successive reflection of the x-ray beam at the channel walls causes a strong reduction of the x-ray intensity contained in the tails of the beam. By choosing a low-absorbing material for the crystal in accordance with the wavelength used only very little attenuation of intensity close to the central wavelength λ_0 can be attained. According to how many reflections the x-ray beam undergoes by traveling through the channel the configurations are denoted as two-bounce, three-bounce, etc., CCC monochromators. In the notation introduced above the setting is denoted as (+ +), (+ + +), etc.

In particular, the Bartels monochromator is mentioned that comprises of two channel-cut crystals each operating in the two-bounce mode as shown in Fig. i8.4 that should be symbolized by (+ − − +). One effect of beam conditioning by this system is the reduction of flank intensity and the other effect is the wavelength selection due to the dispersive part in the beam path. Advantageously, the direction of the outgoing beam is the same as the incoming one and the monochromator may thus be directly inserted into

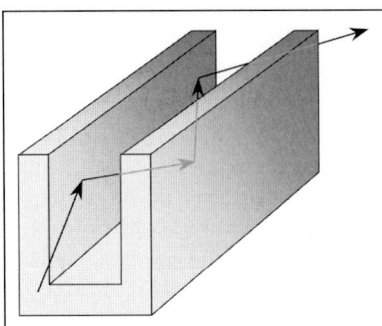

Figure i8.3 Schematic representation of channel cut crystal (CCC) in a three-bounce configuration.

the beam path. In a commercially available version of this concept the channel-cut crystals are fabricated from Ge with 220 lattice planes parallel to the sidewalls. By rotating each crystal the Bragg angle may be changed from that of 220 reflection to that of 440. The 220 reflection is used for increased intensity and the 440 for increased resolution at the expense of lower intensity. The combination of different types of monochromators makes possible advanced diffractometer configurations with extreme high-resolution capabilities. A common type is the so-called multiple-crystal diffractometer, where in addition to the monochromating crystal on the incident beam side a further crystal is inserted on the diffracted beam side (see Fig. 7.3). It should be emphasized again that during the typically performed $\omega/2\theta$ scan or any other scan neither the monochromator nor the analyzer crystal is rotated, but held at fixed Bragg angles to direct the probing beam onto the sample and into the detector. In one widely used multiple-crystal system a Bartels monochromator is used for primary monochromation, while a three- or four-bounce CCC is used as analyzer (see Fig. 7.3).

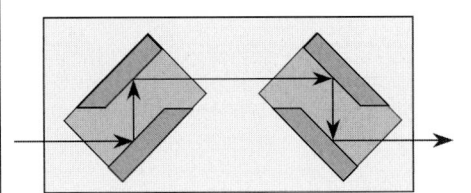

Figure i8.4 Sketch of Bartels monochromator viewed from the top to show the transmitted x-ray beam.

The solution of the Takagi–Taupin equations enabled the simulation of various relevant layer properties like depth dependence of the composition in graded layers and the case of implantation doping of semiconductors.

As a first example the doping of semiconductors by ion implantation is considered [22]. Figure 7.11(a) shows the rocking curve of the 004 reflection from a phosphorus-doped Si wafer into which 500 keV P^+ ions were implanted at a dose of 5 ×

10^{14} cm^{-2} through a 80 nm screening oxide layer. Ion implantation generally leads to a skewed Gaussian concentration profile, where the implanted species has a maximum at a depth that depends on the ion kinetic energy. In the case considered here a concentration profile is obtained as shown by the solid line in Fig. 7.11(b). The maximum P concentration is found at around $z = 650$ nm and amounts on an absolute scale to about 5×10^{19} cm^{-3}. The variation of the electron density associated with the phosphorus content is much too small to be detected by the probing x-ray beam. But the implanted P atoms cause a strain of the Si lattice and the deformation field for the Takagi–Taupin equation was assumed to be purely due to strain. The strained lattice was modeled by a step function assuming a constant deformation field, i.e. strain within each 50 nm thick lamella as displayed by the dash-dot step function in Fig. 7.11(b). The calculated rocking curve is shown as a dashed line in Fig. 7.11(a) and a close agreement between the measured and the simulated curve can be seen.

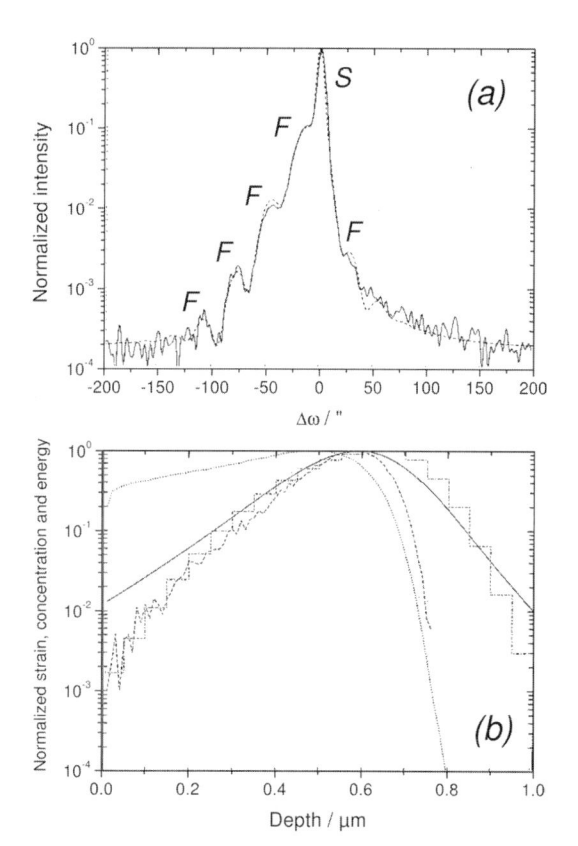

Figure 7.11 (a) Measured (solid curve) and simulated (dashed curve) rocking curves of 004 reflection of P-doped Si wafer. (b) P concentration according to TRIM simulation (solid curve) and optimized strain step distribution to simulate the rocking curve (from Ref. [22]).

The most significant characteristic of the curve is the asymmetry of the Bragg peak. Asymmetry is a universal peak property to be observed with a depth profile of the sample's metric. The second interesting feature is the modulation of intensity – mostly pronounced on the low-angle side of the peak. The fringing can be interpreted in terms of Eq. (7.13) and its application yields a thickness t of 670 nm. Evidently this is the approximate depth at which the maximum P concentration is found. It could be concluded from this investigation that the lattice strain follows the dopant concentrations revealed by previous studies. A careful comparison of the optimum strain profile and the simulated dopant concentration profile revealed that it is not a simple relationship. This is because the strain field indicates the distortions created by the defects caused by the implanted P ions.

The doping of semiconductor wafers is a routinely performed process in the semiconductor industry. Usually, ions are implanted with kinetic energies in the range of some 10 to 100 keV, where they reach implantation depths of the order of 1 nm per keV. In the relevant energy range the ions are mainly decelerated by interaction with the nuclear charges of the crystal lattice. Ion implantation will evidently lead to a degradation of the material electronic properties since it produces interstitials and other lattice defects. The defect concentration might be reduced by postannealing procedures, but the control of the degree of disorder remains an issue after implantation doping. In contrast to other techniques, XRD leaves the sample intact, which turns out of great benefit especially after Si wafer dimensions continuously increase to large sizes and the use of destructive analysis techniques would cause an unacceptable loss of valuable pre-products.

In the next example the rocking curve of a device structure from the group of cubic III–V compounds is considered (see Fig. 7.12). The layer architecture is from

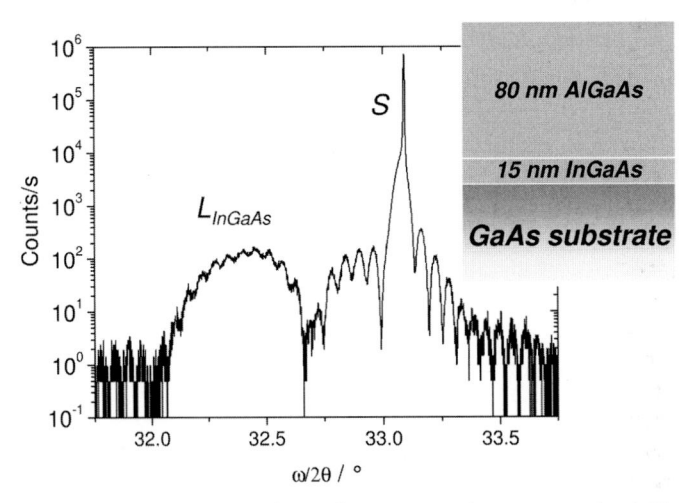

Figure 7.12 Rocking curves of 004 reflection from a InGaAs/GaAs HEMT structure, composed of a 15 nm InGaAs layer on a GaAs substrate, capped with 80 nm AlGaAs. The In concentration was 14% and Al concentration was 25%.

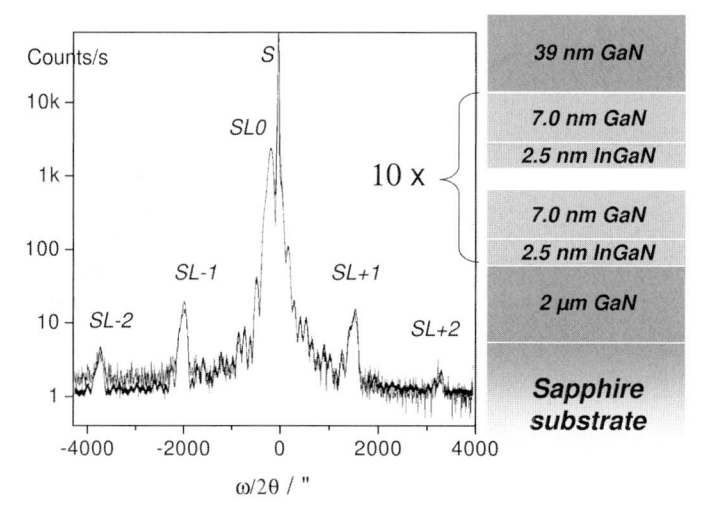

Figure 7.13 Scattering profile of 00.2 reflection from (GaIn)N quantum well structure (left), where dots represent measured intensities and the solid curve the simulation according to the architecture (right) of the layer system (from [23]).

the core of a so-called high-electron-mobility transistor (HEMT) fabricated for high-frequency and high-power electronic circuits. The function is based on the spatial separation of electronic current from the charged donor atoms. The scattering of charge carriers at ionized impurities is the major limiting factor of electronic mobility at room temperature. The basic idea of the HEMT accordingly is to generate the charge carriers in the high-band-gap layer (here AlGaAs) and perform the electronic transport within the low-band-gap layer (here InGaAs). Figure 7.12 shows the 004 rocking curve of such a layer architecture that was measured with Cu Kα radiation on a multiple-crystal diffractometer. The most characteristic features in the pattern are the substrate peak and the broad InGaAs layer peak on the low-angle side. Because the In-containing layer is only 15 nm thick the associated layer peak is rather broad. The pronounced fringing is due to total thickness of 95 nm of both layers. The simulation or fitting of the pattern was performed by assuming a perfect pseudomorphic growth and therefore no relaxation had to be considered. The fit parameters were the composition x of both ternary layers. The thickness of the layers equated to the lamella discussed above. The agreement is very good and the simulated and measured profiles are barely distinguishable. The measurement and simulation of such device structures have become very common in the compound semiconductors industry where the control of thickness t and composition x of individual layers is a necessity at least in device development, but often also in the production process.

The last example in this section is from the field of superlattices the building principles of which are introduced in Chapter 4. Figure 7.13 shows the rocking curve of the 00.2 reflection from an InGaN/GaN multiple quantum well (MQW) structure and its architecture [23]. The structure was grown on a GaN buffer layer deposited on sapphire and capped with a further GaN layer. The quantum well structure was comprised of a stack of ten $In_xGa_{1-x}N$/GaN double layers with an intentional In content of 13%. The pattern exhibits pronounced peaks at multiples of about 1800″ (0.5°) off from the central GaN buffer layer peak. These are caused by the total MQW height and are generally named satellite peaks or superlattice peaks in the literature, indicated by SL. The central superlattice peak is SL0, while those to the low- and high-angle side are indexed with negative and positive integers $SL - n$ and $SL + n$. The angular distance between the satellite peaks can be converted to a thickness, which turns out as the superlattice period t_{per} of the MQW.

In the vicinity of the central peaks certain fringe peaks are observed which are increasingly washed out with increasing distance from $\Delta\omega = 0$. As in XRR the visibility of fringing is countered by surface and interface roughness and to a certain extent the roughness may be deduced from the sharpness of the fringing. Of course, thickness variations of the total stack or of individual layers over the illuminated area of the sample would also reduce the extent of fringing in the measured rocking curve. Although the thickness inhomogeneity might in some cases be revealed by scanning various smaller sections of the sample, this is only achieved at the expense of intensity leading to increased measurement time. It should be emphasized that the combination of XRR and high-resolution rocking curve measurements is a very reliable strategy to elucidate the architecture and metric of superlattice layer structures. The measurement of both enables the determination of the whole set of relevant structural parameters like thickness, composition, relaxation and interface roughness. An example for the case of a metal superlattice is given in Ref. [24].

In the $Si_{1-x-y}Ge_xC_y$ material system HRXRD studies were performed to study the graded Ge concentration in SiGe layers [8] and the out-diffusion of Ge at elevated temperatures [25]. Diffusion coefficients and activation energies could quantitatively be determined by investigating SiGe(C)/Si superlattice structures. The inclusion of a small amount of C into SiGe was initially performed to relax the compressive strain in pseudomorphically grown SiGe layers on Si. The aforementioned studies revealed, moreover, that the addition of carbon suppresses the plastic relaxation of SiGe layers at relevant process temperatures due to an effective dislocation pinning [25]. A general-purpose simulation software for the simulation of epitaxial layer systems according to the lamella approach (Eq. (7.43)) was developed during the course of these investigations [16] that may be applied to any epitaxial layer system.

7.6
Reciprocal Space Mapping

It was already shown in case of a symmetric $\theta/2\theta$ scan that the scattering vector Q has only a nonzero component in the direction of the substrate normal that may be denoted by Q_z in the sample reference frame $\{s_i\}$. During a rocking curve scan, however, there also occurs a nonvanishing in-plane component Q_x for all angular positions ω – except for the one point where $\omega = 2\theta/2$ holds. In thin-film analysis we are mainly interested in the investigation of the surface regions of the sample that are most advantageously investigated by reflection measurements. The latter are restricted to the range between the minimum $\omega_{min} = 0$ and the $\omega_{max} = 2\theta$ condition above which the scattered beam cannot exit the surface.

We can visualize the different scanning techniques in the plane of momentum transfer by the in-plane component Q_x and the out-of-plane component Q_z of the scattering vector (see Fig. 7.14). We are interested in such a representation because the von Laue Eq. (1.11) that determines the condition for a diffraction peak to occur can be visualized in Q space in a simple and instructive form. The relation between the instrumental coordinates ω and 2θ and the more convenient Q_x and Q_z for analysis can be derived from Fig. 5.14. It results from the inspection of this figure (Exercise 10)

$$\begin{aligned} Q_x &= K\left[\cos(\theta - \omega) - \cos(\theta + \omega)\right] \\ Q_z &= K\left[\sin(\theta - \omega) + \sin(\theta + \omega)\right] \end{aligned} \quad (7.47)$$

where $K = 2\pi/\lambda$ is the magnitude of the x-ray wave vector. The Q representation is more directly translatable than the diffraction space angular coordinates for analysis or for comparison with other measurement methods.

The accessible region for data collection can be represented by a hemisphere in Q space having $-2K \le Q_x \le 2K$ and $0 \le Q_z \le 2K$ and obeying $Q_x^2 + Q_z^2 \le 4K^2$. The area of reflection is divided from the areas of transmission by two other hemispheres of radius K that are centered at $Q_x = -K$ and K (Exercise 10). The plane of momentum transfer according to these rules is displayed in Fig. 7.14.

What will the symmetric $\theta/2\theta$ scan and the rocking curve scan now look like in the (Q_x, Q_z) plane? The $\theta/2\theta$ scan will proceeds along the ordinate, which is now the out-of-plane component Q_z. This is visualized in Fig. 7.15(a), where the different configurations during the $\theta/2\theta$ scan are symbolized by varying gray scales of the vectors involved. The Ω scan, however, will look somewhat different. According to Eq. (7.47) the rocking curve scan transforms from a horizontal line to a circular one (see Fig. 7.15(b)). This is evident from the fact that the magnitude of momen-

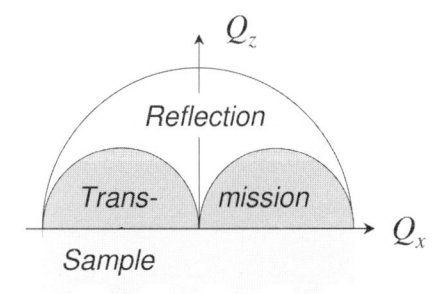

Figure 7.14 Plane of momentum transfer (Q_x, Q_z) in coplanar diffraction geometry.

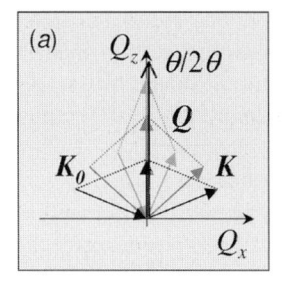

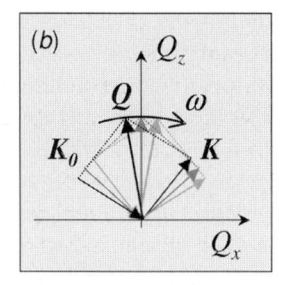

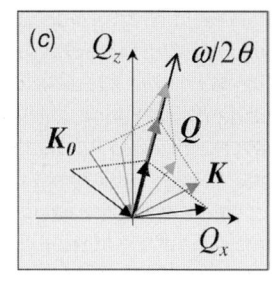

Figure 7.15 Representation of (a) symmetric $\theta/2\theta$ scan, (b) rocking curve scan and (c) so-called radial scan in the plane of momentum transfer.

tum transfer Q remains constant. A further scanning technique is called radial scan, shown in Fig. 7.15(c). In this scan the orientation of the probing Q vector has the same tilt with respect to the surface normal during the whole measurement. The radial scan is performed by starting from any (ω, 2θ) point in the range of reflection and increasing 2θ by $2d\omega$ when ω is increased by $d\omega$. It becomes evident that the $\theta/2\theta$ scan is a specific radial scan.

The usefulness of the Q representation is fully realized when this plot is combined with a reciprocal space map of the sample. Such a combined plot is shown

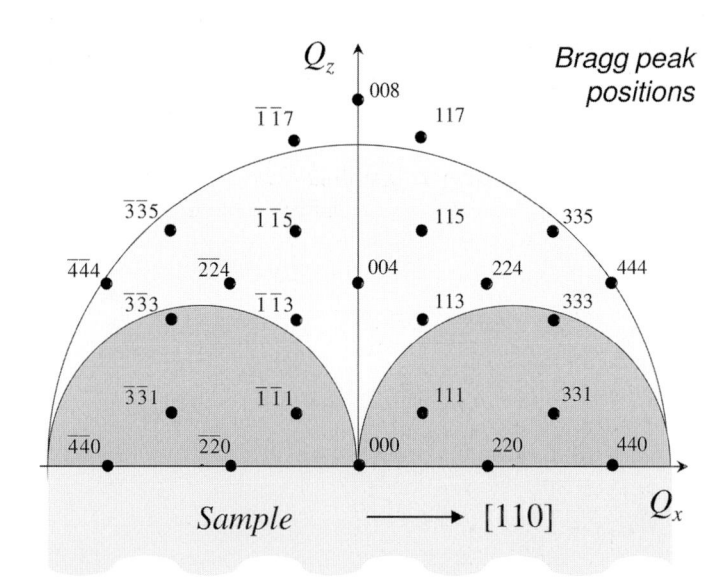

Figure 7.16 Combined plot of (Q_x, Q_z) plane and net of position of Bragg peaks for an epitaxial film on a Si substrate of (001) orientation. Use of Cu Kα radiation is assumed.

in Fig. 7.16 for a silicon wafer of (001) orientation. The [110] direction has been aligned along the sample x axis. In the figure there is an array of points representing the Miller indices of those Bragg reflections that become observable for this sample alignment. The $00l$ reflections allowed in the diamond structure occur all along the Q_z axis and only they would be measured in a symmetric $\theta/2\theta$ scan. This way of plotting also shows very intuitively that the range of measurable Bragg reflections, for instance the number of $00l$, is restricted because of the finiteness of the scannable Q range. This may be restated by saying that Bragg's equation limits the observable reflections to lie within scattering angle $2\theta < 2\pi$ which may also be written as $(h^2 + k^2 + l^2)^{1/2} < 2a/\lambda$. One ends up with three sets of Miller indices: (a) those that are visible in the reflection mode and those that are invisible (b) because they lie in the transmission range or (c) they lie beyond the range accessible with the radiation used. For any substrate of orientation (hkl) the set of Miller indices that show up in this combined plot will depend on the direction of alignment $[uvw]$ of the sample with the x axis of the diffractometer (for another example see Exercise 11).

Reciprocal space mapping is performed such that the Bragg reflection under investigation is fully mapped in a confined area in Q space. This means that the reflection is not only monitored by one rocking curve crossing it, but the whole area in the vicinity of the reflection is included in the measurement. Reciprocal space maps may be obtained by joining together successive one-dimensional scans in Q space for which different procedures are possible. One way of mapping consists in performing various rocking curves with increasing scattering angle 2θ. This is shown schematically in the Q space representation in Fig. 7.17(a). Another technique makes use of successive radial scans (see Fig. 7.17(b)). In both cases an intensity map of the Q area around the Bragg reflection is obtained. A further approach is to use the relationships given in Eqs. (7.47) and map directly in reciprocal space. These maps might deliver further structural information of epilayers to that obtained by rocking curve scans alone.

One may compare reciprocal space mapping with the pole figure measurement as introduced in Chapter 5. The most important distinction between both is that re-

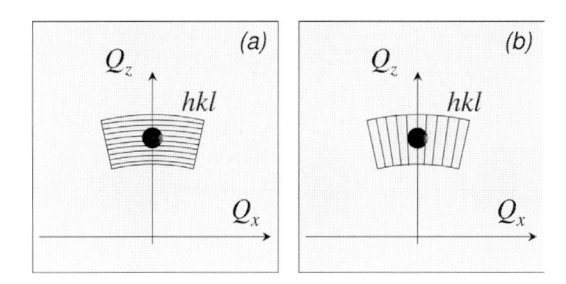

Figure 7.17 Two different ways by which reciprocal space maps may be recorded either by (a) subsequent rocking curve measurements or (b) subsequent radial scans.

336 | *7 High-Resolution X-ray Diffraction*

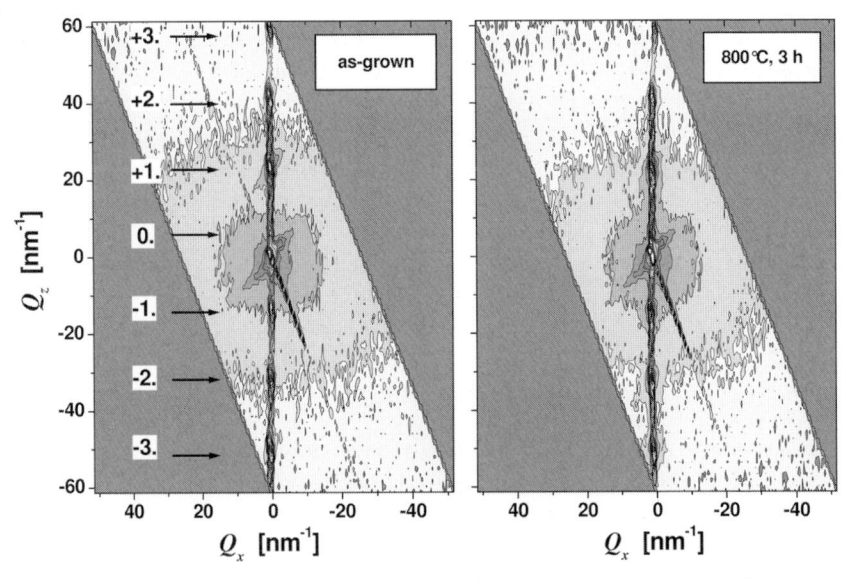

Figure 7.18 Reciprocal space map in the vicinity of 004 reflection from SiGe(C)/Si(C) superlattice structures on Si (001). Numbers and arrows indicate the position of superlattice reflections (compare Ref. [9]).

ciprocal space mapping generally operates at a much higher resolution and in a more restricted Q area. In addition, pole figure measurements are performed in noncoplanar configurations, whereas reciprocal space maps are generally carried out in coplanar configuration (Ω mode). For most experiments on epitaxial layers it is necessary to use a multiple-crystal diffractometer.

As an example Fig. 7.18 displays the reciprocal space maps recorded from an SiGe(C)/Si(C) superlattice structure on Si (001) [9]. The investigation aimed at elucidating the Ge diffusion in the relevant HBT device concentrations of about 20%

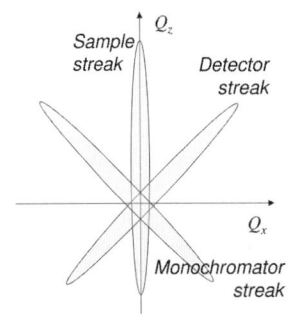

Figure 7.19 Schematic representation of streaking in reciprocal space maps.

for Ge and 1% for C. At relevant processing temperatures between 700 and 950 °C a substantial increase of the Ge diffusion coefficient was identified for small additions of carbon. In order to verify the absence of relaxation and the formation of SiC precipitates a two-dimensional reciprocal space map around the 004 reflection was measured for the sample (a) as grown and (b) after annealing at 800 °C for 3 hours. The measured intensity is displayed in different gray scales in Fig. 7.18. The high intensity for the $Q_x = 0$ line corresponds to the measurable intensity in the symmetric $\theta/2\theta$ configuration. Integer numbers on the left side indicate the positions of the superlattice peaks. The comparison of both reciprocal space maps revealed no increase of diffuse scattering near the substrate or the superlattice peaks that would have been caused by relaxation or formation of precipitates. In combination with *in situ* rocking curve measurement during annealing the investigation thus confirmed the stabilizing effect of C in avoiding the plastic relaxation of SiGe layers [9].

The typical streaking that can be observed in reciprocal space maps due to the influence of the different crystals is shown schematically in Fig. 7.19.

7.7
Diffuse Scattering

Diffuse scattering arises from the imperfections in an otherwise perfect crystal. Hence dislocations, point defects and interface roughness can all contribute to it. The scattering from the defects or dislocation themselves is not very significant; however, these disturb the average lattice by distortion of the crystal planes. With HRXRD the distortion of the average lattice is measurable and occurs close to the Bragg peaks. However, to analyze this scattering reliably a multiple-crystal diffractometer is required. The reason why a multiple-crystal diffractometer is so necessary for mapping reciprocal space is clear from Fig. 7.19. Because the crystal has a surface, a truncation to the crystal lattice, the scattering is extended normal to this surface. This is termed the crystal truncation rod (CTR). Clearly for a conventional triple-crystal diffractometer, the first collimating crystal, the sample and the analyzer crystal all have surfaces that would lead to CTRs producing an almost star-like pattern of intensity associated with each data point. The sample CTR conveys important information; however, the first collimating crystal and the analyzer crystal CTRs do not. The multiple-crystal diffractometer overcomes this by incorporating several repetitive reflections in the monochromator and the multiple-bounce analyzer to remove their CTRs. Thus the diffractometer artifacts are reduced to a minimum. The diffuse scattering close to the Bragg peaks can be analyzed.

Figure 7.20(a) illustrates the diffuse scattering from the dislocation strain fields in an InGaAs layer on a GaAs substrate. With an In content of ~10% and a layer thickness of 70 nm, the structure is just reaching a critical thickness. The diffuse scattering indicates the onset of relaxation before the more characteristic broad layer profile dominates. The latter occurs when the whole layer has undergone macroscopic relaxation, Fig. 7.20(b), i.e. the structure has become "mosaic".

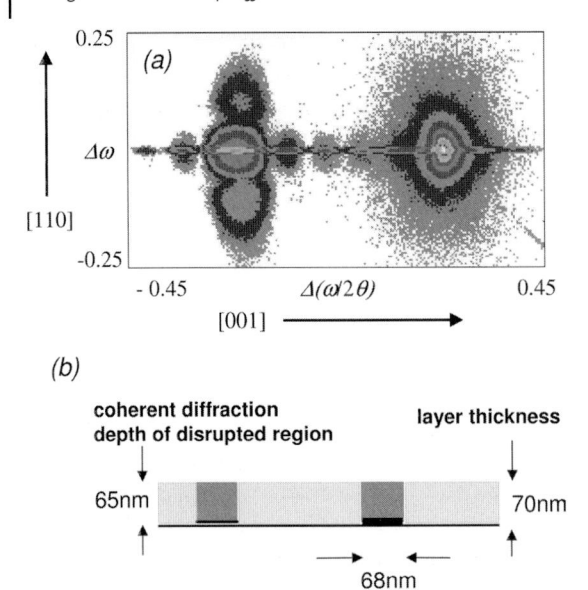

Figure 7.20 (a) Onset of relaxation: GaAs substrate and 70 nm $In_{0.1}Ga_{0.9}As$ layer. Lobes next to the layer peak come from dislocation strain fields. (b) Model of structure giving rise to scattering. Coherent/nearly perfect part is light gray and the darker regions indicate disrupted regions.

7.8
Extensions to High-Resolution Diffraction

What has been outlined above is an indication of the typical analyses that may be undertaken in HRXRD. However, this is not the limit of the methodology, since there are many extensions to these approaches.

X-ray topography is a very useful extension and relies on the idea that within the size of the probing x-ray beam there will be local fluctuations. These fluctuations, for example strain, rotations, etc., will modify the local intensity. So placing an x-ray-sensitive film or camera in the scattered x-ray beam that can measure these changes at the micrometer level can reveal considerable detail. For example, a dislocation has an extended strain field that distorts the crystal planes. So if these planes bow and the divergence of the incident beam is larger than the intrinsic scattering of the perfect structure then there will be enhanced scatter (high intensity) in the regions of these bowed crystal planes. If the divergence of the incident beam is less than or the same as the intrinsic scattering then the intensity is reduced. The strain field of a dislocation extends a considerable distance (many micrometers) and is sufficient for dislocations to be revealed.

So far we have only considered structures that are laterally homogeneous. With x-ray topography it is possible to reveal lateral variations; however, fluctuations on an even finer scale will appear in reciprocal space maps. A lateral periodic structure will create lateral interference fringes in Q_x, and when combined with the vertical fluctuations (layer structures) the reciprocal space maps can become very complex. However, the analysis of these structures is beyond the scope of this chapter.

The complexity of device structures is increasing and their analysis offers a fascinating challenge. At the present state of the art, a full analysis of some of these structures (these could be composed 400–500 layers) does require significant knowledge of the scattering process and the growth process [26].

Exercises

7.1 Calculate the relative bandwidth $\Delta\lambda/\lambda$ for an x-ray beam comprising the Cu $K\alpha_1$–α_2 doublet in order to estimate its influence in Eq. (7.1).

7.2 The diamond and sphalerite structure factors $F(hkl)$ differ for hkl all even and $h + k + l = 2(2n + 1)$ (chapter 2). How does the electron number difference ΔZ between atoms A and B affect the observable intensity?

7.3 Derive Eqs. (7.3) and (7.7).

7.4 What would Eqs. (7.3) and (7.7) look like for (111)-oriented wafers from the cubic system?

7.5 What is the adequate expression of Eq. (7.2) for a (00.l)-oriented hexagonal epilayer?

7.6 In a single-crystalline substrate of orientation (hkl) these lattice planes should be perfectly parallel aligned with the surface. Imagine that a certain misorientation has been introduced by the cutting process that can be quantified by a tilt angle ψ_{mis} between the normal vector $\langle hkl \rangle$ and the substrate normal s_3. How could this misorientation be determined? Under which constraints would a ϕ scan be appropriate?

7.7 Strained layers can cause a bending of the substrate–film composite as was pointed out in Chapter 6 (Stoney's formula). Imagine that you want to quantify the curvature by measuring the difference of the substrate peak position $\Delta\theta = \theta_{sub,1} - \theta_{sub,2}$ at two points on the wafer. (a) How is the curvature r_c derived when both points are a distance l apart? (b) Could you also quantify r_c, when the substrate peak is measured twice at the same point of the waver, but this time with different detector apertures?

7.8 Derive the fundamental equation of the dynamical theory, Eq. (7.30), by inserting the Fourier expansions of Eq. (7.27) into the wave equation of Eq. (7.28).

7.9 Explain on a qualitative basis why the peak intensity in the framework of dynamical theory is higher on the low-angle than on the high-angle side (Fig. 7.10).

7.10 Derive the relation of Eq. (7.46) between the wave vector transfer components (Q_x, Q_z) and the diffractometer rotation angles (ω, 2θ). Make use of Fig. 5.14.

7.11 Draw the reciprocal lattice/sample picture for a thin GaAs layer (sphalerite structure) on a Si (111) wafer (diamond structure) when the [100] direction is in the diffracting plane.

References

Monographs relevant to this chapter

U. Pietsch, V. Holý, T. Baumbach, *High-Resolution X-Ray Scattering – From Thin Films to Lateral Nanostructures*, Springer, Berlin, 2004.
P. F. Fewster, *X-Ray Scattering from Semiconductors*, Imperial College Press, London, 2003.
A. Authier, *Dynamical Theory of X-Ray Diffraction*, Oxford Science Publications, Oxford, 2001.
D. K. Bowen, B. K. Tanner, *High Resolution X-ray Diffractometry and Topography*, Taylor & Francis, London, 1998.
Z. G. Pinsker, *Dynamical Scattering of X-rays in Crystals*, Springer, Berlin, 1978.
M. von Laue (Mitarbeit E. H. Wagner), *Röntgenstrahl-Interferenzen*, Akademische Verlagsgesellschaft, Frankfurt M., 1960.
W. H. Zachariasen, *Theory of X-ray Diffraction in Crystals*, John Wiley & Sons, New York, 1945.

Special papers

[1] W. J. Bartels, Characterization of thin layers on perfect crystals with a multipurpose high resolution x-ray diffractometer, *J. Vac. Sci. Technol.* **B1** (1983) 338.
[2] A. Segmüller, M. Murakami, X-ray diffraction analysis of strain and stresses in thin films, in *Analytical Techniques For Thin Films*, ed. K. N. Tu, R. Rosenberg, Academic Press, Boston, MA, 1988, p. 143.
[3] A. Navarro-Quezada, A. G. Rodriguez, M. A. Vidal, G. Hernández-Sosa, H. Navarro-Contreras, Critical thickness of Ge/GaAs (100) epitaxial films, *Superficies y Vacio* **16** (2003) 42.
[4] D. Knoll, B. Heinemann, H. J. Osten et al., Si/SiGe:C heterojunction bipolar transistor in an epi-free well, single-polycrystalline technology, IEDM 1998, IEEE 1998.
[5] D. Knoll, Industry Examples at The-State-of-the-Art: IHP, in: J. D. Cressler (ed.) *Silicon Heterostructure Handbook*, CRC Press, Boca Raton, 2005.
[6] P. Zaumseil, A fast x-ray method to determine Ge content and relaxation of partly relaxed $Si_{1-x}Ge_x$ layers on silicon substrates, *phys. stat. sol. (a)* **141** (1994) 155.
[7] G. Bhagavannarayana, P. Zaumseil, Diffuse x-ray scattering of misfit dislocations at $Si_{1-x}Ge_x/Si$ interfaces by triple crystal diffractometry, *J. Appl. Phys.* **82** (1997) 1172.
[8] P. Zaumseil, High resolution determination of the Ge depth profile in SiGe heterobipolar transistor structures by X-ray diffractometry, *phys. stat. sol. (a)* **165** (1998) 195.
[9] P. Zaumseil, The influence of substitutional carbon on Si/Ge interdiffusion studied by x-ray diffractometry at superlattice structure, *J. Phys. D: Appl. Phys.* **32** (1999) A75.
[10] J. Zhang, J. H. Neave, P. F. Fewster, H. A. W. El Mubarek, P. Ashburn, I. Z. Mitrovic, O. Buiu, S. Hall, Growth of SiGeC layers by GSMBE and their characterization by x-ray techniques, *Electrochemical Soc. Meeting: SiGe Materials, Processing and Devices*, Honolulu (2004).
[11] C. G. Darwin, The theory of X-ray reflexion, *Phil. Mag.* **27** (1914) 315.
[12] C. G. Darwin, The theory of X-ray reflexion: II, *Phil. Mag.* **27** (1914) 675.
[13] J. Chaudhuri, S. Shah, Thickness measurement of thin films by x-ray absorption, *J. Appl. Phys.* **69** (1991) 499.
[14] T. Metzger, R. Höpler, E. Born et al., Defect structure of epitaxial GaN films determined by transmission electron microscopy and triple-axis X-ray diffractometry, *Phil. Mag.* **A77** (1998) 1013.
[15] P. P. Ewald, Zur Begründung der Kristalloptik: III. Die Kristalloptik der Röntgenstrahlen, *Ann. Phys.* **54** (1917) 519.
[16] P. Zaumseil, RCRefSim (rocking curve and reflectivity simulation), Frankfurt (Oder), 2005, available from: zaumseil@ihp-microelectronics.com.

[17] D. Taupin, Theorie dynamique de la diffraction des rayons X par les cristaux deformes, *Bull. Soc. Franc. Minér. Cryst.* **87** (1964) 469.

[18] S. Takagi, A dynamical theory of diffraction for a distorted crystal, *J. Phys. Soc. Jpn.* **26** (1969) 1239.

[19] J. Burgeat, R. Colella, Effect of alpha irradiation on the X-ray diffraction profiles of silicon single crystals, *J. Appl. Phys.* **40** (1969) 3505.

[20] B. C. Larson, J. F. Barhorst, X-ray study of lattice strain in boron implanted laser annealed silicon, *J. Appl. Phys.* **51** (1980) 3181.

[21] M. A. G. Halliwell, M. H. Lyons, M. J. Hill, The interpretation of x-ray rocking curves from III-V semiconductor device structures, *J. Cryst. Growth* **68** (1984) 523.

[22] J. G. E. Klappe, P. F. Fewster, Fitting of rocking curves from ion-implanted semiconductors, *J. Appl. Cryst.* **27** (1994) 103.

[23] P. F. Fewster, N. L. Andrew, O. H. Hughes et al., X-ray studies of group III-nitride quantum wells with high quality interfaces, *J. Vac. Sci. Technol.* **B18** (2000) 2300.

[24] J. Birch, J.-E. Sundgren, P. F. Fewster, Measurement of the lattice parameters in the individual layers of single-crystal superlattices, *J. Appl. Phys.* **78** (1995) 6582.

[25] P. Zaumseil, G. G. Fischer, K. Brunner, K. Eberl, Comparison of the termal stability of $Si_{0.603}Ge_{0.397}$/Si and $Si_{0.597}Ge_{0.391}C_{0.012}$/Si superlattice structures, *J. Appl. Phys.* **81** (1997) 6134.

[26] P. Kidd, Investigation of the precision in X-ray diffraction analysis of VCSEL structures, *J. Mater. Sci.: Mater. Electr.* **14** (2003) 541.

Index

Page numbers in italics refer to figures.

μt product   28, 60–66, 153, 318

a

abscissa transformation   *103*
absorption   55
absorption coefficient   167, *214*
absorption edges   56
absorption effects   26
absorption factor   26, 28, 57, 196
 – exponent   156
 – GIXRD   153, *154*
 – ω scan   212
absorption index   161
acceleration voltage, x-ray tubes   30
accuracy   79
 – loss   76
 – phase mixture   67
aerosols   118
American Society for Testing and Materials (ASTM)   52
amorphous fraction   74
amorphous phases   70
amorphous silicon   70
amorphous substances   52
amorphous thin films   70–74
angle of rotation   191, 279
angle of the incident beam, GIXRD   144
anisotropy   115
 – growth   124
anode materials, x-ray tubes   30
anode rotation   176
anomalous dispersion corrections   320
anti-site domains   92
apertures   44
ASTM   *see* American Society for Testing and Materials
asymmetric diffraction   278
 – geometry   278
 – penetration depth   *280*
 – scheme   *279*
asymmetry
 – bragg peaks   321

 – *see also* symmetry
atomic form factors   17–18, 161
atomic mass   55
atomic number   50, 55
atomic radius   21
attenuation   55
 – coefficient   67, 143
 – x-rays   277
Aurivillius phases   135
average contrast factors   125
average density   67
average information depth   156, 158
average RS values   270
average size parameter   119
average stress, of sample   261
averaging, over full grain ensemble   119
axis, naming convention   221
axisymmetric texture   204
azimuthal scans   218–220

b

background, x-ray   30, 46
background correction   197
background function   96, 128, 208
ball milling   131
band gap engineering   300
band gaps   48
bandwidth, monochromator   326
barrier coatings   37
Bartels monochromator   313
$Ba_{0.7}Sr_{0.3}TiO_3$ film (BST)   214
bcc   *see* body-centered cubic structure
Be windows, x-ray tubes   34
beam angle   12
beam attenuator   146
beam broadening   194
beam conditioning
 – crystals   325
 – techniques   144

Thin Film Analysis by X-Ray Scattering. M. Birkholz
Copyright © 2006 WILEY-VCH Verlag GmbH & Co. KGaA, Weinheim
ISBN: 3-527-31052-5

beam divergence 69, 144–145, 220, 325–326
- axial 78, 132
- defintion 44–46
- finite 29
- planes 132
beam length 146
beam mask 46
beam optics, parallel 144
beam shaping 46
- elements 44
bending, thin samples 251
Bertaut's column theorem 111
biaxial residual stress state of rotational symmetry 275, 284
biaxial stress model 248
biaxial stress state 274–276
biaxial textures 216–228
blocking layer 173
body-centered cubic structure (bcc) 20, 91
bonding spheres 70
Bragg angle
- aluminum 24
- definition 9
- distinct 151
- with beam refraction 164
Bragg equation 9–10
- film thickness 170, 310
- limits 335
- selection rule 261
- single-layer system 174
- vector form 320
Bragg line profiles, crystallite shapes 114
Bragg mirrors 173
Bragg peaks
- asymmetry 330
- epitaxial film 334
- line profile 109
- monitoring 229
- position shift 165
- splitting see splitting
Bragg reflection 9
- crystallite number 194
- crystallite orientation 189
- enhancement 183
- experimental diffractogram 52
- high scattering angles 249
- intensity 14–37
- lattice orientation 191
- number of 289
- principles 14–37
- reduction 153
- rutile structure 149
- shift 321
- splitting 304
- structural information 297
- width 85
Bragg–Brentano geometry 15
brass substrate 129
Bremsstrahlung 30
BST see $Ba_{0.7}Sr_{0.3}TiO_3$ film
bulk elastic constants 289
Burgers vector 90, 124

c

Caglioti formula 100
Cambridge Crystallographic Data Center (CCDC) 52
Cambridge Structure Database (CSD) 52
carbon 48
carbon layers, amorphous 134
Cauchy function 92, 93, 129, 136
c-axis
- alignment 218
- layer texture 232
CCC see channel-cut crystal monochromator
CCDC see Cambridge Crystallographic Data Center, see charge-coupled devices
CCP14 see Collaborative Computational Project #14
central beam, diffractometer 16, 44
centroid determiniation 250
centroid peak 86
channel-cut crystal (CCC) monochromator 327
characteristic radiation 30
charge-coupled devices (CCD) 256
chemical composition 57, 60
chemical phases, identification 42–84
chemical vapor deposition (CVD) 53, 299
chi-square function 88
cluster, atomic 71–72
coatings
- fluorescing 45
- molybdenum 61
- protective 241
- razor edge 135
- technology 242
- thin hard 287
$CoFe_2O_4$ nanoparticles 136
coherency, cessation 17
coherently diffracting domains 85
coherently scattering domains 107
Collaborative Computational Project #14 (CCP14) 89
column height distribution 111–112
- determination 115–117

column theorem 111
columnar honeycomb structure *316*
complex refractive index, x-rays 161
composition
  – epitaxial electronic films 306
  – epitaxial layers 303
composition-relaxation net 312
Compton scattering 2
configuration factor 151, 196
configurations, diffractometer 325
contaminants 60
contrast factor 132
convolution 136
coplanar geometry 211
  – HRXRD 312
correlation length 117
counting statistics 257
count-mean diameter 119
counts per second (cps) 253
cps *see* counts per second
cracking 239
critical angle 161–165, 280
  – GIXRD 150
  – sequence 168
critical temperatue 51
cross-section 55
  – x-rays 56
crystal truncation rod (CTR) 337
crystalline lattice faults 90–92
crystalline matter, elastic behavior 239
crystallite group method, strong texture 268
crystallite orientations 183
crystallite shapes 112–115
crystallite size 107–120, 134
  – area averaged 120
  – distribution function 118–120
  – volume averaged 119–120
crystallite subsets, selective perception 13
crystallographic lattice planes 9
crystallographic lattive planes 25
crystallographic structures 43
crystallographic unit cell 9, 19, *21*
crystals 325
  – cubic 193, 197, 223
  – imperfections 323, 337
  – lattice 4
  – monochromators 46, 297
  – reference frame 223, 258
  – structure 48
  – symmetry 193
  – tetragonal 149
CSD *see* Cambridge Structure Database
CTR *see* crystal truncation rod

Cu K_α emission line 97
cubic crystals 193, 197, 223
cubic group IV alloys 299
cubic III–V compounds 300
cuprous oxide 134
Curie temperatures 185
CVD *see* chemical vapor deposition
cylinders 113
Czochralski-grown silicon wafers (CZ-Si) 62

d

damping
  – thin-film analysis 36
  – within the crystal 316
damping terms 172
Darwin treatment 314
data reduction 86
database, diffraction patterns 47
Debye scattering equation 70, *72*
Debye–Waller factor 168, 172
DEC *see* diffraction elastic constants
deconvolution 101–107
defect incorporation, thin films 272, 275
defect, lattice 90
defocusing 196, 213, 252
  – correction 196
deposition process 43, 190
  – optimization 43, 226
depth dependence, phase dispersion 80
depth monitoring 229
depth-dependent properties, x-ray beams 158–159
depth-resolved XSA, methods 282
depth-variable 27
detector mode, open 197
detector slits 220
detectors 44
  – x-ray 253–258
deviation parameter, line profiles 322
device structures analysis 331
dial gauge 77
diamond structure 48–49, 299
diamond-like carbon (DLC) 134
dielectric displacement vectors 320
dielectric susceptibility 161
diesel engine technologies 135
differential equilibrium conditions 273, 275
diffracted-beam monochromators 325
diffraction
  – experiment 259
  – high-resolution 297–342, 338
  – x-ray 1–41

diffraction elastic constants (DECs) 239, 258–260
– nitrides 264
diffraction gratings 7
diffraction histogram 47
diffraction pattern 13, 43
– ball milled Ni sample 133
– comparison 226
– epilayer–substrate material systems 297
– GID 175
– nanocrystalline powder sdample 105
– simulation 324
– symmetric 219
– W-C:H films 135
– YBCO 218
– ZNO 188
diffraction-reflection technique 175
diffractometer, θ/2θ 11, 15–17
– Eta 285
– four-circle 193, 286
– triple-crystal 337
diffractometer configuration
– GIXRD 152
– XRR 165
diffuse scattering 337–338
diffusion 37
disarmament 40
dislocation
– contrast factors 124–125, 128
– core 90
– density 91
– edge 124
– glide 91
– lattice 121
– networks 107
– pile-ups 134
– screw 90, 124
dislocation density 124, 126
dislocation-free crystals 323
dispersion correction 160
dispersive elements 46
displacement field 120, 126
distortion broadening 121
d distribution
– definition 247
– measurement 249–258
distribution function
– column height 111–112, 116, 127
– crystallite size 115–116, 118–120
– monodisperse 119
– size coefficients 127
divergence see beam divergence
divergence slits (DS) 44–45

divergent beam optics 44–47
DLC see diamond-like carbon
double-crystal configuration 326
doublet resolution 32
DS see divergence slits
DuMond diagram 326
dynamical theory 37, 62, 319–324

e
edge dislocation 90
edge filters 33
EKK see Eshelby, Kröner and Kneer (EKK)
elastic constants 124
elastic modulus 241
elastic strain 241, 271
electric field vectors 161
electroacoustic devices 188
electrodeposition 134
electron microscopy 62, 85, 107
electron radius, classical 3
electronic diode solid-state detectors 253
electroplating 37
elementary metals 21
emission, x-ray 31–35
emission spectrum 30
– laboratory x-ray tube 98
energy resolution, proportional detector 255–256
epitaxial films
– bragg reflection 305
– composition 303
– structural features 312
– thin 297, 299
epitaxy 303
equal area projection 192
equilibrium conditions 275
error margin 88
error propagation 226
Eshelby, Kröner and Kneer (EKK) 263
Eta diffractometer 285
Euler angles 200
Euler cradle 146, 184, 193, 287
Euler plane 23
Euler space 201, 220
Ewald's dynamical theory 319–324
excitation errors 321
extended x-ray absorption fine structure spectroscopy (EXAFS) 57
extinction 314
– coefficient 319
– conditions 20, 49, 149
– length 319, 323
– mosaic crystal 317
extrinsic stresses 271

Index

f

face-centered cubic structure (fcc)   20, 91
Fachinformationszentrum (FIZ)   52
fcc   *see* face-centered cubic structure
FeRAM   *see* ferroelectric random access memory
ferroelectric random access memory (FeRAM)   198
fiber axis   184
fiber textures   184, 200, 204–215
  – ODF   201, 205
  – volume fraction   *207*
films
  – aluminum   13, 178
  – amorphous   70–74
  – CrN   266
  – deposition   275
  – epitaxal   297, 299, 303, 305, 312
  – magnetic   177
  – metal   37, 80
  – nanocrystalline   102
  – optical   177
  – piezoelectric   188
  – polycrystalline   36, 54
  – polycrystalline   183
  – Pr2O3   177
  – silicon   61
  – thickness   156, 170–171, 275, 309, 318
  – TiN   *247*
  – TiO2   154
  – titanium   38
  – W-C:H   135
film–substrate composite, stress   *245*
filters   97
  – materials   30
fine structure   57
first scattering peak (FSP)   72
fitting parameters   95
FIZ   *see* Fachinformationszentrum
flat sample   77
fluctuations, x-ray beam   338
fluorescence, x-ray   30
focus modes, x-ray tubes   34
food packaging   37
four-circle diffractometer   286
Fourier coefficients   117
Fourier length   103
Fourier transform   89, 102, 127
  – discrete   23, 104
Fourier-conjugated variable   103
Fraunhofer diffraction   6
Fresnel diffraction   6
fringes   170, 310, 330

FSP   *see* first scattering peak
full width at half maximum (FWHM)   86
fundamental parameter approach   100, *132*, 136
FWHM   *see* full width at half maximum

g

GaAs (001) substrates   318
GaN buffer layer   332
gases, scattering equation   70
gate oxides   176
Gauss function   92, *93*, 129, 136
  – three-dimensional   221
Gaussian concentration profile   329
general orientation distribution   220
general textures   216–228
geometry factor   25
  – ω scan   212
GIABD   *see* grazing incidence asymmetric Bragg diffraction
giant magnetoresistance effect (GMR)   177
GID   *see* grazing incidence diffraction
GIXRD   *see* grazing incidence x-ray diffraction
glass
  – fiber   161
  – substrate   53, 61, 71
GMR   *see* giant magnetoresistance effect
goniometer   15, 184
  – radius   45
graded multilayer mirrors   77
grain boundary migration   241
grain growth   241
grain interaction models   261–264
grain orientation, preferred   126
graphite   48
grazing incidence   143–182
grazing incidence asymmetric Bragg diffraction (GIABD)   154
grazing incidence diffraction (GID)   175, 282
grazing incidence x-ray diffraction (GIXRD)   143, 148–155
growth parameters, thin films   65, 275

h

H Mode   286
half population tilt   207
hard nitride coatings   233
harmonic analysis   222
harmonic method, texture factors   204–207
harmonic oscillators, damped   160
harmonic reflections   63–64
HBT   *see* heterojunction bipolar transistor

hcp   *see* hexagonal close-packed structure
height misalignment   76
HEMT   *see* high-electron-mobility transistor, *see* homogeneous elastic matrix models
heteroepitaxy   303
heterojunction bipolar transistor (HBT)   299, *300*, 310
hexagonal close-packed structure (hcp)   20
hexagonal Group III nitrides   301
hidden elastic stresses   268
high-electron-mobility transistor (HEMT)   331
highly oriented pyrolytic graphite (HOPG)   46
high-resolution rocking curve   306–313
– Ge epilayers   *308*
high-resolution x-ray diffraction (HRXRD)   297–339
high-temperature superconducting films   231, 233
high-temperature superconductors (HTSC)   184, 187
histogram based techniques   43–55
homogeneous elastic matrix (HEM) models   261
Hooke's law   241
HOPG   *see* highly oriented pyrolytic graphite
HRXRD   *see* high-resolution x-ray diffraction
HTSC   *see* high-temperature superconductors
hybrid technique, GID   175
hydrogen concentration   70
hydrogenated amorphous carbon layers   134

i
IBAD   *see* ion beam-assisted deposition
ICDD   *see* International Center for Diffraction Data
ideal components, method of   228
impact ionization   31
incidence angle, grazing incidence   148
incident beam, multilayer system   171
information depth   155–158
– variation   229
INLT   *see* inverse numerical Laplace transform
Inorganic Structure Data Base (ISDB)   52–53
in-plane stress   247

instrumental calibration, stoichiometry determinations   75
instrumental effects, XRR scan   167
instrumental errors   76
instrumental line profile   97–101
instrumental parameters   79
instrumental peak shifts   79
integral breadth   110
– diffraction line   92–93
– model functions   96
– sample broadened profile   130
– total   123
integral intensities, x-ray diffraction   29, 35, 61, 64, 110
intensity damping   155
intensity loss, incoming beam   314
intensity oscillations   168
interactions between x-rays and matter   1
interatomic distances   5
interface roughness   174
interference
– constructive   1
– destructive   1, 20, 50
interference function   7, 29–37, 108, 112, 116
interlayer stresses   275
International Center for Diffraction Data (ICDD)   52
International Union of Crystallography (IUCr)   53
intralayer stresses   275
intrinsic stresses   271
inverse numerical Laplace transform (INLT)   277
inverse pole figure   202
– thin CrN film   *266*
inverse problem, thin-film analysis   87
ion beam-assisted deposition (IBAD)   217, 289
iron powder
– reflections   *121*
– Williamson-Hall plot   *123*
iron pyrite   80
ISDB   *see* Inorganic Structure Data Base
IUCr   *see* International Union of Crystallography

j
JCPDS   *see* Joint Committee on Powder Diffraction Standards
Joint Committee on Powder Diffraction Standards (JCPDS)   52

k

Kiessig oscillations 168
kinematical theory 37, 314, 319
kinetic energy, particles 275

l

LaB$_6$ standard 136
laboratory reference frame 24, 260
ladder of complexity 240
Lambert-Beer law 26
La-modified PbTiO$_3$ films 203
Laplace transform 89, 159, 164, 277
large area electronic 61
laser diodes (LDs) 301
laterally graded multilayer mirror (LGMM) 145, 151
lattice
 – bcc 20, 91
 – crystal 4
 – fcc 20, 91
 – simple cubic 9
lattice constants, YBCO 219
lattice distortion 120
lattice faults 90–92
lattice mismatch 305
 – strain 271
lattice parameters
 – determination 74
 – dislocation 132
lattice planes 191
 – angular relations 208
 – crystallographic 9
 – vertically tilted 147
lattice vector 4, 319
Laue conditions 8, 10, 108
Laue equation 74
Laue function 110
layer peak, interpretation 309
layer period 145
layer properties 158
layer structure profiles 324–332
layer textures 204–215
layers
 – blocking 173
 – carbon 134
 – epitaxial 303, 307
 – fully strained 305
 – GaN buffer 332
 – Ge epi- 308
 – III–V compound 298
 – inter- 275
 – multi- see multilayers
 – nickel 129
 – semiconducting 232
 – silver top 228
 – single 168–171, 174
 – superlattice 332
 – ZNO:Al 232
 – see also thin films
LD see laser diodes
least square minimization 96
LED see light-emitting diodes
Levenberg–Marquardt algorithm 88
LGMM see laterally graded multilayer mirrors
light-emitting diodes (LEDs) 301
line broadening 74
line focus 45, 66, 146
line profile analysis (LPA) 85–142
 – standard sample 100
 – x-ray 134
line profiles
 – instrumental 97–101, 133
 – standard 101
linear absorption coefficient 60, 278
linear attenuation coefficient 27, 55–60, 58
liquids, model systems 70
load test 259
logarithmic normal function 118
Lorentz factor 26, 35, 249
Loschmidt's number 3
LPA see line profile analysis

m

macroresidual stress fields, variations 273
macroscopic equilibrium conditions 275
macroscopic relaxation 337
macrostress 240, 268, 270
magnetic thin films 177
March function 208–209
masks, diffractometer 44
mass absorption coefficient 27, 57–58
 – chemical elements 59
mass attenuation coefficient, average 66
mass density 55
mathematical fit 96
MAUD program, Rietveld refinement 132
Maxwell's equation 320
MBE see molecular beam epitaxy
mean quadratic deviation 23
measurement geometry, x-ray diffractometer 15
mechanical anisotropy 260
mechanical relaxation phenomena 304
mechanical stress 241
 – thin films 239
medium-range order (MRO) 72

Index

metal lattices, atomic form factor 23
metallic atoms, form factors 18
metallic films 80
metallic interconnect 37, 230
- copper 38
metallic ribbons 227
metal-organic chemical vapor deposition (MOCVD) 301
metals 37
- full texture analysis 227
- hexagonal close-packed (hcp) 21
microcavities 64
microscopic structural features 313
microstrain 85–86, 127, 134
microstress 240, 269–270
microstructural reorganization 241
microstructure
- single crystals 315
- solids 107
- thin films 134, 136
Miller indices 9, 262
- pole figures 225
mirrors
- Bragg 173
- graded 77
- multilayer 145, 151
- x-ray 175
misalignment, superconducting material 217
misfit dislocations 299
mismatch, epitaxial layers 303
misorientation, epitaxial thin films 315
MOCVD see metal-organic chemical vapor deposition
model functions 208
- profile analysis 86–97
model patterns 54
model simulation 96
modeling, structural 87
molecular beam epitaxy (MBE) 299
molecules, in gases and liquids 71
moments
- distribution function 116
- statistical 117, 127
momentum transfer
- GID 175
- plane of 333
- vector 6
monochromators 97, 325
- α_1-α_2 99
- crystal 46, 144, 297
- graphite 47
- secondary 44, 47, 98
monodisperse distribution 111

mosaicity, single crystals 314–319
MRO see medium-range order
multilayer mirrors
- graded 77
- laterally graded 145, 151
multilayers 62, 68
- magnetic 177
- periodic 173
- reflectivity 171–175
multiphase materials, residual stresses 270
multiple-crystal diffractometer 307, 328
multiplicity 24

n

nanocrystallites, palladium 117
nanoparticles, $CoFe_2O_4$ 136
National Institute of Standards and Technology (NIST) 53
Neerfeld–Hill approximation 263
nickel layer 129
NIST see National Institute of Standards and Technology
nitride coatings 242
- hard 233
non-coplanar configuration 195
Non-Proliferation Treaty, nuclear arms 40
normal function, logarithmic 118
nuclear fission bombs 40
numerical data analysis 87–89

o

octahedrons 113
ODF see orientation distribution function
one-elemental metals, crystal structures 22
optical coatings, properties 169
optical films 177
optoelectronic devices 301
optoelectronic thin-film applications, texture 231
orientation, crystallites 183–238, 200
orientation densities, representation 202
orientation distribution function (ODF) 25, 183, 201, 220
- coefficients 223
- epitaxial films 315
- expansion 205
- normalization 225
- sections 228
- XSA 267
Ostwald's rule of stages 80
outer cut-off radius 90, 132
overlap 102, 106
- bragg peaks 68
- peak 128

p

pair distribution function 73
pair production 56
palladium nanocrystallites 117
parafocusing geometry 194, 252
parafocusing mode 16
parallel beam configurations 176
parallel beam geometry 77, 194
parallel beam mode, diffractometer 151
parallel beam optics 144
parallel plate collimator (PPC) 144
pattern decomposition 131, *132*
$Pb_{0.88}La_{0.08}TiO_3$ (PLT) 203
PDF *see* powder diffraction file
PSD *see* position-sensitive detectors
peak asymmetries 97
peak intensity measurement *198*
peak parameters, profile analysis 86–97
peak shift 244
 – GID 176
peak splitting 98
Pearson function 95, 130
PECVD *see* plasma-enhanced CVD
peeling 239
penetration depth 27, 59, 155–158, 162, *164*, 280
 – solids 60
 – variations 253
 – x-ray 278
perovskite structure 135, 185
 – notations 187
perpendicular projection *192*
phase composition 43
phase difference
 – beams 169
 – derivation *170*
phase factor 162
phase mixtures 66–69
phase problem 23
phase shift 2, 6
phase system comparison 199
phases
 – amorphous 70
 – Aurivillius 135
 – chemical 42–84
 – metastable 53
 – nonequilibrium 53
PHD *see* pulse height distribution
photoelectrons 56
photoionization 1, 55
physical vapor deposition (PVD) 37, 60
piezoelectrical fields 302
piezoelectricity 188

plane stacking 22
plane wave propagation 162
plasma-enhanced CVD (PECVD) 60
plastic deformation 91, 271, 309
PLT *see* $Pb_{0.88}La_{0.08}TiO_3$ (PLT)
PMS *see* pseudo macrostresses
point focus 66
 – x-ray tubes *31*
Poisson ratio 241, 248, 304
polarizability, spatially dependent 319
polarization 3
polarization factor 26, 321
pole figures 184, 191–195, *193*
 – Ag *228*
 – diamond *233*
 – incomplete 197
 – inverse 200–203
 – measurements 195–200
 – visualization 192
 – YBCO *219*
polonium 40
polycapillary lenses *146*, 147
polycrystalline films 81, 216
 – residual stress 271
 – residual stress distribution 282
polycrystals
 – elastic constants 263
 – textured 325
porosity 60, 65, 239
position-sensitive detectors (PSDs) 257
postoxidation 241
powder cell 53
powder diffraction 47
powder diffraction file (PDF) 52
powder diffractometers 15
powder pattern 125
powder reflection, intensity 189
PPC *see* parallel plate collimator (PPC)
precision
 – cell edge determination 76
 – difference to acccuracy 79
 – instrumental 249
preferred orientation 25, 68
 – derivation 207
 – second axis 216
primary extinction correction 317
profilometry 62
property profiles, thin films 159
proportional counter 253–254
protective coatings 241
pseudo macrostresses (PMS) 270
pseudomorphic growth 303
 – breakdown 305
pseudo-Voigt function 95, 130

pulse height distribution (PHD) 255
pulse height distribution (PHD) discrimination 255
PVD see physical vapor deposition
pyroelectricity 186

q
quartz glass 82, 166
quartz rocks, deformed 208

r
R values 88
Rachinger correction 99, 104
radial scan 334
rank of harmonic expansion, ODF 226
RCRefSim see rocking curve and reflectivity simulations
real-time multiple strip (RTMS) 258
reciprocal space map 298, 332–337
 – SiGe(C)/Si(C) superlattice structure 336
reference frame, crystal 111, 120
 – laboratory 111
 – transformation matrices 17
reflection broadening 107–120, 107–120
reflection centroids, shift 164
reflection coefficients 172
 – multilayer system 171
reflection intensity 67
reflection profile 89
reflection splitting, Ge films 308
reflective coatings 37
reflectivity 171–175
 – single layer 168–171
 – substrate 166–168
refractive index 147, 160–161, 165
 – imaginary part 280
regression, nonlinear 88, 96
relaxation
 – determination 311
 – onset 338
 – strain 303
relaxation parameter 304
residual stress 134, 270
 – origin 271
residual stress analysis (RSA) 239–296
 – depth-dependant properties 159
 – with texture 265
residual stress evaluation, scattering vector method 286
residual stress fields, nonuniform 278
residual stress gradients 273–276
residual stress measurement, x-ray diffraction 240

residual stress state, biaxial 275, 284
residual stress tensor, off-diagonal components 251
resistivity, ZNO:Al layers 232
resolution
 – divergent beam optics 45–46
 – energy 255–256
 – epitaxial layer investigations 307
 – high 297–339
 – instrumental line profile 97–98
 – spectrum 103
Reuss approach 262
Rietveld fitting schemes 208
Rietveld programs 210
Rietveld refinement 131, *132*
rock salt structure 242
rocking curve and reflectivity simulations (RCRefSim) 166, 322
rocking curves 184, 210, 297
 – corrected 214
 – high-resolution 306
 – measurement *211*
rotating anode 35
rotational symmetry
 – biaxial residual stress state 275, 284
 – thin films 204
roughness, surface 168, 171
RTMS see real-time multiple strip
Rutherford backscattering 65
rutile films 154
rutile structure 149–150

s
sample
 – adjustment 146
 – chemical composition 290
 – depth *153*
 – length 45
 – reference frame 11, 69, 221
 – rotation 69, 151, 279, 284, 286, 306
 – tilt 196
 – transparency 77
 – XRR 165
satellite peaks 110
 – parallelepipeds 113
SAW see surface acoustic wave
SAXS see small-angle x-ray scattering
SBT see $SrBi_2(Ta_{1-x}Nb_x)O_1$ (SBT)
scaling factor 36
scan
 – azimuthal 218–220
 – radial 334
 – θ/2θ 1, 11–13, 211
 – ω 211

scattering 55
– coherent 1
– diffuse 337–338
– from atomic nuclei 3
– inelastic 1
– Rutherford backscattering 65
– Thomson 2
scattering angle 2
scattering factors, anomalous 18
scattering plane 2
scattering vector 6, 102, 151, 284, 332
– construction 7
– GIXRD 150
Scherrer constants 115
Scherrer equation 108–110
scintillation counter 253, 255, 256
screw dislocation 90
secondary extinction 317
selective perception, subsets of crystallites 13
semiconducting layers, texture 232
semiconductors 48, 76
– doping 328
– high dielectric constants 176
shadowing optics 44
shape parameter 89
sharpness, texture 224
signal-to-noise ratio, improvement 69
silane (SiH_4) 70
silicon 75
– amorphous 70
simple cubic lattice 9, 19
single-crystal heteroepitaxy 271
single-crystalline film 217
single crystals 36, 325
– microstructure 315
single layer, reflectivity 168–171
single-line analysis 129–130
size broadening 101, 120–134
size distribution function 107
size parameter 110
– average 119
slip system 91, 125
– fcc lattices 92
slits 44, 144
– detector 45
– divergence 45, 97
– receiving 78, 97, 132
– Soller 46, 78, 97
small-angle grain boundaries, mosaic block 315
small-angle x-ray scattering (SAXS) 135
Snell's law 161–162
solar cells 70, 80

Soller slits 46, 78, 97
sphalerite structure 50–52, 299–300
spherical harmonic functions 204
splitting
– Bragg peaks 13, 33, 45, 98–99
– tetragonal 185–187
square root intensity plot 13
$SrBi_2Nb_2O_9$ (SBNO) 135
$SrBi_2(Ta_{1-x}Nb_x)O_1$ (SBT) 198
stacking faults 92, 107, 121, 135
standard data 106
standard profile 128
standard sample 75
– LPA 100
steep residual strain, detection 284
stereographic projection *192*
stick pattern 47
stoichiometry determination 74–75
stoichiometry deviation 81
Stoney's formula 253
strain 121, 244, 303–306
– root mean square 122
strain broadening 101, 120–134
strain distribution, thin films 265
strain field 85, 90
strain tensor 246
strain-free direction, determination 250
strain-free tilt angle 248
stress
– classification of 268–277
– extrinsic 271
– first kind 240
– intrinsic 75, 271
stress components 260, 274
stress model, biaxial 244
stress states, formation 289
stress-free tilt angle 260
stress-strain relation 261
structure
– body-centered cubic (bcc) 20–21, 91
– chalcopyrite 82
– close-packed *21*
– columnar honeycomb *316*
– crystallographic 43
– diamond 48–49, 299
– face-centered cubic (fcc) 20–21, 91
– fine 57
– hexagonal 22
– hexagonal close-packed (hcp) 20
– micro- *see* microstructure
– perovskite *see* perovskite structure
– polonium 4
– profiles 324–332
– rock salt 242

- rutile   149–150
- simple cubic   4, 5
- sphalerite   see sphalerite structure
- wurtzite   see wurtzite structure
- zinc blende-type   242

structure factor   19–24
- diamond lattice   48
- perovskite structure   185
- rock salt structure   242
- rutile structure   149
- zinc blende structure   50

substrate, reflectivity   166–168
substrate normal   204
substrate peak   61
substrate temperatures   53
superlattices   173
- layer structures   332
- reflectivity   171–175

surface, ideally smooth   168
surface acoustic wave (SAW)   189
surface layers, thin   176
surface oxidation, analysis   178
surface roughness   168, 171
surface sensitivity   171
- GID   176

surface-induced disorder phase transitions   178
symmetry
- crystals   193
- perovskite   185
- rotational   see rotational symmetry
- see also asymmetry

symmetry constraints, ODF   223
symmetry-adapted functions, cubic crystal class   223
synchrotron GID study   177
synchrotron radiation sources   229

t

Takagi–Taupin equations   323–324, 328
take-off angle, anode focus   34
tensor transformation   246
tetragonal crystals   149
tetragonal distortion, heteroepitaxy   303
tetragonal splitting   185–187
tetrahedral coordination   48
tetrahedrons   113
texture analysis   201, 211
- Ca-doped $PbTiO_3$ film   209
- computer programs   206

texture factors   25, 63, 184, 188–190, 208
- derivation   204
texture influence, minimization   266
texture profile, normalized   214

texture states, formation   289
textures   25, 134, 183–238
- axisymmetric   204
- biaxial   184, 216–228
- effect   265–268
- fiber   see fiber textures
- fully general   225–228
- general   216–228
- high-temperature superconducting composite ribbons   231
- layer   204–215
- metals   227
- optoelectronic thin-film applications   231
- polycrystals   325
- residual stress analysis (RSA)   265
- semiconducting layers   232
- sharpness   224
- thin-film   228–230

thermal barrier coatings   172, 173
thermal expansion coefficients   299
thermal stresses   271
thermal vibrations   23
thickness   60, 80
- film   27
- geometrical   64
- mass   65

thickness factor, metals   28
thickness homogeneity   46
thickness inhomogenites   65
thin films   52
- absorption effect   27
- aluminum   13, 178
- amorphous   70–74
- analysis   143, 162, 333
- CrN   266
- defect incorporation   272, 275
- depth dependency   80, 228–230
- diamond   233
- diffraction pattern   189
- elasticity   289
- epitaxial   315
- ferroelectric   203
- GID measurement   178
- growth   55, 124, 129, 155, 230
- macroresidual stress   273
- magnetic   177
- material parameters   305
- mechanical stress   239
- metal   37
- microstructure   134, 136
- nanocrystalline   102
- piezoelectric   188
- pole figure measurement   198

- polycrystalline 36, 54
- polycrystalline 183
- powder pattern 250
- Pr_2O_3 177
- property profiles 159
- refractive index 161
- residual stress analysis (RSA) 159
- residual stress gradients 276
- rotational symmetry 204
- silicon 61
- strain distribution 265
- strains 244
- stress gradient analysis 281
- structural properties 313
- thickness 156
- TiN *247*
- TiO_2 154
- titanium 38
- W-C:H 135
- see also layers

Thomson scattering 2
three-bounce configuration, CCC *328*
tilt angle 191
 - strain-free 248
total absorption 163
total external reflection 161–165
triaxial residual stress 276
triple-crystal diffractometer 337
tube window 44
twin planes 92, 107
twist and tilt, minimized 217
two-dimensional wire detector *257*

u
ultrahigh density recordings 107
unit cell 48
 - columns 111
 - crystallographic 9
 - edge 74
 - perovskite *186*
 - see also structure
unit vectors 200
unmeasured reflections, computation 206

v
VCSEL see vertical-cavity surface-emitting laser
Vegard's rule 75, 80, 306
vertical-cavity surface-emitting laser (VCSEL) 301
voids
 - mosaic blocks 315–316
 - thin-film growth 64
Voigt approach 261–262

Voigt function 95, 129
volume
 - sample 35
 - unit cell 35

w
Warren-Averbach analysis 126, 134
wave field 319
wave vector 2, *161*, 172
 - GID 175
whole pattern fitting (WPF) 207–210, 289, 130–133
Williamson-Hall analysis 122–126
Williamson-Hall plot 123, 134
 - modified 126
WPF see whole pattern fitting
wurtzite compounds 188
wurtzite structure 50–52, 242, 299, 301
 - ZnO 231

x
XANES see x-ray absorption near-edge spectroscopy
XAS see x-ray absorption spectroscopy
x-ray absorption near-edge spectroscopy (XANES) 57
x-ray absorption spectroscopy (XAS) 57
x-ray attenuation coefficient 143
x-ray beam conditioners 325
x-ray beam divergence 213
x-ray diffraction 1–41
 - high-resolution 297–342
 - perovskites *186*
x-ray emission 31–35
x-ray fluorescence 30
x-ray mirrors 175
x-ray penetration depth 278
x-ray reflectivity (XRR) 143, 165–175
x-ray reflectometry (XRR) 143, 257
 - curve 166–167
 - fringing 171
 - patterns *173*
 - Ta_2O_5 *169*
x-ray refractive index 44
x-ray residual stress analysis (XSA) 240, 247
 - fundamental equations 246
x-ray stress analysis 243
x-ray stress gradient analysis 281–282
x-ray topography 338
x-ray tubes 30–35, 249
 - ceramic *31*
 - copper 13
 - focus modes 34

– laboratory *31*
– mode 146
XRR *see* x-ray reflectometry
XSA *see* x-ray residual stress analysis

y
YBCO film 216
– critical current density *217*
Young's modulus 260

z
zero shift 76
zinc blende-type structures 242
– structure factor 50
zinc sulfide (ZnS) 50
ZnO:Al layers, resistivity 232
ZnSe films 53
– diffraction pattern 54